# 11   GEOGRAPHICAL CHANGE AND INDUSTRIAL REVOLUTION

Coalmining in South West Lancashire, 1590-1799

CAMBRIDGE GEOGRAPHICAL STUDIES

1   *Urban analysis*, B. T. Robson
2   *The urban mosaic*, D. W. G. Timms
3   *Hillslope form and process*, M. A. Carson and M. J. Kirkby
4   *Freight flows and spatial aspects of the British economy*, M. Chisholm
    and P. O'Sullivan
5   *The agricultural systems of the world: an evolutionary approach*,
    D. B. Grigg
6   *Elements of spatial structure: a quantitative approach*, A. D. Cliff,
    P. Haggett, J. K. Ord, K. A. Bassett and R. B. Davies
7   *Housing and the spatial structure of the city: residential mobility and the
    housing market in an English city since the Industrial Revolution*,
    R. M. Pritchard
8   *Models of spatial processes. An approach to the study of point, line and
    area patterns*, A. Getis and B. N. Boots
9   *Tropical soils and soil survey*, Anthony Young
10  *Water Management in England and Wales*, Elizabeth Porter

# GEOGRAPHICAL CHANGE AND INDUSTRIAL REVOLUTION

Coalmining in South West Lancashire,
1590-1799

## JOHN LANGTON

*Lecturer in Geography at the*
*University of Liverpool*

CAMBRIDGE UNIVERSITY PRESS
CAMBRIDGE
LONDON · NEW YORK · NEW ROCHELLE
MELBOURNE · SYDNEY

CAMBRIDGE UNIVERSITY PRESS
Cambridge, New York, Melbourne, Madrid, Cape Town, Singapore, São Paulo, Delhi

Cambridge University Press
The Edinburgh Building, Cambridge CB2 8RU, UK

Published in the United States of America by Cambridge University Press, New York

www.cambridge.org
Information on this title: www.cambridge.org/9780521103923

First published 1979
This digitally printed version 2009

*A catalogue record for this publication is available from the British Library*

*Library of Congress Cataloguing in Publication data*
Langton, John.
Geographical change and industrial revolution.
(Cambridge geographical studies; 11)
Bibliography: p.
Includes index.
1. Coal trade – England – Lancashire – History.
2. Coal mines and mining – England – Lancashire – History.
3. Industries, Location of – England – Lancashire – History.
4. Lancashire, Eng. – Industries – History. I. Title.
II. Series.
HD9551.7.L36L36    338.2'7'2094273    78-67428

ISBN 978-0-521-22490-1 hardback
ISBN 978-0-521-10392-3 paperback

# CONTENTS

| | |
|---|---|
| *List of figures* | ix |
| *List of tables* | xi |
| *Acknowledgements* | xiii |

**1 Introduction**
Geographical change and industrial growth 1
In search of a methodology
  *The statement of an historical geographical problem* 3
  *The construction of a scheme of explanation* 5
  *The basic market area theory* 7
  *A non-operational model* 9
  *Theoretical concepts and empirical variables* 13
  *A scheme of enquiry* 26
Concluding methodological remarks 30

PART ONE: SEVENTEENTH-CENTURY STABILITY

**2 The colliery location pattern, 1590-1689** 35
The numbers and distribution of collieries 35
The duration of collieries 39
The size of collieries 40

**3 Factors affecting the location pattern of collieries, 1590-1689**
Demand 45
  *Transport* 45
  *The distribution of coal markets* 46
  *Alternative sources of fuel* 55
  *Coal markets and the colliery location pattern* 57
Supply
  *Coal seams* 61
  *Colliery sites* 64
  *Labour supplies* 67
  *Capital and entrepreneurship* 72
  *Competition between collieries* 78
The geography of the South West Lancashire mining
industry, 1590-1689 79

PART TWO: THE BEGINNINGS OF GROWTH

**4   The colliery location pattern, 1690-1739**
The numbers and distribution of collieries                              83
The duration of collieries                                              85
The size of collieries                                                  85
Supplementary evidence                                                  89

**5   Factors affecting the location pattern of collieries, 1690-1739**
Demand
    *The distribution of coal markets*                                  94
    *Alternative sources of fuel*                                      101
    *Coal markets and the colliery location pattern*                   102
    *Variable transport costs and methods*                             105
Supply
    *Coal seams*                                                       107
    *Colliery sites*                                                   110
    *Labour supplies*                                                  113
    *Capital and entrepreneurship*                                     120
    *Competition between collieries*                                   128
The geography of the South West Lancashire mining
industry, 1690-1739                                                    130

PART THREE: THE EARLY WATERWAY ERA

**6   The colliery location pattern, 1740-99**
1740-56
    *The numbers and distribution of collieries*                      136
    *The duration of collieries*                                      136
    *The size of collieries*                                          137
1757-73
    *The numbers and distribution of collieries*                      139
    *The duration of collieries*                                      140
    *The size of collieries*                                          141
1774-99
    *The numbers and distribution of collieries*                      145
    *The duration of collieries*                                      147
    *The size of collieries*                                          148
Summary of main trends, 1740-99                                        153
Supplementary evidence                                                 155

**7   Factors affecting the location pattern of collieries, 1740-99**
Demand
    *The waterways*                                                    161
    *Other developments in transportation*                            174
    *Local industries*                                                 176

| | |
|---|---|
| *Local households* | 181 |
| *Coal markets and the colliery location pattern* | 183 |
| Supply | |
| *Coal seams* | 189 |
| *Colliery sites* | 193 |
| *Labour supplies* | 194 |
| *Capital and entrepreneurship* | 212 |
| *Competition between collieries* | 224 |
| The geography of the South West Lancashire mining industry, 1740-99 | 234 |
| **8 Conclusion** | 237 |
| **Appendix 1.** Measures of weight and volume used in the South West Lancashire coal industry during the seventeenth and eighteenth centuries | 243 |
| **Appendix 2.** The collieries of South West Lancashire, c.1590-1799 | 249 |
| **Appendix 3.** The output of Haigh cannel colliery (2), 1725-73 | 253 |
| **Appendix 4.** The Land Tax Assessment Returns as a source of colliery data | 255 |
| *Notes* | 260 |
| *Bibliography* | 306 |
| *Index* | 313 |

# FIGURES

| | | |
|---|---|---|
| 1 | Model of a Löschian mining system | 10 |
| 2 | Costs, prices and outputs with different techniques | 20 |
| 3 | Relationships between the variables of a mining system | 27 |
| 4 | Numbers of collieries, 1590-1709 | 36 |
| 5 | Shifts in the colliery centroid, 1590-1689 | 37 |
| 6 | Distribution, duration and size of collieries, 1590-1689 | 38 |
| 7 | Recorded and estimated colliery outputs, 1590-1689 | 41 |
| 8 | Population in 1673 and the distribution of collieries 1590-1689 | 47 |
| 9 | Origins of coal bought by the Shuttleworths at Smithils, 1588-99 | 49 |
| 10 | Distribution of coal-using tradesmen who left wills, 1590-1705 | 51 |
| 11 | Pithead coal prices, 1550-1690 | 59 |
| 12 | Relief and the distribution of collieries, 1590-1689 | 63 |
| 13 | Sources of capital and entrepreneurship, 1590-1689 | 75 |
| 14 | Coal royalties, 1590-1689 | 76 |
| 15 | Numbers of collieries, 1680-1749 | 83 |
| 16 | Shifts in the colliery centroid, 1680-1749 | 84 |
| 17 | Distribution, duration and size of collieries, 1690-1739 | 86 |
| 18 | Recorded and estimated colliery outputs, 1690-1739 | 87 |
| 19 | Distribution by township of fathers recorded as colliers in parish baptism registers of the 1720s | 91 |
| 20 | Numbers of fathers recorded as colliers in parish baptism registers of certain townships, 1720-45 | 92 |
| 21 | Trend of baptisms in selected boroughs and hundreds, 1700-99 | 95 |
| 22 | Growth of coal-consuming trades during the seventeenth and eighteenth centuries according to the evidence of wills | 97 |
| 23 | Distribution of coal-using tradesmen who left wills, 1690-1755 | 97 |
| 24 | Pithead coal prices, 1690-1739 | 103 |
| 25 | Coal prices in the South and South Western areas, 1690-1756 | 104 |
| 26 | Relief and the distribution of collieries, 1690-1739 | 111 |
| 27 | Pauper migrants to and from Parr, 1694-1739 | 116 |
| 28 | Sources of capital and entrepreneurship, 1690-1739 | 124 |
| 29 | Coal royalties, 1690-1739 | 126 |
| 30 | Numbers of collieries, 1740-56 | 136 |
| 31 | Distribution, duration and size of collieries, 1740-56 | 138 |

| | | |
|---|---|---|
| 32 | Numbers of collieries, 1756-73 | 139 |
| 33 | Distribution, duration and size of collieries, 1757-73 | 142 |
| 34 | Numbers of collieries, 1774-99 | 145 |
| 35 | Distribution, duration and size of collieries, 1774-99 | 149 |
| 36 | The Orrell collieries as recorded in Land Tax Assessment Returns, 1782-99 | 153 |
| 37 | Distribution by township of fathers recorded as colliers in parish baptism registers of the 1780s | 156 |
| 38 | Distribution by township of fathers recorded as colliers in parish baptism registers of the 1790s | 157 |
| 39 | Numbers of fathers recorded as colliers in parish baptism registers of certain coalfield townships, 1740-99 | 159 |
| 40 | Coal sales from the higher and lower coalpits at Winstanley (6) in 1766 | 163 |
| 41 | Coal tonnages hauled on the Weaver and Douglas Navigations, the Leeds and Liverpool Canal, and exported in the late eighteenth century | 166 |
| 42 | Liverpool coal prices, 1752-1810 | 168 |
| 43 | Distribution of coal-using tradesmen who left wills, 1741-80 | 177 |
| 44 | Distribution of population in 1801 and collieries of the 1790s | 182 |
| 45 | Pithead coal prices, 1740-73 | 185 |
| 46 | Pithead coal prices, 1774-99 | 186 |
| 47 | The spread of colliery steam engines | 192 |
| 48 | Three eighteenth-century collier family trees | 196 |
| 49 | Pauper migrants to and from Parr, 1740-99 | 198 |
| 50 | Variations in colliers' weekly earnings at three pits in the late eighteenth century | 208 |
| 51 | Sources of capital and entrepreneurship, 1740-99 | 218 |
| 52 | Sources of capital and entrepreneurship at collieries serving the Leeds and Liverpool Canal, 1774-99 | 222 |
| 53 | Output of the South West Lancashire coalfield, 1590-1799 | 237 |
| 54 | The South West Lancashire coalfield transformed in area according to the time taken to travel a mile in the eighteenth century | 241 |
| 55 | Locations of all collieries worked 1590-1799 | 250 |
| 56 | Haigh (2) receipts and outputs, 1725-73 | 254 |
| 57 | Colliery data from Land Tax Assessment Returns, 1782-99 | 255 |

# TABLES

| | | |
|---|---|---|
| 1 | Numbers of hewers and productivity per pit, 1600-87 | 42 |
| 2 | Urban populations in 1664 | 48 |
| 3 | Colliery profits, 1590-1689 | 73 |
| 4 | Productivity per pit, 1690-1739 | 88 |
| 5 | Estimates of the annual outputs of coalfield areas, 1690-1739 | 93 |
| 6 | Population estimates for Lancashire in 1700 and 1750 | 94 |
| 7 | Registration of vital events of sometime colliers in Up Holland and St Helens, 1690-1739 | 114 |
| 8 | Occupational mobility of sometime colliers registered in Up Holland and St Helens, 1690-1739 | 114 |
| 9 | Residential mobility of sometime colliers registered in Up Holland and St Helens, 1690-1739 | 114 |
| 10 | Wage payments, 1676-1748 | 118 |
| 11 | Some demographic characteristics of collier families registered in Up Holland, St Helens, Eccleston and Douglas Chapel in the early eighteenth century | 120 |
| 12 | Colliery profits, 1690-1739 | 121 |
| 13 | Ratios of total recorded collieries and average numbers working in each year, 1757-73 | 140 |
| 14 | Numbers of hewers and productivity at cannel pits, 1725-1801 | 150 |
| 15 | Numbers of hewers and productivity at coal pits, 1737-99 | 150 |
| 16 | Output in the three main areas of the coalfield, 1740-99 | 154 |
| 17 | Registration of vital events of sometime colliers in Up Holland and St Helens, 1740-99 | 195 |
| 18 | Residential mobility of sometime colliers registered in Up Holland and St Helens, 1740-99 | 195 |
| 19 | Occupations of colliers' fathers and other occupations followed by sometime colliers in the eighteenth century | 197 |
| 20 | The residential ascriptions of three colliery engineers in parish registers, 1758-96 | 199 |
| 21 | Occupational mobility of sometime colliers registered in Up Holland and St Helens, 1740-99 | 199 |
| 22 | Some demographic characteristics of collier families registered in Up Holland and St Helens, 1740-99 | 200 |

| | | |
|---|---|---|
| 23 | Wage payments, 1746-98 | 202 |
| 24 | Day wage payments at Standish colliery (41) during sixteen weeks of 1745-6 | 205 |
| 25 | Wage payments at Skelmersdale colliery (105) in 1798 | 206 |
| 26 | Colliery profits, 1750-99 | 213 |
| 27 | A comparison of the auditors' cash payments to the steward and the value of output at Haigh cannel colliery (2), 1748-55 | 253 |
| 28 | Colliery assessments in the Land Tax Assessment Returns, 1783-99 | 257 |
| 29 | Output, receipts, profit and land tax at Skelmersdale colliery (105), 1793-9 | 258 |

# ACKNOWLEDGEMENTS

Figures 13, 28, 51 and 52 are reproduced here with the permission of the Institute of British Geographers and figure 53 with that of the Economic History Society.

The clarity of the figures is due to Michael Young of the University of Cambridge and Alan Hodgkiss, Joan Treasure and Sandra Mather of the University of Liverpool. Alan Hodgkiss gave unstinted advice and help in solving the technical problems involved in drafting the maps and graphs so that they were suitable for different levels of reduction. The photographic reduction of the figures to final page size was done by Douglas Birch and Harold Taylor of the Arts and Social Sciences Joint Photographic Unit at the University of Liverpool. Maureen Bryce and Leonora Sandman deserve thanks and congratulations for producing a presentable manuscript from original drafts which were often barely intelligible to anyone but themselves and me. I am deeply grateful to (and was for a long time considerably in awe of) Elizabeth O'Beirne-Ranelagh of Cambridge University Press, who systematically exposed and removed inconsistencies and errors of style and content: they were more numerous than might be expected even in a draft written over such a long period and containing as many numerical items and catalogue references as the one presented to her.

Without the enormous help given to me by the staffs of the Lancashire Record Office, the John Rylands Library and Wigan Public Library between 1963 and 1966, when most of the material on coalmining itself was collected, I would not have become aware of many of the manuscripts quoted. I owe a great debt of gratitude to their kindness, enthusiasm and knowledge.

The form of the monograph and the concepts and methods used in it were devised fitfully, over more years than I care to remember, whilst I was a postgraduate student at the University College of Wales at Aberystwyth and a member of staff at the Universities of Manchester, Cambridge and Liverpool. A general acknowledgement must suffice for most of the many conversations which, often unknown to the victims, contributed to the way I approached and performed the work. The theoretical section was conceived and written and much of the data processing was performed in the late 1960s and early 1970s when I was an assistant lecturer in geography at the

University of Cambridge. Without the stimuli provided by my colleagues and students of that time my approach and method would have been considerably different.

Particular thanks are due to E. A. Wrigley, B. H. Farmer and John Carney, who read drafts of the first and some of the later chapters. Perhaps the finished version would have been better if I had been able to incorporate more of their suggestions into it.

CHAPTER 1

# INTRODUCTION

## Geographical change and industrial growth

The work of geographers derives whatever distinctiveness it possesses from a preoccupation with patterns of spatial distribution and spatial contact. Of course, the two kinds of pattern are intimately related. Differences in human activities from place to place are in part a cause of and in part caused by differences in the kinds of contacts made from those places. To understand one set of differences it is necessary to examine the other set. If the activities carried on in particular places change in different ways, so must their connections with other places. Geography defines its characteristic problems in a search for knowledge about patterns of distribution and contact, about the relationships between them, and about their reciprocal changes through time.

Economic historians and development economists have recently been paying more and more attention to the geography of economic growth.[1] It is commonplace, now, to consider national economic growth as something which can only be made intelligible at a subnational level. Regional differences and regional inter-relationships are an increasingly common focus of attention for those who study economic change in the long run because national economies do not surge forward as a whole. Different parts of the national territory grow economically at different rates. Typically, growth is initiated in one or a few regions and gradually permeates to others. The initiating regions derive stimulus from some particular concatenation of economic circumstances existing within them and from linkages with other growing regions. Growth is eventually stimulated elsewhere as laggard areas are bound into the matrices of flows which link and feed the growth regions. Spatial economic differences and particular patterns of economic flows are linked and change together; each state of national economic development has its geography of regional differences and contacts and each phase of growth is associated with economic geographical changes.

In the seventeenth and eighteenth centuries difficult, slow and costly communications ensured a high degree of regional economic autonomy. Indeed, the increasing integration of previously loosely articulated regional economies was one of the processes which stimulated growth in the

1

industrial revolution. It has been widely recognised for a very long time that the loci of rapid economic growth were tightly circumscribed in the eighteenth century and that variations in resource endowment, acted upon by technological change, were responsible for spatial disparities in rates and patterns of growth. But knowledge of the economic geography of the industrial revolution has expanded little further. Were there, in fact, cohesive regional economies or merely spatially scattered points of growth? If they existed, where, exactly, were the growth regions? What was their nature: did polarisation occur around metropolitan growth points, or did expansion permeate from congeries of originally scattered economic specialisms which dynamised the regions in which they lay? How was growth communicated to different economic sectors within the growth regions and to other parts of Britain? How far was the growth which occurred in particular areas based upon an endogenous dynamic, how far upon linkages with other small areas of growth; how far was it due to the character of the national economy within which the areas were embedded, how far to the nature of the particular international trade systems in which they operated? If the industrial revolution is to be seen as a process requiring elucidation, rather than as a national economic event to be explained as simply as possible – and this is a viewpoint which is now advocated by at least one economic historian[2] – questions such as these are important. Economic dynamics have a spatial as well as a sectoral dimension.

Such broad questions as these cannot be pursued in all their variety in a research monograph, the primary aim of which is to convey the results of work in archival repositories on the mining industry of South West Lancashire in the two centuries before 1800. However, one can attempt to convey the information gathered in such a way that some understanding is gained of the geography of change in the industrial revolution: of the spatial variations of output, of the ways these changed and of the relationships between such changes and shifting patterns of contact over the coalfield and its surroundings. Viewed in the broadest terms, the subject of enquiry is the emergence of spatial integration within the hinterland of Liverpool, the welding together of scattered economic pockets into a functionally coherent economic region.

The traditional methodology of historical geography is not suited to such a task. Something beyond a series of cross-sectional reconstructions of the distribution pattern of the mining industry of the region is obviously required to reveal the processes of geographical change within the region. The consensus of recent criticism from outside and of methodological discussion inside historical geography seems to be that distribution maps of the phenomenon being studied should be considered not as the conclusion of research effort, but as the definition of a problem requiring explanation.[3] The benefits which should accrue as a result of this prescription can be convincingly demonstrated in terms of methodological generalities. But arguments couched in such terms do not, and cannot, demonstrate how far what Fischer has called 'explanatory paradigms' – that is, 'interactive structures of workable questions and the factual statements which are adduced to answer them'[4] – can be articulated around the kinds of problems which are typically defined in empirical work.

In the rest of this introduction an attempt is made to devise such an analytical structure, a framework within which the disparate bits and pieces of factual information can be organised so as to illuminate general problems concerning the geography of industrial revolution. Emphasis will be given to the difficulties which crop up in the process of trying to practise the precepts of what might be called 'analytical historical geography', an academic genus which so far has been mainly evident only vicariously in critical reviews and methodological treatises. If these difficulties result only from my own failure to grasp the real nature of what has been promulgated, to understand fully the economic theory that it is necessary to consult or the intricacies of the historical situation being analysed, they may still be worthy of rehearsal. At least, such a presentation will display the problems which materialise when an attempt is made to emerge from the intellectual barricade behind which historical geography sheltered for quarter of a century or more.[5]

## In search of a methodology

### *The statement of an historical geographical problem*

Before searching for ways in which a structure of questions can be devised so that something is explained, the nature of that something must be defined. This definition should state a problem of some kind, but the statement of an historical geographical research problem in such a way that an adequate explanation can be devised is not at all easy. The more narrowly specific the stated problem, the easier it will normally be to devise an adequate explanatory scheme. 'The historical geography of Radnorshire', 'geographical change in the Lake District', or, for that matter, 'the geography of the South West Lancashire coalmining industry, 1590-1799', are not specific problems. Some particular aspect must be drawn out of such portmanteaux for explanation. To argue about what exactly ought to be drawn out is undoubtedly to sink into that 'sterile navel contemplation' castigated by Olsson and contemptuously ignored by Harvey.[6] But, unfortunately, it is a difficulty which confronts all empirical workers at the very outset of their studies. And it *is* a difficulty. As soon as the problem, whatever it might be, is defined, the totality of the actual empirical situation is left behind and the distance of this departure is directly proportional to the degree of specificity in the definition of the problem. Even without adherence to the view that geography, like history and, some would argue, economic history, is concerned with the synthesis of 'total' empirical situations, it might still be felt that to emerge with something resembling the sorts of problems used by theorists to illustrate the value of their products is to limit the scope of enquiry so far that it ignores most of the interesting aspects of the situation from which the problem has been abstracted.[7]

An historical geographer can rarely define his problem with precise brevity. The problem at present under investigation comprises more than the total output of the coalfield and the way that it changed. Its specification must incorporate the spatial distribution of that output and the changes in this pattern, and these cannot easily be specified. In fact it is not even possible to describe a static surface of this kind in terms of a simple mathematical function which might serve as a precise definition of events to be explained. If data were fully available, then perhaps a count of the output raised in each cell of a grid superimposed over the coalfield could be used as an approximation to such a specification, explanatory variables being incorporated into the analysis through techniques of spatial correlation. Even in this unreal world of perfect data availability (just how unreal it is will transpire later), these methods of describing the spatial distribution of output would ignore such aspects of the problem as colliery size and duration. These must be incorporated in any description of the colliery distribution pattern because fluctuations in the way that total output was apportioned over the coalfield could have been due to differing rates of change in colliery size between areas and/or to different rates of change in colliery numbers. The latter, in turn, may have resulted from differences in the rates at which collieries were opened, or from differing ratios between varying rates of opening and closure. Which of these was occurring, and to what extent, must obviously be known before any explanation of aggregate spatial change can be attempted.

The simple point at issue is that any statement of this locational problem must incorporate information about colliery numbers, size and duration. Generalised statements of the number of units of output raised over arbitrarily defined areas are an insufficiently detailed specification of the problem. And to present the required information three separate quantitative series must be used, each broken down subregionally in some way.

Thus, even after accepting that such holisms as 'geographical change' are not sufficiently specific problems for explanatory analysis and that particular elements of such an imprecise compound must be abstracted if explanation is to be successful, there is a limit to how specifically and simply the abstractions can be defined. To a greater or lesser extent, their definitions must comprise messy amalgams of a number of elements. And this, in turn, has repercussions on the degree of rigour with which a scheme of analysis can be defined, and thus on its explanatory power. Simply because historical geographers are, almost by definition, interested in relatively complex problems, even when they pitch their studies on a level below that of synthetic empirical holisms such as 'geographical landscapes', their explanations must be relatively weak and the devising of explanatory schemes must be relatively difficult.

## The construction of a scheme of explanation

If geographical explanation, in its simplest sense, comprises a structured enquiry into the functioning of the objects which make up the pattern being studied, then what is required in this particular case is a set of interlocking questions about the way that a mining system operates. This set of questions must obviously be as comprehensive as possible: it must cover all the subsystems which contribute to the functioning of a mining system, and it must be concerned with the ways in which the various subsystems interact. A rigorous statement of the hypothetical possibilities is provided by economic location theory.

Economic location theory is founded upon the assumption 'that economic activities in general and locational choice in particular are governed by the desire to maximise nominal profits'.[8] These represent the difference between money receipts and money costs, and the size of this difference varies through space as the costs of operation and receipts revenue vary. Both costs and receipts are determined by a number of factors and the interrelation of all these factors makes the analysis of location difficult.[9] This complexity can only be penetrated if further assumptions are made, and different theoretical schemes have been devised by making different sets of simplifying assumptions. Although much recent work has been directed at integrating them, there are, basically, two categories of location theory: that based on the least cost approach of Alfred Weber, and that based on the market area approach of August Lösch.[10] Each of these sets of theoretical propositions is more suited to the analysis of certain types of empirical situation than others.

Weber's least cost theory is predicated on the assumption that markets exist at a particular point where demand is constant or unlimited. Given knowledge about the relative quantities of the resources required in production, their prices at source, the cost of hauling them and the cost of transporting a unit quantity of finished goods to the market, the point of minimum cost can be determined. This point represents, with a single market and therefore a single market price, the site where maximum profits would be made by a firm contemplating entry into an industry. But this approach to the problem of location deals with the case of one firm only and it cannot accommodate variations in demand and finished goods prices.

Lösch's market area theory expresses the influence of variations in prices and demand and, unlike Weber's theory, it encompasses all the firms in an industry simultaneously so that optimum location patterns, not just the optimum site for the next firm to enter an industry, can be determined for an equilibrium situation. But this theory only applies under the assumptions that firms serve discrete market areas which surround their plant (which holds given certain stipulated conditions) and that costs are spatially uniform on a featureless plain where population is scattered and, like raw materials, evenly distributed.

Thus, neither theory can accommodate both spatially variable costs and spatially variable demand. Even when the industry concerned is as economically uncomplicated as mining, and given full data on all relevant variables, no theory exists which will allow the ascertainment of the optimum location pattern of the industry – and therefore provide an explanation of that pattern under the working hypothesis of profit maximisation. Theory indicates most forcibly, then, not how explanations may be devised, but that explanations are impossible because of the inherent complexity of the situation being studied. Yet theory can be pursued beyond this chastening point for indications of the exact nature of this complexity, and therefore as a potentially useful aid in structuring a scheme of enquiry which is as adequate as the nature of the problem allows.

Lösch's scheme is the best starting point for the present analysis. Generally, situations in which producers are located at points and consumers are extended over the area occupied by the points are most amenable to analysis through his market area theory. The necessarily punctiform nature of mining, in which agglomeration economies are relatively unimportant, the predominance of domestic demand and the absence of overweeningly large towns in the region, at least in the seventeenth century, all suggest that this situation would obtain in the case in hand. Moreover, the greatest merit of Weber's approach is that it can accommodate variations in raw material quantities and prices and in the distances of supply points from potential places of production. Such considerations are irrelevant to mining because mineral producers obtain their inputs on the site of production without paying the costs of assembling them there. Coal can, at least as a preliminary assumption, be regarded as a ubiquity on a coalfield and labour and capital are to a great extent mobile without cost to the producer of coal. These cost conditions applied *a fortiori* on an almost unexploited coalfield at a time when the labour and capital requirements of individual collieries were generally small. Furthermore, because coal is heavy and bulky, and because weight loss occurs *in extremis* during use, mining is generally likely to be closely tied to market locations. Again, this was especially true of the industry in the seventeenth century when haulage costs were very high.

Clearly, then, the market area model provides a better starting point for this particular analysis than the least cost scheme. This is not to argue that costs can be ignored, but rather that in this particular case the necessary analytical groundwork is better provided by a theoretical scheme which assumes that monopolistic competition exists in space and goes on to elucidate the effects of variations in demand, rather than by one which is primarily aimed at demonstrating the effects of variable haulage costs for raw materials and finished goods.

## The basic market area theory

It will be apparent from what has been written so far that the standard theoretical case of the market areas scheme is one of an industry which (1) produces a single good from resources which are (2) available everywhere at (3) equal cost.[11] In the elementary version of the theory demand is assumed to be constant through space, with (4) an evenly spread population of which (5) each member has the same demand curve for the product. In a study through time such as this, (6) temporal invariance of demand and (7) inexhaustible raw materials must also be assumed to generate the simplest possible case. If these assumptions held, then there would be no temporal variation in mining from colliery to colliery, each being permanent, and the number, size and spatial distribution of collieries would be dependent upon the shapes and sizes of the market areas which they served.

With uniform demand curves, the size of these sales areas would depend upon the prices that consumers had to pay at the points of consumption. Consumer prices, in turn, would depend upon the pricing system used in the industry. 'Three possibilities are open to the entrepreneur . . . He can adapt his prices to the individual case; or keep them so rigidly fixed that all buyers pay the same f.o.b. price [i.e. mill or pithead price] or the same c.i.f. [or delivered] price. Depending on his policy, his prices will differ on the average from place to place.'[12] Differences in the size of market areas and therefore in the size of collieries and/or their number would thus obtain under different pricing systems. A constant pricing system must therefore be assumed, and that which is nearest to the South West Lancashire situation was (8) equal pithead pricing. With such a policy, the prices at the points of consumption would depend on the pithead price plus the cost of haulage from the pithead. For simplicity it must, therefore, also be assumed that (9) transport costs were everywhere proportional to the weight moved and the distance over which it was moved.

Certain deductions can be made about the size and shape of the market areas that would develop around collieries for the simple situation in which conditions (1) to (9) are satisfied. The first is that 'adjacent market areas . . . are separated by boundaries along which delivered prices from the two supply points are equal . . . The boundary is a straight line, the normal bisector of the line connecting the two supply points.'[13] Thus, all collieries with market areas which lay wholly on the coalfield, and which therefore directly impinged upon those of other collieries, must be surrounded by market areas of equal size and all such collieries would thus have equal outputs. Collieries located near to the edge of the coalfield would have market areas that were 'open' off the coalfield, and their extra-coalfield extent would be limited only by the eventuality that at some point transport cost plus pithead price reached a limit beyond which demand ceased to exist. The shape of the market area boundary in this direction would describe an arc centred on the colliery, its radius depending upon transport costs, pithead price and the demand curve. The high transport costs of the seventeenth and eighteenth centuries would keep this radius small, but it might still exceed half the distance to the nearest colliery, so that works on the edge of the coalfield could have larger market areas, and therefore higher outputs, than those with market areas wholly on the coalfield. It can be demonstrated that the wholly intra-coalfield markets areas would take the form of hexagons, the collieries located centrally within them, so that the colliery location pattern would be one of points located at the apices of a regular triangular lattice.[14]

It can, then, be concluded that if assumptions (1) to (9) were satisfied, collieries would have been permanent; they would all have produced the same outputs, except those near the edge of the coalfield which may have been larger than the rest, and they would have been located in a regularly spaced pattern. They must, too, if the assumptions were satisfied, have sold coal at equal and constant pithead prices.

Statements about the size of the collieries which had market areas wholly on the coalfield, and therefore about the exact number of collieries and the distances between them, can only be made for the equilibrium state of such a system, and this would only occur if five further conditions obtained. The first is that (10) *'the location for an individual must be as advantageous as possible.* An entrepreneur therefore makes his choice within the whole district and within his market area in such a way that his profit . . . shall be greatest.'[15] (This assumption is, of course, the basic 'working hypothesis' of the whole preceding discussion, but it is restated here because it has not so far appeared in the enumeration of assumptions.) The second is that (11) *'the locations must be so numerous that the entire space is occupied'* (ie all demand must be supplied), and the third is that (12) all *'abnormal profits must disappear'.*[16] This last condition requires that prices must everywhere equal average costs. This would only occur if the number of collieries was maximised, so that (13), *'the areas of . . . sales* [and therefore the size of collieries] *must be as small as possible'.* The fifth of the conditions necessary for the achievement of equilibrium is that (14) *'at the boundaries of economic areas it must be a matter of indifference to which of two neighbouring locations a consumer belongs'.*[18]

The conditions (10) to (14) can be expressed as five equations from which, if assumptions (1) to (9) hold, 'everything can be obtained . . .; size and limits of market areas, the situation of production locations within them and within the entire area and f.o.b. prices'.[19] The solution of these equations requires, *inter alia*, the specification of the assumed constants of demand and supply schedules and it obviously cannot be attempted here. All that can be said about the area under study, further to the earlier conclusions, from the addition of conditions (10) to (14) is about the number and size of collieries. Given the requirements of conditions (11) to (13) to allow the maximisation of the number of firms, that collieries actually existed which produced less than 100 tons per year, and that, therefore, profit could be made on this output, and given that total output in, say, the late seventeenth century was c.13,000 tons per year, then there would have been c.130 collieries producing c.100 tons per year (with, possibly, larger works near the coalfield edge), equally spaced in a triangular lattice, selling at constant pithead prices.

Reality was far from congruent with these theoretical expectations: no more than twenty-one collieries existed at any one time before 1700 and some of them produced over 2000 tons per year. The theory is not 'valid' in this particular set of empirical circumstances; it does not provide an explanation of the location pattern which existed because this differed from that which the theory predicts. But what the theory does provide is an enumeration of the variables, expressed in assumptions (1) to (14), which must be taken into account in a locational analysis. It demonstrates what must be closed within an analytical system aimed at explaining a colliery distribution pattern, and it describes each of the parts of that system. Furthermore, it provides information about how these various elements would be inter-related if they assumed certain states.

*A non-operational model*

The assumptions (1) to (14) fall into two groups: those concerning resources and those concerning the ways in which people behave. The latter group can be further split into those concerning the behaviour of consumers and those concerning the behaviour of producers. The assumptions of the theory, then, define a system in which resources are operated upon by two control subsystems. A simple model of this system and its components can be constructed, as in figure 1. The transposition of the assumptions into the terms of this model will aid their further discussion.

Theory always represents statements of an if–then variety. In the present case, if elements A–G and K–M assume certain values, then subsystems I–III attain certain states. If these states obtain, then the supply and demand curves J and N will be spatially constant, and if this occurs then deductions can be made about the geographical distribution of collieries and the prices at which they sold coal. Transposing assumptions (1) to (14) into the terms of the diagram, and progressing from subsystems I and II to J and from subsystem III to N, the theory states that if all necessary resources were inexhaustible and available everywhere at equal cost (assumptions (2) to (7) concerning elements A–D of subsystem I) and if they could be combined

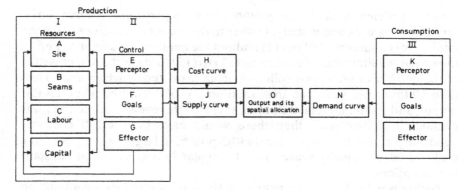

1. Model of a Löschian mining system.

only into undifferentiated products (assumption (1) concerning elements A–D of subsystem I); if entrepreneurs had (i) perfect knowledge of these resources (an implicit assumption about IIE), (ii) aimed at maximising their profits (assumption (9) about IIF) and (iii) were able to enter the industry freely using best practice technology (required by assumptions (1), (10) and (12) about IIE), then the cost curves of all producers would be the same and would equate with their supply curves. To the right of the diagram, if consumers had perfect knowledge about possible sources of supply (an implicit assumption about IIIK subsumed under (14)); if their tastes were the same (assumption (5) about IIIL); if they were all able to travel unit distances at equal costs to obtain supplies (assumption (9) about IIIM), and if they were evenly distributed (assumption (4) about the whole of subsystem III), then a common demand curve would also exist. The interaction of these supply and demand curves would fix price, total output and the spatial distribution of output.

All the explicit and implicit assumptions of the theory can thus be conveniently expressed in terms of this simple model. From the theory it can be asserted with confidence that the analysis can be closed around the elements A–G and K–M. Knowledge about nothing else is required to explain O. If this model could be put into operation with known values of A–G and K–M, it *must* produce a pattern resembling the output of the system being modelled. But it cannot be put into operation and when the theory is cast into the form of figure 1 the reason why it cannot is obvious. Deductions about the spatial allocation of production can *only* be made from variables A–G and K–M when they assume values which ensure the existence of spatially constant supply and demand curves. It is to ensure this eventuality that assumptions must be made about them. As soon as these assumptions are relaxed, supply and demand curves may vary spatially and the model falls to pieces. With variable supply and demand a number of patterns which are optimal could theoretically occur – the model need no longer have a determinate solution. Worse than this, the places where profits would be maximised can only be ascertained by running the model an infinite number of times to test, in an *ad hoc* way, various combinations of locations. But this logistical extremity is also impossible, even in principle, because demand at any point could only be calculated if the demand and costs at all neighbouring points were known. The demand curve at a point represents an aggregation over an area surrounding that point and to estimate the demand curve for any one point the demand at all surrounding points must therefore already be known. Moreover, the costs of all surrounding points must also be known, compared with the cost at the point being considered, because these will determine the prices asked elsewhere and therefore the exact quantity demanded at the point being considered at the price which is fixed by costs at that point. Thus, what is unknown at one point must, to be made known, already be known for other points; if it is unknown for all points it can be discovered for none of them.

Lösch's theory cannot, then, be put into operation. It simply demonstrates that under certain conditions an equilibrium location pattern is possible. And this is all that it was intended to demonstrate: 'in my opinion, systems of general equations of equilibrium like this one have no other significance. I consider it utopian to assume that they can be gradually improved, and employed to solve practical problems.'[20]

The conclusion which is reached through a consultation of location theory is that because of the inherently complicated nature of locational problems, it is impossible to build an operational model to explain a colliery distribution pattern. However, this excursion into theory was begun in the same spirit as Mrs Huntington prepared for her escape from Grassdale: 'brilliant success, of course, I did not look for, but some degree of security from positive failure was indispensable'.[21] A number of such securities have been derived incidentally from this analytical circular tour, besides a realisation of why successful explanation is not possible. The theory has yielded a definitive list of the sorts of elements which must be considered and, moreover, these elements have been derived within a structure, described in figure 1, which might greatly aid their articulation into a coherent scheme of analysis.

From here the theory, or the model derived from it, can be used as a point of departure for two quite different analytical destinations. Either a certain minimum number of assumptions (which would necessarily be large) could be maintained and a model built to aim at providing definite explanatory statements within the limits set by the assumptions. Or each element of the theoretical system as defined in figure 1 could be examined in an analysis which was more nearly reflective of fully ramified historical reality, but which could not provide any reliable causal statements: which could not explain, in any rigorous sense of the term, the pattern of colliery distribution. Neither of these analytical extremes could, of course, be reached owing to the problems involved in the empirical specification of the conceptual terms of the theory and in the measurement of the variables defined in such a specification. Nonetheless, one of the two divergent paths which lead towards these analytically distant ends must be chosen: the explanatory scheme which is constructed must be aimed towards either econometric rigour or empirical verisimilitude. No scheme of analysis can be produced to home in on both of these targets simultaneously.

In the first kind of scheme two of the subsystems, or the major parts of them, would be assumed to have existed in the states required by the theory. For example, the whole of subsystem II and K and L from subsystem III could be held to the theoretical requirements, and a least cost analysis could then be performed to discover how variations in the costs of production and transport affected the pattern, which, except for any residual discrepancies between it and the pattern produced by the model, would then have been 'explained' in these terms. The causal statements derived would be definite and incontrovertible. The progression from theory to theoretical model to operational model to rigorous explanatory conclusions, as advocated by the methodological sophisticates of geography and economic history, would have been achieved. But what, exactly, would this achievement represent? The high degree of hypothetical-deductive rigour would have been obtained only at the expense of gross oversimplification. What would emerge is an explanation which would definitely have been correct if the situation had been of a nature which it quite probably was not. What is more, how far the real causal situation lay from that which emerged from the assumed context, and therefore the amount of credence which the results possessed, would remain unknown.

A scheme of the second kind could be devised in which the causes of deviations of the actual location pattern away from a Löschian equilibrium pattern could be sought in terms of discrepancies between the actual states of each of the elements defined on figure 1 and the states they would have to achieve to produce the Löschian pattern. Such a scheme of analysis would comprise what might be termed a 'factor model' – one in which each factor which could have influenced the pattern is examined separately, in an order and manner that the theory suggests is appropriate. The theory would thus be used merely, to use David Harvey's words, 'to codify our expectations about reality'.[22] Such models possess a very low status on the scale of explanatory rigour. Because each factor is examined separately it is only possible to conclude that a certain one *might* have influenced the pattern in a particular way. There is no means of definitely finding out which one of a number of factors that might have skewed the pattern away from a Löschian equilibrium in a similar particular way was, in fact, responsible for the discrepancy. Any such evaluations can be made only in an *ad hoc* manner. Such loosely articulated models as these, then, sacrifice much in analytical power and certitude to gain their greater comprehensiveness. But, at least to someone with my own academic inclinations, their flaws are not as objectionable as those which are innate within the more rigorous kinds of formulation which were briefly discussed earlier.

When it has been decided to employ a 'factor model' in order to jettison the least possible amount of empirical referability, using theory only to help to order the analysis and suggest the kinds of hypotheses which might be usefully examined, then a problem is immediately apparent which is perhaps not so important in the formulation of models of the first type, given their less comprehensive treatment of reality and the academic ethos which they reflect. This is the problem of how far it is possible to specify the abstract conceptual terms of the theory in such a way that they are empirically recognisable and measurable. The value of theory in any particular piece of historical geographical analysis, and the degree to which it is possible for the latter to utilise properly the former, obviously depends upon the success with which this translation can be made.

## Theoretical concepts and empirical variables

Lösch's theory is concerned with market areas, and the heaviness, bulk, low value and weight-loss-in-use of coal all suggest that the location of collieries must be strongly influenced by the location of coal markets. It is, therefore, appropriate to examine the concept of demand first.

*Demand.* Lösch's theory assumes that demand is spatially constant. This condition is, to say the least, extremely unlikely, and two sets of questions arise in consequence. First how far is it possible to specify what the distribution of demand was? Second, because the other elements of the model only operate in the manner described by the theory when demand *is* constant, what would be the implications of variable demand in terms of the possible effects of other factors on the colliery location pattern?

The specification of the way in which demand for coal varied over South West Lancashire, or any other commodity over any other area, is extremely difficult. Lösch used a partial equilibrium formulation in his treatment of this problem: he assumed that the demand for a good is the same function of the price of that good at every point on the plain which provided the assumed basis for his calculations. But Long has pointed to the serious inadequacy of such partial equilibrium models in spatial analysis.[23] The amount of coal demanded at any point is not only a function of the price of coal at that point, even if income is held constant. The prices of other goods besides coal at the point in question also act to determine the amount of coal which will be demanded at a certain coal price. In other words, the problem should be formulated as one of general rather than partial equilibrium. Of course, this is an ideal requirement of all economic history, but often it can be dispensed with and a simpler partial equilibrium model can be used instead without too much loss of realism. But when spatial variations are being studied this is not the case because attention is focused upon how the demand for one commodity varies as its price varies away from the point at which it is produced, and any such variation is obviously intimately related to the way in which the prices of other goods vary away from the points at which they are produced. It is therefore inconsistent to hold all other prices constant over space whilst at the same time using the fact that coal prices will vary over space according to distance from pitheads as a pivot of the analysis of coal demand.[24]

Thus an economically rigorous calculation of the geographical variations of coal demand must involve an examination not only of coal prices but also of all other prices and the ways in which they, too, vary through space. Clearly this is a completely impracticable assignment. This being so, some simple descriptive notions derived from the concepts depicted in figure 1 will have to suffice as an attainable half-way house between an unrealistic assumption that demand was spatially constant and an unachievable formulation of what it actually was.

As the diagram on figure 1 indicates, even a simple approximation of demand must include some consideration of the different ways in which different kinds of consumers might behave and of the variations in the spatial distribution of these different kinds of consumers. This is not merely to allow an aggregation which can provide a picture of 'total' market conditions: differences in the behaviour patterns of consumer groups might have had different repercussions on colliery location, and these possibilities must be explored.

It is desirable to distinguish between domestic, industrial and export markets, at least. It seems reasonable to hypothesise that the domestic consumer had the least knowledge of the range of possible purchasing places, the least stringent requirements in terms of the type of coal purchased, and the greatest degree of indifference between coal and other types of fuel. When these characteristics are combined with the fact that mining had begun in some parts of the coalfield as a component of a manorial economy, the net result would be a high level of 'consumer inertia' in the domestic market, with coal being bought almost habitually at a particular colliery as long as it continued to produce at a constant price, little thought being given to alternative sources of supply or other kinds of fuel. But because of their high degree of indifference between coal and other fuels, price rises might cause a wholesale switching out of coal by domestic consumers if an alternative was locally and easily available, even though the presence of cheaper coal at a non-habitually frequented colliery in another manor might not have the same effect if no price change occurred to trigger off the shift. Thus, the amount by which prices could be raised by colliery owners might be severely constrained in the domestic coal market, whilst price cuts would not certainly bring commensurate increases in sales. Simply because demand was not a continuous function over a range of prices the colliery owner would more than usually be encouraged to keep prices stable and this, in turn, and perhaps irrespective of the pricing policies of other colliery owners who hazarded to change, would ensure a reasonably constant output.

There were, then, in the characteristics of the domestic coal market, forces which would act strongly to maintain a *status quo*. On the other hand, if no colliery existed in an area it did not mean that a sufficiently large market was not available to sustain a colliery selling at a certain price which could wean householders away from other fuels — either by virtue of its being run by a landowner to whom they were beholden, or because other fuel sources were becoming increasingly difficult to procure. This kind of situation could produce many experimental incursions into production, some of which would prove abortive, and thus a degree of flux would be created in the location pattern. Moreover, if price changes were made necessary by rising costs, then the coal market in an area might evaporate and mining might completely cease there. Whether the nature of the domestic market engendered stability or flux thus depended on the way in which costs changed as collieries continued to be operated, on the availability of alternative fuels and on the way in which the cost of their procurement changed with continued exploitation of the sources of their supply.

The industrial consumer might have differed considerably from the householder whose behaviour would produce the effects just described. The coal needs of the industrial consumer were larger and they were part and parcel of his making a living. Usually there was no substitute for his coal and, especially for metalworkers, coals of different types produced radically different qualities of finished goods. As a result, the industrial consumers' searches for information about alternative sources of supply of coal, according to criteria of price and type, were probably much more thorough. Furthermore, their demand probably described a more nearly continuous function, in the aggregate, over a range of prices than did the demand of the domestic consumers. It is, therefore, likely that the industrial market was more stable than the domestic market in the face of changing prices and colliery openings and closures in an area. On the other hand, switches of custom between different local collieries and competition between coalmasters for industrial markets would, one would imagine, have been much more common. The stabilising and unstabilising effects on the colliery location pattern that were generated by the nature of the industrial market would, thus, operate on a different scale from those same effects operating as a result of the character of the domestic market. The former encouraged the continuance of demand over an area but not at particular collieries within that area, whilst the latter encouraged the stability of demand at particular collieries, once established in an area, but not over the area as a whole in the face of rising prices or a turnover in the colliery population. This could affect entrepreneurial behaviour profoundly: price cutting for industrial consumers would be an attractive proposition. Moreover, the direction of causality between mining and markets could be completely reversed. Whereas the prospective coalmaster who intended to serve domestic markets responded to given location patterns of demand and chose to locate his works accordingly, the possible supply points of higher grade or very cheap coal would be relatively rare and the industrial consumer, with low capital costs, might move towards the collieries which could best supply his needs. Householders drew domestic coal works towards them; high grade or cheap coal drew industrial consumers towards the collieries.

Export markets simply represented points of high consumption available to those parts of the coalfield nearest to them. The consumers in these markets were, of course, domestic or industrial and the only additional effect on the colliery location pattern which export markets would evince would be upon the size and/or number of collieries on the coalfield edge, and this effect would be mediated through the pricing or other competitive strategies of the coalmasters there. The coalfield boundary is irregular and export markets are by definition situated at some distance from it. Differences in the distances between various collieries and such markets could be insignificant as a proportion of the total distances. Consequently, the possibility of rivalry between a number of collieries for single large markets – of oligopoly – rather than spatially monopolistic competition emerges. This, as it will transpire later in the discussion of entrepreneurial behaviour, is of potentially great importance as an influence on the pricing policies of coalmasters and therefore on the outputs of their collieries given fixed supply and demand schedules.

So far, only two of the three determinants of consumer behaviour (their 'perception' and 'goals') have been dealt with. Their 'effector' capabilities can, without too much artificiality, be considered as the availability of means of getting coal from the pithead to the place of consumption. Generally in conditions of f.o.b. pricing, transport costs are a major influence on the price paid at the point of consumption, so that it is usual to consider transport costs as one of the major determinants of location patterns. This is not always a realistic convention, but in the pre-industrial revolution mining industry it is most appropriate. Prices usually doubled at a few miles from the pithead, and the inelasticity of domestic demand, plus large export markets where a large proportion of the price would comprise haulage costs and where coal from other coalfields might compete, would all mean that coal transport costs would be an important control on the aggregate demand available to particular supply points and therefore on the size and density of collieries. Any spatial variations in transport costs could obviously have an enormous impact on the geography of mining, and the opening of a cheap transport route would be a most powerful stimulus to changing it.

A number of points have emerged clearly from this examination of the make-up of demand. Although demand cannot be calculated, and although the various facets of consumer behaviour depicted on figure 1 cannot be specified in such a way that they can be examined explicitly and fully, it is apparent that a consideration of coal markets should as far as is possible include a treatment of transport costs and separate examinations of domestic, industrial and export markets. Moreover, because of the nature of the domestic coal market it is necessary also to study the availability of alternative fuels. Because of differences in the amount of information available to and the degree to which coal was substitutable by domestic and industrial consumers, and because of the possible existence of large export markets at some roughly equal distance from various places on the coalfield, the pricing policies of coalmasters need not have been the automatic responses to irrefragable market conditions which exist in Lösch's theory. Entrepreneurial behaviour thus becomes, even under the analytical aegis of a profit-maximising assumption, potentially more variable and therefore a potentially more important determinant of the shape of the colliery distribution pattern.

So far, the spatial distribution of the various kinds of coal purchaser has been ignored, but it is obviously of considerable importance. Buyers must, to a greater or lesser extent, have been clustered in villages and towns and this would create irregularly shaped market areas around collieries. Suppliers too would become relatively concentrated in the areas of clustered demand and the distance between collieries could thus shrink sufficiently to allow oligopolistic competition. But price competition would not necessarily result. These conditions might, in fact, reinforce the already observed tendency for price stability within the middle section of the demand curves. The demand curves of domestic consumers were, to use jargon, kinked, and it might well be that price competition would be less prevalent than the existence of oligopoly alone would suggest because in such a market it is impossible to predict the outcome of price cuts and counter-cuts.[25]

Just as important is the emergence of the possibility of collusion amongst sellers in order to keep up profits at the expense of the customers, if a number of collieries was located close to a large market which was inaccessible to more distant coalmasters. Such collusion would be equally likely if the group of sellers had advantageous access to a cheap transport route. The tendencies towards locational and price stability produced by uncertainty and kinked demand curves alone would, of course, be reinforced by restrictive practices of this kind.

*Resources.* Coal seams were not ubiquitous in the South West Lancashire coalfield. Certain measures were completely barren and others contained few seams separated by great thicknesses of barren rock.[26] A small southern fringe of upper coalmeasures can be ignored on the former grounds, and the lower (or *Lenisulcata*) coalmeasures can be distinguished from the middle or main productive coalmeasures on the latter. However, some coal was available even in the *Lenisulcata* rocks, and on the main productive coalmeasures the cost of winning coal and the quality of the coal must have varied from place to place. Each of these factors – presence/absence, varying costs and varying quality – might have had an important effect on the geography of coal production.

The complete absence of coal beneath an area, or its presence only at depths which were unfathomable by the then current mining techniques, would simply create 'holes' in the colliery location pattern and extend the potential market areas of the nearest workable sites.

It became obvious in the discussion of coal markets that variations in coal quality could be of great significance in moulding the geography of mining if industrial consumers were relatively numerous. It is important, therefore, to discover how far, and in what ways, the quality of coal varied over the coalfield and whether particular coals were especially suitable for particular purposes. If this was the case, and if the coals in question were sufficiently rare, a significant skewing of the pattern away from a Löschian equilibrium could occur: production would be greatly stimulated at certain sites, irrespective of considerations of price and distance from markets, and the collieries thus fattened would grow to a larger than 'normal' size and cast a shadow over mining in the areas around them.

The nature of the coal seams could also cause variations in the cost of mining, the results of which would be much more complex kinds of warping of the Löschian location pattern. Such cost variations could cause the complete absence of mining in some areas; consequent price variations could produce market areas of different radii and therefore collieries of different sizes or, if entry into the industry was not free, the cost variations could be absorbed in profit differences. Moreover, given the nature of domestic and industrial demands, variations in cost through time at particular mining sites might be a potent cause of instability in the colliery location pattern. It is easy, then, to suggest ways in which cost variations attendant upon differences in the resource base from place to place might have been important contributors to the geography of mining. But unfortunately this is not a variable which is easy to define operationally. Not only are cost data few, but the separation of that component of costs which reflected the nature of coal resources from that which reflected the prices of other factors of production is not at all easy. Moreover, if the influence of variability in these costs *per se* is to be adduced, then their level at sites which were not mined, as well as those which were, should ideally be known. Some proxies can be sought by examining how mining costs were comprised, by separating out those effects which might have been due to the cost of the mining site, the coal itself and the work involved in getting it out and bringing it to the pit bank for sale.

It seems that mining sites were rarely costed separately from the coal itself. In leases, rents – or perhaps more properly, royalties – were assessed on the coal only. The surface property used during the course of mining was automatically provided, free of all costs except those involved in restoring it to the condition it was in before mining began. But it should not be inferred from this arrangement that site conditions can be safely ignored. They could have affected the location pattern of collieries in two ways. First, the returns to the site itself on alternative uses to mining may have been sufficient to preclude the digging of pits, spreading of spoil and opening of access routes there. Second, site characteristics of a particular kind may have been either necessary or attractive. For example, if water or the depth of seams were important causes of increased expense, then steeply sloping ground and/or valley bottom sites might yield conditions which appreciably reduced it. This would represent, in effect (and insofar as the value of the site was not inexplicitly included in the coal royalty), the substitution of the free resource represented by the propitious site conditions for the cost of deeper mining, longer soughs or pump drainage. This possibility posed a potential geomorphological constraint on colliery distribution and must, therefore, be examined.

Freeholders did not pay a royalty and mining could prove cheaper for them than for lessees. In principle, I suppose, the freeholder should require greater profits to compensate him for the irrecoverable loss of personal property on the depletion of his coal stocks. Again, it can be supposed that this increment should, theoretically, equal the royalty he would require, so that the prices charged by freeholders and lessees for the same coal worked with the same techniques should be equal. However, if the freeholder was not enough of an economic sophist to think in this way, then his cheaper mining could allow him to reduce his prices, and this competitive advantage would feed his colliery output at the expense of the production of leaseholder rivals. Certainly, the leaseholder/freeholder distinction, reified in the size of royalties, is an interesting line of enquiry into possible causes of any observed skewdness from a Löschian mining geography. Furthermore, any differences in royalty charges from place to place could have the same effect amongst the collieries of leaseholders.[27]

The cost of actually digging down to, extracting, and raising the coal could vary according to a legion of factors such as its hardness, depth, wetness, gassiness, the roof and floor conditions and so on. It was, of course, the cost surface defined by these factors plus royalties which interacted with that of demand to produce the location pattern of collieries: if demand was spatially constant, low cost sites would grow in production at the expense of higher cost sites with neighbouring market areas. It is necessary to discover not only how the cost of mining varied in space, but also through time, because shifts in the cost surface alone could cause the location pattern of collieries to change.

The possible lumpiness of costs, as well as their aggregate level, was also of importance. If large sums had to be invested in fixed capital installations or in wages over short periods during which they were not nearly covered by receipts, then the degree to which capital was available could be a major determinant of mining geography. Of course, if other things were equal, then mining would not occur at such sites if others were available where large lumpy investment was not necessary. But some fixed investment, if only in pits, was required everywhere and, more important, it is highly likely that other things were not equal. Because transport costs were high, the spatial variability of demand could have ensured that high cost sites were profitable, and the more concentrated the demand the greater would be the possibility of absorbing high costs in high pithead prices at collieries near large markets. Variability in coal quality could have had the same effect, particularly in relation to the industrial market, where high prices might be available for particularly choice coals. High cost sites could, then, have been attractive to mining, but only if capital was available and not inordinately expensive.

Thus, spatial variability in the availability of capital could instigate discrepancies between the demand and cost surfaces. But the degree to which capital was available could also have other effects on the location pattern of collieries by affecting the size of collieries and the way that size varied from area to area and from time to time. In a Löschian equilibrium all collieries are equal in size: free entry ensures that those quantities are produced at which costs are minimised. But if costs varied, as noted earlier with respect to cost variations due to physical conditions, so could prices and therefore outputs. Costs could also vary according to the techniques used to mine the coal, and techniques could vary in their capital requirements and in the level of output at which they could be most profitably employed. In short, indivisible economies of scale might be available and this would bring into being what might, for symmetry's sake, be called 'kinked supply curves'. As a non-economist, I suspect that the hypothetical implications of this possibility are a multitude: some of the obvious ones which have implications for this discussion are illustrated graphically on figure 2.

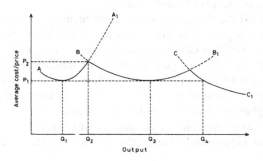

2. Costs, prices and outputs with different techniques.

If the curve A–A$_1$ represents the 'normal' cost curve of the industry, with cost rising steeply as output is increased, Q$_1$ is the output at which cost and price are minimised and would be the size of all collieries in a Löschian equilibrium. Price and output could only be increased beyond P$_1$ and Q$_1$ if the entry of competitors selling at P$_1$ could be blocked in some way. With the existence of oligopolistic competition resulting from a clumping of demand this might be possible, but one can surmise that the extent to which colliery outputs and f.o.b. prices moved up in unison in this way would be relatively insignificant given the nature of the domestic demand schedule. The use of a technique which required larger capital investment produces the cost curve B–B$_1$. This technique would obviously not be used at outputs of less than Q$_3$ except insofar as restrictions on entry into the industry allowed the coalmaster to charge prices between P$_1$ and P$_2$. If this was possible, then the production of outputs greater than Q$_2$ would allow the coalmaster to reduce prices until they were back to P$_1$ at Q$_3$ (or increase his profits). If there was completely free entry all collieries would produce Q$_1$ at P$_1$ in equilibrium using technique A. But, given that Q$_3$ could be sold, the greater gross profits available at this output would prove more attractive than those made on Q$_1$ and there would be an incentive for coalmasters to attempt to restrict the entry of competitors so that Q$_3$ could be produced.

Moreover, Q$_3$ could be an even more popular option if temporal considerations are taken into account. If costs increased more rapidly using the more primitive technique, then the continuance of mining at a site might require the innovation of technique B. A coalmaster whose works had reached this point before that of his rivals, or under the threat of the local opening of fresh collieries using technique A, might still innovate B because he could be confident that competition from the smaller collieries could only be short lived. Given this confidence and his heavy fixed capital investment, he might well be willing to sell smaller quantities than Q$_3$ at less profit or even at a loss for short periods until A's costs increased. Thus, quite naturally, the greater the investment the more stable the larger colliery locations even in a non-equilibrium situation.

A third technique, C, might allow cheaper production than either A or B, but only at outputs greater than or equal to Q$_4$ and only after greater fixed investment. Whether or not C was innovated would depend upon the size of the available market, and very high transport costs, for instance, could completely preclude its use. If a cheap transport route existed, the use of C might be possible on that route but nowhere else, so that the introduction of a cheap transport route could skew the colliery location pattern not only in terms of the spatial distribution of collieries but also in terms of their size range.

Thus, if a variety of productive techniques was available, the options open to coalmasters in their formulation of competitive strategies would become numerous despite the existence of kinked demand curves and oligopoly, and the size of collieries could vary very markedly in such circumstances. But the extent to which the blocking of free entry, the selling at a loss over the short term and the heavy investment near cheap transport routes – all of which produced large and long-lived collieries – occurred obviously depended upon the availability of capital. Conversely, a small size range, free entry and instability, 'natural' products of the demand and haulage cost situation, would only automatically occur in a varied technical environment if capital was scarce.

The availability of capital was, therefore, potentially an extremely important variable in determining the shape of the colliery location pattern and the manner and speed of its response to changing transport technology. But the degree to which capital was available to actual or putative coalmasters is not at all easy to discover. Hence, perhaps, the frequency in most economic analysis of the assumption that it is freely available everywhere. But this side-step is obviously not an acceptable means of escaping from the difficulty in this particular case. The least that is satisfactory – and perhaps the most that can be done – is to try to discover how much capital was necessary and whether the possibility of scale economies did exist; whether any difficulties occurred in the accumulation of sufficient capital at colliery sites to allow their exploitation at the level desired by the entrepreneur, and where capital came from, geographically, socially, and economically, how mobile it was, and therefore the ease with which shortages could be rectified. From this knowledge it might be possible to see if the shortage and immobility of capital had the hypothetically possible effects described above; if it did prevent mining at high cost sites or the continuance of mining at long exploited sites, prevent size variation in propitious technological and market conditions, prevent the kind of competitive behaviour which would encourage the development and stability of large collieries, and prevent full and speedy responses to developments in transport or markets which would otherwise be conducive to the growth of large collieries.

In principle, the absence or high or variable cost of labour could skew the pattern away from a Löschian equilibrium in exactly the same way as the absence or high cost of reserves, sites or capital. The labour requirements of mining must be discovered to ascertain whether or not this was an empirical possibility, and so must the mobility of labour, occupationally and geographically, to see whether or not movement could militate against it. Variations in wage rates, too, may have been important, especially if labour immobility allowed their long-continued existence. To the extent that wages were a cost of production, wage differences over space could skew the cost surface significantly. Moreover, to the extent that labour and capital were substitutable, cheap and abundant labour could compensate for some of the possible effects of capital shortage, or dear labour could intensify them.

Each of these facets of labour – the amount required, mobility, and wages – can all, in principle, be measured (although whether or not data survive to provide these measures is, of course, another matter).

*Entrepreneurial behaviour.* Which of the many possible effects of non-uniform demand and resource costs and characteristics actually exercised the colliery location pattern would to a large extent be determined by the manner in which entrepreneurs responded to their existence. That this particular factor should assume such significance will come as no surprise to those with even a passing familiarity with modern historical geographical writing, geographical work on location theory, or locational economics. But neither the bald assertions which stem from the obvious pragmatism of historical geographers,[28] nor the increasingly complex, abstract and sophisticated theoretical elaborations of the behavioural economists, which seem to collapse the whole problem into one of 'uncertainty',[29] help very much in a search for ways in which this element of the clutch of location determining variables can be analysed in a particular historical instance. The tripartite framework of the 'entrepreneurial subsystem' given on figure 1 is not, of course, definitive or theoretically comprehensive. It is much less sophisticated than that set of components described by Isard.[30] But, on the other hand, it is more realistic than the almost exclusive preoccupation with environmental perception which is what most historical geographers seem to have distilled from the produce of their expeditions to the wells of behavioural science. It can usefully serve as a basis for a discussion of some of the notions about entrepreneurial behaviour which seem to be relevant to the present analysis.

Like all classical location theory, that of Lösch makes assumptions about the nature of the three components which, according to my scheme, determine behaviour. To recapitulate: knowledge must be perfect, all information which is relevant to 'rational' behaviour must be fully and instantaneously perceived; behaviour based on that knowledge must be geared solely towards maximising the producer's profits, and the producer must possess fully the wherewithal necessary to allow him to enter the industry and to maximise profits in it.

What kinds of knowledge are required before a profit-maximising decision can be made? The entrepreneur must know, at least, the disposition of seams and other necessary resources, about markets and, if in imperfect competition, he must know what responses competing coalmasters would make to his beginning mining or changing output or prices. It is patently obvious that none of these pieces of information can be known with any certainty and imperfect knowledge alone could prevent mining occurring at optimum sites and at optimum outputs. However, it is equally obvious that it is impossible to discover how much the eighteenth century entrepreneur did know about these aspects of the local mining economy. Some foray into 'behavioural historical geography' seems necessary, then, in pursuit of explanation, but it is easier to preach than to practise.

Even if, *à la mode*, attention is focused only on entrepreneurial perception of the physical environment, insoluble problems are immediately apparent. It is clear that the regular spatial patterning of collieries in a Löschian equilibrium would occur only over the *known* coalfield, and not over the full extent of the coalfield as it was fixed by the Geological Survey and as it appears on my maps. But how can one discover how much of the coalfield was known at any one time and how this knowledge unfolded during the two centuries following 1590? The only commonly available evidence that people knew of the existence of reserves is evidence that they were mined, but it is not very productive to measure one of the explanatory variables solely in terms of that variable which is to be explained. Similarly with markets: the only way that information about entrepreneurs' knowledge of markets can be obtained is by reference to the markets which they *did* serve. Details of markets assessed and rejected in a search towards an optimum marketing strategy, even if this was ever performed, remained in the experience of the coalmaster or agent and died with them.

All that the entrepreneur knew cannot be discovered, but only some part of what he did on the basis of what he knew. If the object of the analytical exercise was merely to reconstruct a 'perception surface' as an esoteric and independent exercise of the kind which seems to be advocated in much of the behavioural writings of historical geographers, then it would undoubtedly be useful to draw inferences about what entrepreneurs knew from evidence of what they did. But in the present case, knowledge of what they knew is required to explain what they did, and such inferences would just represent unproductive circular reasoning. The best that can be done with the problem of entrepreneurial perception is to assemble those bits and pieces of evidence which might exist independently of the acts of mining and marketing themselves to see whether any conclusions can be drawn about the extent of the knowledge of entrepreneurs. Certainly, whatever is scratched together, it will be a far cry from a definition of the 'perception surfaces' upon which the collieries were, in fact, located.

In a similar way, it is clear that it is analytically and theoretically important to discover whether entrepreneurs were trying to maximise profits and, if they were not, what, if anything, they were attempting to maximise. But it is again equally clear that this is just not the sort of information that is available, even if 'economic rationality' and 'optimisation' are concepts with any real meaning in pre-industrial society. One suspects strongly, of course, that they are not and that any profit was acceptable, particularly to small-time lessee coalmasters; that the continuance of an economic enterprise which had existed for many years, especially one which made for seasonally harmonious labour organisation on an agricultural estate, was reason enough for the larger gentry to mine their coal; that to the largest estate owners of all, who consumed nearly all that their mines produced in their own and their kinsmen's mansions, carting it from pit to hearth with boon labour, the concept of profit was completely irrelevant; and that personal rivalries, hatreds, jealousies and friendships were motives enough for many of the entrepreneurs of the time (indeed, all these possibilities will become demonstrable actualities later). The only conceptual rug beneath which this motley of reasons can be swept is 'satisficing behaviour'. But, as Harvey has trenchantly demonstrated, this is a concept which is devoid of empirical usefulness.[31] The best that can be done is, again, to hold implicitly onto the assumptions of the theory whilst searching for information which might get one nearer to the empirical truth without really helping to solve the analytical, explanatory, puzzle.

The free entry into the industry of competitors is a requirement of Lösch's theory. If the wherewithal to practise mining was not completely available this alone could skew the pattern away from a Löschian equilibrium, the direction being determined by the precise nature of the restrictions. A number of ways in which the 'capacity for effective action' of entrepreneurs may have been rendered less than perfect, even given unsullied perception, are subsumable under a lack of resources and have already been dealt with. But, even given markets and physically available resources, a lack of managerial or technical expertise or institutional constraints could have barred entry into the industry. Duckham has demonstrated the importance of the former in the Scottish coalfields[32] and in principle it is possible to measure whether this was also so in South West Lancashire: whether entrepreneurial expertise varied over the coalfield and whether any such variations were nullified by mobility.

Two institutional constraints on free entry might have exerted an important influence: the requirement of non-freeholders for leases and the system of competition, most particularly entrepreneurial combinations, which might or might not have had their effect through the former agency. The first of these has already been partially dealt with. If only estate owners mined coal then the cost of resources might, in effect, be less, and variations in rents might cause mining costs to vary for leaseholders. Furthermore, estate owners need not pay restitution costs or soughing rights and need not assemble a mining labour force from scratch and solely and specifically for that purpose. In such circumstances, too, insofar as coalmasters ran their own works, the mobility of mining and managerial expertise would be completely dampened.

The kind of competition which prevailed must also have exerted some control over the colliery location pattern. Löschian theory assumes perfect monopolistic competition, with active competition for markets in all parts of the space occupied by the collieries. The concept and the situation it describes are obviously theoretical abstractions, but it is possible to pursue the hares let fly when the assumption is relaxed over a short distance. It is at least hypothetically possible to discover whether any kind of competition over space existed by searching for market area overlaps and price cutting between neighbouring collieries, which could have occurred for few other reasons. The degree to which any spatial competition that existed was imperfect can be sought in two directions. First, it is possible that there was product differentiation, that coals from different collieries were of different qualities. This would not only have the effect that the produce of different collieries might serve markets with different demand characteristics, but also that the produce of one colliery might be able to compete over a larger area than that of other pits at the same price in the same market sector, with obvious repercussions on colliery size variations or pricing policies. Second, the degree to which competition was prevented by the operation of cartels or other price fixing agreements, or agreements between freeholder-coalmasters to limit the supply of reserves to putative competitors, could also have obvious locational repercussions. The spatial distribution, size and duration of collieries could be radically affected if particular individuals or combinations of them owned the monopoly rights to coal over wide areas, or if locational clusters of collieries, brought into being by the geography of demand, led to cartel agreements, which would then tend to crystallise the mining pattern that existed when they were instituted.

*A scheme of enquiry*

The discussion has now moved a long way from a search for a means of 'explaining' a colliery location pattern. The model which might have provided an explanation consistent with the canons of econometrics is patently inoperable unless grossly unrealistic assumptions are made about some of the factors which must have had causal influence. An examination of the components of the mining industry which are defined by Löschian theory, of the degree to which these concepts are empirically recognisable, and of what might happen to the colliery location pattern when the assumptions made about some of them in Lösch's theory are relaxed, has dissolved the rigorous and elegant structure of abstract theoretical certainties into a multitude of hypothetically possible sets of relationships. An operational analysis is not simply a matter of testing these hypotheses, and it would not be even if data were available on all the variables in these relationships. Each set of hypotheses about the ways in which the variables subsumed beneath the terms 'demand', 'resources' and 'entrepreneurship' could have affected the geography of mining is not independent of the remainder. Some of the possible ways in which mining could be affected by particular variables are mutually exclusive and some are contingent upon the operation of other variables in particular ways. What is now required, therefore, is some framework which provides a reasonably systematic progression through this mass of possibilities, in which as many subsequent possibilities are eliminated at each step as possible, and in which as many as possible of the fragments of data which have survived can be integrated.

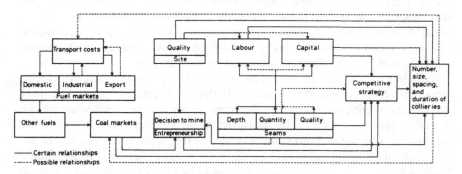

3. Relationships between the variables of a mining system.

Figure 3 is a diagram of the major conclusions drawn in this chapter.[33] The conceptual elements of the theoretical system which can be empirically recognised are expressed as boxes and the links between them as arrows. The solid lines represent necessary or certain relationships, the broken lines possible ones. The reciprocal of the number of arrows feeding into a box provides a rough and ready indication of the degree of independence of that element of the system, although the number of arrows feeding out of a box gives a less sure measure of the importance of that element of the system in influencing the rest. An example might make this mode of expression clearer. The clump of boxes and arrows in the top left-hand corner of the diagram simply state that transport costs definitely influence the size of the domestic and export markets available to any point on the coalfield; that they possibly influence the size of the available industrial market – only possibly because industrial consumers might move their plant to colliery sites; that the size of the industrial and export markets might have influenced the cost of transport (if they were large enough to stimulate the building of a cheap transport route) and that the sheer size of the mining industry itself could have had the same effect.

Reading across the diagram from the top left-hand corner in this way provides a brief summary of this chapter so far and an ordering of the various relationships in the system which provides a framework for their empirical analysis. Given the amount of interaction and the number of possible feedback loops within the system, any examination of its parts is bound to render a replication which is considerably frayed at its edges. Clearly, the best place to begin is with that variable or group of variables which is most independent of the remainder of the mining system in its operation. Here lie the most likely sources of causal influence and stimuli to change. In other words, the examination can best start with those elements which are largely 'given', in the sense of the term as it is used in a geometrical theorem, and which therefore provide the rigid constraints within which the choices of entrepreneurs and consumers must operate. If the diagram is crudely split into demand and supply factors, linked by the entrepreneurs' decisions to mine, then the left-hand, demand, side is seen to be both most independent and least complex. Within that group of elements transport costs clearly exerted an important blanket constraint on market size.

Examination of the stable pattern of colliery distribution which (to pre-empt a later conclusion) existed until the end of the seventeenth century will be begun with an attempt to determine how far the constraints on the size of market areas exerted by transport costs might have been responsible for the size, spacing and duration of collieries that were characteristic of the period. If all other elements of the system existed as in Lösch's theory, then if haulage costs were high close spacing would be expected, and variations in transport costs alone could have caused variations in colliery size and/or density. Then demand will be analysed to see how far it was spatially non-uniform and how far any variations in demand might have accounted for variations in colliery spacing. Some estimate of variations in aggregate market size will be sought by summing the domestic, industrial and export markets available to different parts of the coalfield and subtracting that part of this aggregate which might have been satisfied by other kinds of fuel. In Lösch's system, with all other things equal, demand variation alone could not be responsible for variations in colliery size. But certain kinds of variation in demand might provide conditions in which variations in supply factors could, and in which different expectations about pricing behaviour and colliery duration would occur. This is one of the main reasons why different kinds of market must be considered separately.

The examination of transport costs and markets will be concluded with a summary enumeration of those aspects of the location pattern which cannot completely be attributed to the influence of these factors alone. The discussion of the supply variables will be aimed at explaining these residual characteristics. Because of the degree of interaction and possible substitution between the numerous variables which make up even the simple idealisation of what might be termed the supply subsystem as described on figure 3, any analysis of it must be more untidy than an examination of demand. Again, however, it is most logical to begin with those elements which were physically given determinants of costs and therefore of supply schedules: that is, with the availability of seams and sites. The unavailability of either could, of course, alone cause gaps to appear in the colliery location pattern. But, equally important, variations in the quality of either of these resources could cause spatial variations in mining costs which could, in turn, affect the pattern of colliery size. Differences in the amount of fixed capital available, especially if combined with restricted entry, could also affect colliery size, and variations in the quality of coal would, in certain market conditions, do the same. Changes in costs over time, resulting from limitations in the quantity of coal available at given depths, could, alone or in combination with primarily domestic markets, be an important control on the duration of mining at particular sites.

Spatial variations in the availability and/or cost of labour might have also had some effect in skewing the pattern of colliery location away from that which would be expected on the basis of complete demand orientation. *A priori,* it might be surmised that this is improbable; that wages could not vary as widely as geomorphologically and geologically controlled costs because labour availability would to a large extent reflect demand if the domestic hearth was the main market, and that labour mobility would equalise any marked discrepancies. Nonetheless, labour conditions are of great intrinsic interest and the propositions in the above list must be examined.

Thus far, some notion of the way that costs varied over the coalfield will have been produced from the examination of supply variables. This material, together with that relating to demand, should indicate the ways in which opportunities to profit from mining varied. These sets of information would form the basis for decisions to sink capital into mining. A lack of either entrepreneurial initiative or capital, or spatial variations in their incidence, could prevent the emergence of a location pattern which reflected the culmination of the operation of demand and supply stimuli. Entrepreneurship and capital are almost impossible to separate in this period and they must be dealt with as a single variable. Who provided them and whether different social groups were more active investors in different areas were potentially important controls, in certain demand conditions, on the size of collieries and, most elementally, on their presence or absence.

Finally, an examination of the degree of spatial competition which existed, whether combinations occurred, and whether products were differentiated according to quality or price will be undertaken in an attempt to discover whether institutional factors skewed the pattern away from that which would be expected on the basis of demand, costs and mining agencies acting according to separate individual considerations to sell indentical products.

The examination of the changes which occurred during the late seventeenth and eighteenth centuries will follow the same format. Those variables most independent of the mining industry as a whole, and therefore most amenable to stimulus from outside it, were transport and demand. The physical resource base could not change and independent changes in the labour supply situation would be unlikely to engender growth in the mining industry as would, all other things remaining equal, the independent action of entrepreneurs to inject capital or institute new competitive strategies. Changes on the supply side can most realistically and readily be considered, initially, as responses which were triggered off by changing demand. Observed changes in the colliery location pattern will first be related to changes in transport networks and the geography of markets. The ways in which these changed can be used to sketch the outline of an optimum demand oriented colliery location pattern, and then the differences between this and reality and any temporal lags between them, will be examined in terms of the supply variables.

## Concluding methodological remarks

The causal models which are implicit in these schemes for enquiring into the reasons for stability and change are extremely simple. In no sense can they be described as rigorously explanatory. Considered as causal models, the first is simply an exercise in non-mathematical stepwise regression, the variables being treated roughly in their order of causal importance, with the residuals left after the examination of one variable forming the basis of what is to be explained by the next, and so on, through the sequence from transport to competition. Some concessions will be made to the joint operation of variables by trying to show, through references back at each stage, how a particular variable might have operated in a particular way only because one which was considered earlier operated in another particular way. The references to theory have allowed some attempt to order the variables in terms of their causal importance and to adduce ways in which they might have operated jointly. But the rigour which mathematical expression gives to a true stepwise regression is entirely lost and in causal terms all that can emerge is a tissue of might-have-beens.

The scheme for the analysis of change is even more imperfect as an explanation. The causal model implicit within it is one of a single stimulus (changing transport and/or demand) bringing about a change in the colliery location pattern, this change being modified from that which would occur if other things were equal by the checks on its direction and speed imposed by the supply variables. As a causal model this is extremely naive. It ignores the possibility that changes in supply variables – notably capital supply and in the techniques of mining, and therefore in the costs and the constraints applied by depth, wetness and so on – could occur independently of changes in demand. It also ignores the possibility of an interaction between the supply and demand factors to cause cumulative change. In jargon, the system defined is one of 'simple action', and feedback between its parts cannot be accommodated. If what I have written elsewhere (in one of those methodological flights of fancy which are easily launched when the bonds imposed by the need to consider a particular empirical problem are loosed) is to be taken seriously, then this neglect of the action of feedback around – rather than the progress of a stimulus through – a system means that one of the most powerful causes of change is ignored.[34] However, it is very difficult to see how feedback could be handled in the absence of adequate data and in a system whose operation is more complex than one in which the progression of causal influence moves from variable to variable around a loop, one which can be represented as a series of regression equations and solved in a difference equation. Of course, suggestions about the possibility of such causal behaviour can be made. After the variables have been considered as respondents their possible roles as instigators and reciprocators of change can be speculated upon. But this is hardly a demonstration of the operation or an evaluation of the extent of the more complex causal processes.

The general conclusion towards which this introduction has led is a simple one. Those who lament, from the sidelines, about the lack of firm and definitive causal analysis in historical geography, of explanations of the geographical past and the way that it changed, are crying for the moon. A search through the relevant theory does not provide a means of injecting rigorously derived causal statements into historical geographical analysis, even though colliery location is a relatively simple geographical problem and industrial location has been given continuous attention by theorists for over fifty years. Those who lament the lack of causal rigour in historical geography either have an oversimplified view of what explanation comprises or they have never tried to accomplish it.

But to draw this conclusion is not to call into question the value of theory in historical geographical work. The only alternative destination to a set of incontrovertible causal statements is not a descriptive treatise on the maps yielded up by a particular tax return. The only product of a scheme of enquiry which is theoretically based is not meant to be explanatory certitude. At a less rarefied intellectual level such a scheme should also provide an ordered examination of all those aspects of historical reality which might have impinged upon the way the mining industry developed, a framework over which the historical fabric can be stretched so that it can be displayed coherently and comprehensively. Simply because the fabric's pattern is massively complicated, much of it obliterated by the holes caused by the shortage of data, it is more than ever necessary to use whatever help theory can give to pull it into a firm outline so that the bits of pattern that are left can be recognised and interpolations made between them.

It is, of course, true that this kind of exercise is merely an 'arduous preliminary to historical research',[35] and this particular example has perhaps been made more arduous and protracted than was absolutely necessary in order that the usefulness of theory in the examination of a particular problem of historical geography could be thoroughly examined and displayed, without glossing over difficulties and without taking the short-cuts around some of the more thorny theoretical thickets that intuition alone might have provided. But such 'preliminaries' are important conditioners of the whole of the subsequent empirical analysis. They are not some kind of ritual activity which is performed before getting down to the real business: they define what the business is to be. It cannot, surely, be doubted that the agenda outlined above contains a more varied range of items concerning the geographical patterns and processes of the industrial revolution, and therefore about one aspect of its dynamics, than would the bill of fare presented by historical geography's stereotype?

# PART ONE:
# SEVENTEENTH-CENTURY STABILITY

CHAPTER 2

# THE COLLIERY LOCATION PATTERN, 1590-1689

The information that can be culled from account books, records of law suits, correspondence, title deeds, topographical writings, and so on, does not amount to much. Neither the colliery location patterns of particular years nor the detailed histories of individual collieries can be discovered. The first of these deficiencies can be overcome reasonably effectively by using, as did Nef,[1] decennial units of time and assuming, for the purposes of calculating colliery numbers and outputs, that all collieries recorded in a decade worked continuously through it. The second problem is more difficult to circumvent because the locational descriptions given in the sources of information are often very inexact and because references to collieries are often widely scattered in time. In consequence, it is often impossible to decide with reasonable certainty whether a number of contemporaneous pieces of evidence relate to one or more collieries, or whether successive references through time plot the progress of a single works or not. If rough estimates are to be made of the numbers and outputs of the collieries that worked the coalfield, and if maps of their distribution are to be plotted, decisions have to be made about such uncertainties in order to compile the necessary colliery biographies. This would have to be done anyway, implicitly, if more than an ill-assorted rag-bag of information was to be produced. I think it best to do it explicitly, and so each colliery has been given a reference number under which that information has been tabulated which seems most likely to refer to a single works. The resulting compilation obviously contains the results of some arbitrary decisions as well as conclusions based on the reasonably sound criteria which are outlined in Appendix 2. But it should be substantially accurate within the limits set by the nature of the information. The tables, graphs and maps of this chapter, and of later chapters in which colliery location patterns are dealt with, have been derived from this compilation and from a series of decennial distribution maps based upon it.

## The numbers and distribution of collieries

There were, according to the evidence available, never fewer than eleven collieries at work on the coalfield before 1700 and never more than sixteen, eighteen if collieries working small patches of urban copyhold ground are

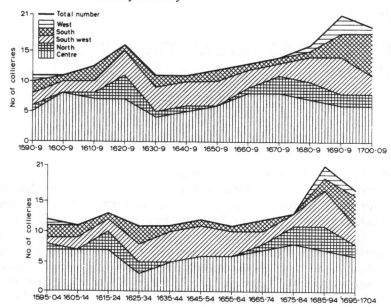

4. Numbers of collieries, 1590-1709. Totals given per decades as listed on the horizontal axes.

counted. The graphs of figure 4 show the numbers of collieries recorded in each decade from 1590 to 1709 in each area of the coalfield, excluding workings on urban copyholds which were recorded in relatively large numbers on one or two occasions. The period was divided into decades beginning in 1590 (top graph) and 1595 (bottom graph) so that any trends in the series which might solely be due to the process of chronological classification can be recognised. The two graphs differ in detail. Whether the century before 1690 was one of gentle oscillations in colliery numbers, or one of slow growth punctuated by a peak in the 1620s, cannot be discovered from the evidence: neither graph is inherently more reliable than the other.[2] But there is no doubt that a significant upsurge occurred in about 1690 to end a period of relative numerical stability and initiate steady growth, the trend of which was interrupted only by a small hiccough just after it began.

The coalfield has been split up into five areas so that subregional trends can be monitored. Their boundaries are somewhat arbitrary, of course, but they do run through those tracts which were most consistently empty of collieries and each area has some geomorphological coherence. The North largely comprises the basin of the Yarrow; the Centre, the basin of the middle Douglas; the South, the basin of the Sankey Brook; the South West, the basin of the Ditton Brook, and the West the mossland area to the west of the 400 foot downfault which created the imposing western flank of the

Billinge–Ashurst ridge. (The geomorphological build of the coalfield is illustrated on figure 12.) Of these areas, the West was almost continuously unexploited and the North and South were usually of only slight importance before 1690. The graphs show clearly that the small South Western area and the much larger Central one were peculiar in their continuous exploitation, with the latter usually containing about half of the total colliery population of the coalfield. The subregional balance only changed markedly when the numerical trend changed, for it was expansion in the South and South West alone which pushed up the coalfield total in the 1690s, after four decades during which numbers in these two areas had declined slowly but steadily.

The continual fluctuations in the numbers of collieries in each area, even before 1690, meant that there was considerable short-term change within the colliery location pattern. This is summarised on figure 5. Each solid point on the grid to the left represents the 'average position' – or centroid[3] – of the collieries which worked in a decade in the series from 1590 to 1690, and the open circles mark the centroids of the collieries working in each decade of the series beginning in 1595. The way that the centroids calculated to a 1590 base shifted around the coalfield is indicated on the diagram. They tracked northwards as the South and South West gradually slumped in importance after the 1640s, then lurched southwards again in the 1690s. But

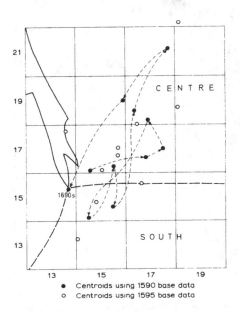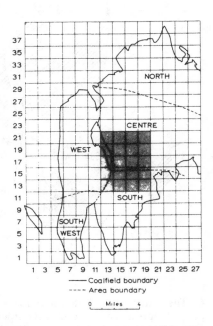

5. Shifts in the colliery centroid, 1590-1689. The area of the left-hand diagram represents the shaded area of the map to the right. The arrowed pecked lines show the successive shifts from decade to decade of the centroids calculated to a 1590 base.

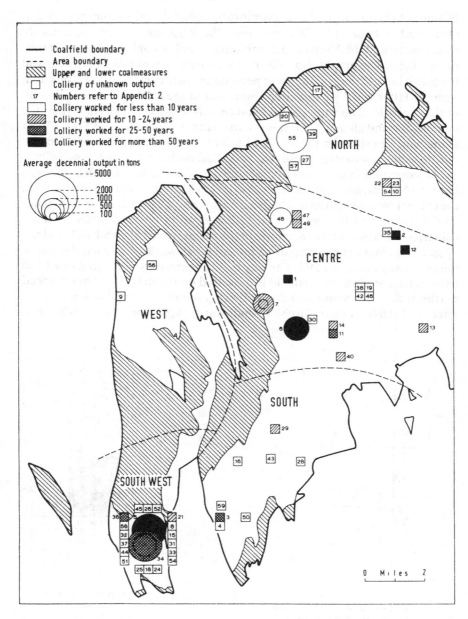

6. The distribution, duration and size of collieries, 1590-1689. The colliery sizes plotted represent average annual outputs in decades in the series beginning in 1590. Two symbols are plotted for a colliery if records of output are available during two such decades.

superimposed upon and preceding this strong directional movement were numerous patternless shifts. These resulted from small changes in the relative importance of the different areas and from some quite marked changes of locational patterns within the areas. Although the subregional apportionment of the total colliery population was reasonably constant before 1690, there was continual rearrangement on a smaller scale.[4]

Figure 6 is an attempt to portray, *inter alia,* this local variability. Only in and around Prescot in the South West (where the large number of workings requires that their distribution is conventionalised) and in the discontinuous ring of collieries which circled Wigan (collieries 19, 38, 42 and 46 were located in the town) was there continuous activity throughout the century. Elsewhere, sites were mined sporadically, often by two collieries working immediately adjacent reserves.[5]

## The duration of collieries

It seems, then, that the industry was characterised by a high entry and exit ratio before 1690, despite the relative stability of the total number of collieries. Of course, it is impossible to tell how far this impression is merely a product of the patchy nature of the source materials. Put baldly, only five of the fifty-six or fifty-seven collieries recorded in the century before 1690 had documented working lives of over fifty years, and seventy per cent of them had recorded durations of less than ten years. Much of this impression must be due to the discontinuous nature of most of the records, but there is evidence that it is not a completely warped reflection of reality.

John Rigby's Shevington colliery (49) opened in 1672 and closed before 1679.[6] Two Whiston collieries (24, 25) opened in 1627 were closed by 1636,[7] and the Gerards' Aspull colliery (12) was opened originally for only one year in 1601 before it was revived with greater success ten years later.[8] Similarly, Prescot Hall colliery (5) experienced a fitful prelude to a long life, working for only about eight years in its initial period of activity.[9] It is unlikely that workings in the small urban plots of Prescot and Wigan (8, 15, 21, 26, 31, 32, 33, 38, 42, 46 and 53) functioned continuously for very long. In 1606 John Goodicar covenanted to sink three pits in three years to get the coal to which he had the right in a Prescot copyhold,[10] and the colliery of Nicholas Pennington in Wigan seems to have operated successfully for only one night in 1653.[11] Peter Plat's pit in Millgate, Wigan, was opened in November, 1619, closed nine months later and re-opened in 1621 for an unknown period.[12] In the 1630s the lords of both these urban manors passed injunctions which forbade or severely limited mining on their demesnes.[13]

This evidence of actual closure can be supplemented by records of litigation, often the agency through which pits were shut. Twenty-two of the fifty-odd collieries that operated during the century had suits filed against them, four more than once.[14] The opening up of neighbouring reserves was common, and so were the apparently inevitable sequels of attacks on workings or owners followed by attempts to get the law to intervene and forbid the continuance of one of the collieries. Owners of neighbouring properties often filed suits simultaneously or in quick succession and some of these disputes dragged on for many years before one of the adversaries withdrew, forced out by the courts or exhausted.

It can be concluded, at least tentatively, that exploitation was short-lived in many parts of the coalfield and that the collieries at Orrell (1), Haigh (2), Winstanley (6), Aspull (12), Pemberton (11), Sutton Heath (3), Prescot (5) and Whiston (34 and 36) which operated for more than twenty-five years were exceptional. All but one of these long-lived collieries were situated in the ring around Wigan or near Prescot.

### The size of collieries

Tonnage accounts and statements of rents received at known royalties are rare for this period. Only eight have been discovered, and even this apparently sound evidence must be treated with some caution because different systems of coal measurement were used in different parts of the coalfield. The basket, the basic unit of measurement throughout the coalfield in this period, varied quite widely in content. Of course, it is not usually known which particular system of measurement was in use at a colliery and this makes standardisation difficult and liable to error. Nonetheless, a reasonably firm pattern of subregional variations in measure contents seems to have existed, and an attempt has been made to equate each of the different measurement practices in Appendix 1. All statements of output have been converted through the ratios suggested there to tons of twenty baskets of 120 lb. The eight surviving accounted outputs are mapped on figure 6 and plotted as solid symbols on figure 7. Averages are depicted if the accounts run for more than one year and two symbols are plotted if the accounts overlap two decades. This slender evidence suggests that production ranged up to 1500 tons a year from minuscule amounts of a few hundred tons. There seems to have been no noticeable upward trend in colliery outputs over the century before 1690 and the works in the South West seem to have reached larger sizes than elsewhere.

These accounted figures can be supplemented by quantitative statements which give a less precise indication of the size of collieries, such as the number of pits open at a colliery or the number of hewers working there. The productive existence of individual pits was ephemeral. New ones were sunk at Winstanley colliery (6) about every three years between 1676 and 1695,[15] and the pits of Whiston colliery (34) shifted over six fields, tapping two different seams in their progress, between 1636 and 1645 or 1646.[16] Only in the Cannel seam of Haigh and Aspull were longer-lived pits recorded. There, the average working lives of pits seem to have been about five years

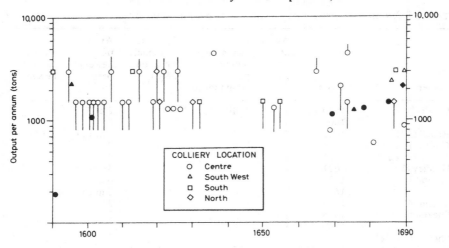

7. Recorded and estimated colliery outputs, 1590-1689. Solid symbols represent recorded outputs, open symbols estimates. Vertical lines show possible ranges of error of estimates.

and one pit at Haigh (2), known, with due deference, as 'the greate Stone low Pitt', was used for about twenty-six years.[17] The usual complement of hewers was three per pit, fewer if demand warranted it,[18] but at the Cannel pits twice this number seems to have been normal.[19]

The recorded relationships between the numbers of pits and hewers and annual output are presented on table 1. This evidence, too, is flimsy, but the relationships don't seem to have changed before about 1750,[20] and it seems reasonably safe to conclude that the norm in ordinary bituminous coal was a colliery comprising one pit, employing three hewers and producing somewhere between 1000 and 1500 tons a year. Sometimes only two hewers were employed in a single-pit colliery[21] and sometimes, too, more than one pit was operated simultaneously.[22] Certainly, when statements of pit numbers and manpower are converted to estimates of annual output, together with other evidence such as accounts for very short periods and rents payments for which royalty rates are not definitely known, the picture presented by the accounts is not radically changed (see figure 7). There is still no indication that collieries grew in size over the century and outputs of less than 1500 tons a year seem to be the norm. However, these estimates contain larger outputs than 2500 tons a year. Each of those which exceed this figure relates to the colliery at Haigh (2), the only one which definitely worked without interruption right through the seventeenth century. Three pits worked the cannel there on at least two occasions and when only two were recorded they employed twelve hewers between them.[23] In 1687 it was ordered that each hewer at Haigh (2) must send up thirty baskets each day of

TABLE 1. *Numbers of hewers and productivity per pit, 1600-87*

| Colliery | Years | Av. no. of pits p.a. | Av. annual output per pit in tons (o) | Coefficient of annual variation of (o) | Av. no. of hewers per pit | Av. output per hewer |
|---|---|---|---|---|---|---|
| Orrell (7) | 1600/1 | 1 | 1040 | | 2 or 3 | |
| | 1605 | 1 | | | 3 | |
| Aspull (12) | 1622-4 | 2 | | | 6 in 1 pit | 200* |
| Goose Green (11) | 1624 | 1 | | | 2 | |
| Goose Green (14) | 1624 | 1 | | | 2 | |
| Hindley (13) | 1624 | 1 | | | 3 | |
| Haigh (2) | 1636 | 2 | | | 6 | |
| | 1645 | 2 | | | 5 in 1 pit | |
| Wigan (42) | 1653 | 1 | | | 2 | |
| Shevington (47) | 1666 | 1 | | | 3 | |
| | 1669 | 1 | 1100 | | | |
| Shevington (48) | 1672 | 1 | | | 3 | |
| Winstanley (6) | 1676-95 | 1.05 | 1334 | 18% | | |
| Eccleston (52) | 1686 | | | | 2 in 1 pit | |
| Haigh (2) | 1687 | | | | | 390† |

* Scowcroft pit employed six hewers on average, and produced, on average, 1000 wain-loads per year.
† Specified in the colliery orders, but not definitely achieved.

a five day week and this,[24] like the pit and manpower information, indicates that the annual output of Haigh (2) was nearer to 4500 tons than to 2500. Estimates from these rough and ready indexes of colliery size also suggest that the urban works were small as well as transient. Nicholas Pennington's colliery at Wigan employed two casual hewers in its one pit in 1653,[25] and Ralph Fletcher's pit in a Prescot copyhold probably produced less than 300 tons in 1583.[26] Peter Plat's Wigan colliery had only one working pit in the 1620s,[27] and if the five shillings fine imposed on one of the six Prescot copyholders against whose mining Henry Ogle, the lessee of Prescot Hall, complained in 1636 was commensurate with the magnitude of their transgression, then their production, too, must have been small, despite Ogle's assertion to the contrary.[28]

Output varied from week to week and from year to year at particular collieries as well as from colliery to colliery. Output series are very scarce, but there are enough to give a glimpse of this short-term variability. Annual production at Winstanley colliery (6), 1674-95, ranged between a minimum of 649 tons and a maximum of 2352 tons and the coefficient of annual variation was just over thirty per cent.[29] At Aspull colliery (12) in the 1620s, 'some times as much is gotten in one yeare as is in other times gotten in two',[30] and the usual rental income from a Whiston colliery (24) between 1636 and 1646 (at the constant royalty of twenty-five per cent of output) was c.£80, although £140 was received in one year and, generally, output was higher in the six years before 1642 than afterwards.[31] There were also marked variations from week to week at Winstanley (6) between 1676 and 1695. In 1678, which was not an abnormal year, weekly production ranged from two to fifty-four tons and the coefficient of weekly variation was forty-three per cent. In some years there were long periods of complete idleness, as in 1690 and in 1683, when a stoppage of at least three months entailed a loss of about half the normal annual output.[32] At Aspull in the 1620s, £40 worth of coal was raised in some weeks, £30 worth in others, and nothing in excess of wages in yet others. Complete closure for periods of up to three months was not uncommon.[33] It is difficult to determine whether a regular seasonal pattern was usually present in these fluctuations. There was certainly none at Winstanley (6) between 1676 and 1695, when between thirty-five and fifty-seven per cent of the annual output was raised during the winter months of October to March inclusive. Only in 1679 was a significantly small proportion of the year's get – twenty-five per cent – raised in winter, and this was clearly exceptional.[34] On the other hand, the vicar of Prescot reported in the middle of the seventeenth century that the people of the 'Prescot side' of his parish were 'very poore, a greate parte of them liveing in summer tyme by digginge and windinge coals, and in the winter by beggeinge'.[35] There are inconclusive suggestions that the same seasonal pattern prevailed at Shevington colliery (48) in 1675 and at Aspull (12) between 1622 and 1643.[36]

These various strands of evidence point in the same direction: the coal industry of South West Lancashire was in an economically primitive state throughout the century before 1690. A large proportion of the coalfield in the West was almost completely unexploited. Elsewhere, mining was usually intermittent, and when collieries were worked their outputs fluctuated randomly from week to week, from month to month and from year to year. In only two areas was this sporadic small-scale pattern not characteristic of the industry. In and around Prescot in the South West and in the ring of collieries around Wigan in the Centre there was continuous mining and some of the workings there raised over the 'norm' of 1500 tons a year and stayed open for periods upwards of twenty-five years. Even in these areas, colliery outputs fluctuated through time and openings and closures were common. Nonetheless, the number of working collieries remained roughly constant in these two areas and, on the reasonable assumption that short-term fluctuations in output were out of phase at different collieries, output probably remained roughly stable in each area, perhaps falling slightly in the South West and perhaps rising slightly in the Centre over the century as a whole. Altogether, about 13,650 tons were raised from the coalfield in the 1590s, and about 21,850 tons a year in the 1690s, far smaller and more slowly increasing quantities than Nef estimated for this coalfield.[37] But Nef's conclusion that this was one of the least productive of British coalfields in the seventeenth century is probably correct. Although collieries as small as the typical South West Lancashire workings, and with similarly intermittent or fluctuating outputs, were common on other coalfields,[38] much larger collieries were recorded elsewhere. Haigh (2) was dwarfed by collieries in Warwickshire,[39] Nottinghamshire,[40] Cumberland,[41] Northumberland and Durham[42] and even Somersetshire[43] in the sixteenth and seventeenth centuries.

# CHAPTER 3

# FACTORS AFFECTING THE LOCATION PATTERN OF COLLIERIES, 1590-1689

## Demand

### Transport

The coalfield is land-locked and penetrated by no naturally navigable rivers, so that road haulage was the only means of moving coal until the Douglas was opened up for navigation in c.1742. The cost of coal rose very steeply away from the pithead. Nef suggested that it doubled at just over two miles;[1] in Lancashire, where pithead prices seem to have been low, it required a haul of nearer four miles to double the cost.[2] Of course, the exact rate of increase would depend upon the kind of vehicle in which the coal was carried. Occasionally horses and panniers were used,[3] but the majority of contemporary references are to carts or wains, in which haulage was about thirty per cent cheaper.[4] These vehicles must have weighed at least 10 cwt when laden, and the Lancashire roads, built mainly over soggy boulder clay, do not seem to have been well suited to their accommodation. Visitors ranging in temperament and purpose from Cromwell to Fiennes complained about the state of the roads on the coalfield, and the inhabitants, too, were well aware of their deficiencies:

> Our wayes are gulphes of durte and mire, which none
> Scarce ever pass in summer without moane.[5]

It was easiest to carry coal over such roads, whether in panniers or wagons, in the summer months when they were firmest, and the domestic consumer probably purchased most of his coal then, storing supplies in readiness for winter. Between 1583 and 1618, the Shuttleworth family made twenty-nine of their forty-three coal purchases in the months between April and September, inclusive, and four of the five occasions on which wains were definitely used fell into this season.[6] The Eltonheads were to take the coal they were allowed from Sutton Heath colliery (3) within sixteen days of July 25,[7] and the coal supplies of Tarbock Hall were all bought in July and August between 1652 and 1654 and four purchases out of five were made in April and May between 1659 and 1661.[8]

There must, then, have been a strongly seasonal pattern to demand. This explains, to some extent, the seasonality of mining which occurred in at least the Prescot area. But it is not a complete explanation. Coal extraction was not seasonal at Winstanley colliery (6), but coal sales were, extra workers being hired 'in sail time' to clear the banks of coal that had been accumulated through the winter.[9] Clearly, seasonality of output was not a simple function of a seasonal demand created by bad road conditions. Even so, one would expect that the spatial constraints, if not the temporal ones, imposed by expensive and difficult haulage would be strong and that mining would be very tightly linked to the location of its markets.

## The distribution of coal markets

*Households.* Nef concluded that 'it was undoubtedly the use of coal in domestic heating and cooking that occasioned the chief demand for it',[10] and Files claimed that only domestic coal was sold in Lancashire before the eighteenth century.[11] Households on the coalfield seem to have taken advantage of their location and consumed coal with relative prodigality. Nef postulated that the consumption of coal per head of the population ranged from about 9 to about 16 cwt a year, which is equal to between 45 and 80 cwt per household.[12] The annual allowances made to demesne tenants from the colliery at Haigh (2) varied between about 100 and 260 cwt.[13] The most common sum expended on coal at Winstanley (6) was ten shillings, which would have purchased 80 cwt, and the annual allowance given to a neighbour by the owner of Winstanley (6) in recompense for trespass was eleven loads, or about 88 cwt.[14] The gentry and nobility consumed considerably greater quantities in their halls and mansions. The Shuttleworths of Smithils bought 30.5 tons in 1597[15] and the Bankeses of Winstanley took 47 tons 3 cwt from their pit bank (6) 'for Hall use' in 1679.[16] The grander household of Viscount Molyneux consumed 60 tons of coal in 1653[17] and the capacity of the seventy-two hearths of the Earl of Derby at Knowsley must have been vastly greater than that.[18]

Population figures are notoriously difficult to obtain for this period, but all the indications are that South West Lancashire was relatively thickly peopled. The county as a whole was the fourth most densely populated in Britain by 1700[19] and it is likely that the portion south of the Ribble, where the coalfield lies, was already more heavily settled than that to the north. This market was not uniformly distributed. There were six towns in the region, none of them very large on the evidence collected by the surveyors for the Hearth Tax of 1664 (which is the only year in which those exempt from the tax are commonly enumerated in Lancashire).[20] A summary of the information on the towns given in the 1664 schedule is presented in table 2 and the density of taxed households for all areas of the region in 1673, the only complete schedule for the county, is mapped on figure 8.[21] It is clear that the high overall density of population was not produced by the size of the towns. Wigan, the largest, was one of Gregory King's great towns, but it contained less than one-third as many hearths as such places as Worcester,

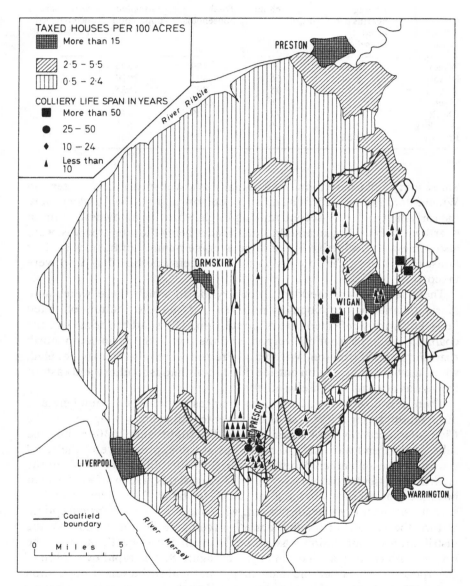

8. Population in 1673 and the distribution of collieries, 1590-1689. Numbers of houses per township were derived from LRO, Hearth Tax Returns, 1673 and township acreages from *VCH*.

TABLE 2. *Urban populations in 1664*

| Town | Average hearths per taxed household | % households taxed | Total households | Estimated population | % increase in taxed hearths, 1664-73 |
|---|---|---|---|---|---|
| Wigan | 2.54 | 59 | 459 | 2295 | 6 |
| Preston | 2.74 | 66 | 402 | 2010 | 11 |
| Warrington | 2.10 | 75 | 327 | 1635 | 18 |
| Liverpool | 2.12 | 78 | 289 | 1445 | 11 |
| Chorley | 1.78 | 71 | 149 | 745 | 5 |
| Ormskirk | 2.12 | 96 | 108 | 540 | 14 |
| Prescot | 2.56 | 90 | 70 | 350 | 11 |

Shrewsbury and Chester, one-half as many as towns such as Gloucester and Winchester, and two-thirds as many as Ludlow and Warwick, which were the forty-first and forty-second ranking English provincial towns in terms of hearth numbers.[22] Wigan and Preston, besides being the largest towns, were also somewhat distinctive in the large average number of hearths in their households and in the low proportions of their householders who were taxed.

These small towns studded a countryside in which there were quite marked variations in population density, if the distribution of taxed households truly reflected that of population. The southern fringes of the coalfield, the area around Liverpool and the line of the main south–north road through Warrington, Wigan and Preston were most thickly peopled, whilst the bulk of the region outside this discontinuous southern and eastern fringe was not half so densely settled.

There must, of course, have been changes in population numbers and distribution during the course of the seventeenth century. Plagues and the Civil War certainly took their toll in the 1630s and 1640s,[23] but the subsequent trend was likely to be an upward one, precursive of the rapid growth of the seventeenth century. Certainly, the difference between the Hearth Tax assessments of 1664 and 1673 suggest this, if such differences can be used to suggest anything at all (see table 2). Given that the seventeenth-century growth did not, with one most significant exception, disturb the pattern of relative densities (compare figures 8 and 44), the distribution of population was probably much the same throughout the seventeenth century as that portrayed on figure 8. The rapid expansion of Liverpool, which caused the main change in the distribution of population after 1690 had not, on the evidence of the Hearth Tax figures, begun before 1673, but in 1680 the port was 'a large, handsome, well built and increasing, or thriving town', and in 1690 it was 'more than twice as big as it was twenty years before that'.[24] Even in 1690, however, Liverpool's population cannot have exceeded 3000 by very much, although it had by then probably outstripped Wigan, where opulent new houses were also being built by the end of the period,[25] as the major concentration of domestic hearths in the region.

The aggregate domestic market for fuel was thus, despite the smallness of the towns, quite a large one, and one which probably increased steadily after c.1650. On the coalfield alone, the fuel capacity of household hearths was probably well in excess of the output of the collieries.[26] And there is no reason to suppose that any of the domestic hearths of the plains to the north, west and south of the coalmeasures were outside a reasonable haulage distance. The Shuttleworths, as figure 9 shows, carted coal over distances which were greater than that between the coalfield and the furthest point of the coast, and in the midlands hauls over the twenty-six miles between

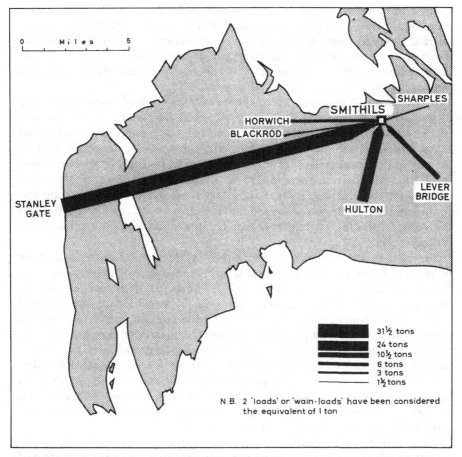

9. Origins of coal bought by the Shuttleworths at Smithils, 1588-99. Data abstracted from Harland, 'Shuttleworth accounts'.

Wednesbury and Coventry were not uncommon despite the expense of cartage.[27] This being so, the households of the region could, potentially, have offered a market of about four times the capacity of its collieries. There was a discrepancy between the domestic market and coal production in spatial terms, as well as in capacity. It is quite clear from figure 8 that the areas of heaviest and most continuous exploitation did not correspond exactly with those of densest population. Thus, the mining industry did not simply and directly reflect the size and distribution of the domestic fuel market.

*Industries.* Coal undoubtedly found its way into industrial markets in the region right from the outset of the seventeenth century. In 1601 Thomas Gerard claimed that cannel from his Aspull colliery (12) 'was necessary for fyringe and fuele & w[th]out the w.[ch] diverse tradesmen & other handycrafte men dwellinge . . . nere unto yo[r]. saide subiects myne & other psons. in the saide Countye of Lanc. & elsewhere cannot convenyently use their trades & handycrafts'.[28] In the following year, the owner of an Orrell colliery (7) made a similar claim: without her coal 'divse. Artifycers and other trades men cannot use their trades & occupacions'.[29]

It is difficult to discover which these trades were and where they were located. In the coalfield manor of Up Holland eleven kilns and five blacksmiths augmented domestic demand in 1653[30] and lime burning and blacksmithying must have topped up the demands of domestic hearths throughout the region. But these activities served severely constricted local markets, and their distribution must have reflected that of the rural population they served, so that these activities cannot have warped the spatial pattern of demand away from that shown by figure 8. A comparison of figure 8 with the distribution of blacksmiths shown on figure 10 supports this assertion.[31]

Just as there were no large towns in the region, so there were no deposits of raw materials which could form a large concentrated market for coal. South West Lancashire contained negligible reserves of iron ore, no copper, tin or salt, and the inland location of the coalfield precluded the use of its fuel in the evaporation of brine. Nor did the region contain, in the seventeenth century, significant alum, sugar, soap, glass, copperas or starch producing industries, which were, nationally, the remaining large industrial consumers of coal. But an industrial demand which was small in comparison with that of other coalfields could be significant for the small-scale mining industry of South West Lancashire, and small concentrations of coal-using tradesmen did exist in various parts of the coalfield, although their exact significance as part of the coal market is difficult to determine.

Figure 10 shows the distribution of those craftsmen who required fuel and who left wills. The evidence is undoubtedly grossly under-representative of the numbers of craftsmen in any particular trade, but there is no reason to

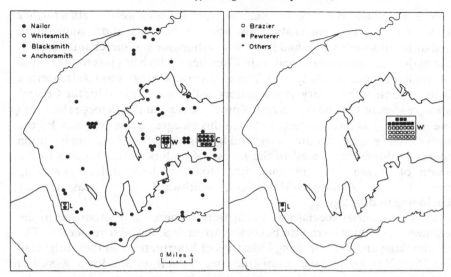

10. Distribution of coal-using tradesmen who left wills, 1590-1705. Tradesmen who died between 1690 and 1705 are included to represent those working before 1690 who died subsequently. Based on 'Indexes to wills'.

suppose that it does not provide a sample which reflects proportions and distribution patterns fairly. Nailors were second in number to blacksmiths. They flourished despite the absence of local iron, their smithies fed by metal brought in from as far afield as Staffordshire in 1590. By 1637 the iron supplies of the region were organised by Warrington ironmongers, who drew on the Cheshire furnaces, and in the 1650s and 1660s South Yorkshire iron was imported into Wigan.[32] The small deposits of iron ore at Haigh do not seem to have been worked at that time, although Sir Roger Bradshaigh of Haigh sold 663 lb of iron to a local smith in 1664 and 1665 and Spencer and Co. opened a slitting mill on the swiftly flowing Douglas in Haigh at about the same time. A forge was added later and by 1673, when they were abandoned, these works supplied large quantities of iron to nailors in the hamlets of Chowbent, Standish, Blackrod, Hindley, Billinge and Winstanley.[33]

Their market was evidently quite large, if scattered (see figure 10), for the nailing industry was producing in excess of local requirements before 1613.[34] The inventory of Nicholas Withington of Chowbent,[35] taken in 1679, illustrates the size to which the nailing industry had grown by the end of the period and the complexity of its organisation. Withington was owed £111. 6. 9 in 'book debts' by twelve persons, two from Wales and Liverpool, and at least fifteen workmen owed him sums ranging up to £6. 10. 0. His 'Goods in the shopp & Smithy Iron Nayles & work tooles' were valued at

£69. 5. 3, and goods in Chester, Skipton and Liverpool were worth a further £64. 14. 4. This is a remarkable illustration of the range and volume of the activitites of a nailor who had become a putter-out and merchant, based in the major concentration of nailors in Chowbent, which lies just to the east of the region defined for study here. The area around Wigan was less important in the seventeenth century. A few nailors' wills have survived for the Central area (see figure 10) and the record of purchasers from Haigh forge show that the industry was more widespread than this meagre source suggests. By the turn of the century, 'hammers worked by hand and foot were ringing upon scores of domestic stithies' in the homes of nailors in the hamlets to the south of Wigan[36] and by that time, too, the lock and hingemaking specialism of Ashton-in-Makerfield, southward again, was already beginning to develop.

Other industrial specialisms which were to become pronounced in the eighteenth century were also beginning to emerge in the seventeenth. The glassmaking and wire-drawing industries of Warrington had their origins in the 1640s[37] and so had the pottery industry of the Rainford–Prescot–Sutton district, which was already exporting its crude wares through Liverpool by 1667.[38] The port itself was not yet a manufacturing centre of any importance, but anchorsmiths and blacksmiths were relatively numerous and the erection of a sugar refinery in the 1660s marked the beginnings of the import processing industries.[39] An interesting development which was not precursive of later expansion was the growth at Haigh of industries besides iron processing attracted there by its coal. During the second half of the seventeenth century John Dwight, who later invented a porcelain-making process at Fulham, conducted experiments at a Haigh pottery, and at about the same time John Blackburne, who was not a local man either, opened 'glass-houses' next to the Bradshaigh's colliery (2), from which he obtained all his fuel at the concessionary rate of 1d per basket, half its normal price.[40]

The ancient borough of Wigan, to the south west of Haigh, was by far the most important manufacturing centre of the region, an interesting contrast to Preston, where manufacturing was of little importance and social and service functions were dominant.[41] Its manufactures were varied. Like all towns of its size, it contained congeries of victuallers and clothing-makers and by the 1670s there were Companies of websters, tailors, shoemakers, smiths and butchers.[42] As early as 1630 there were four water mills in the manor and the power of the Douglas had been supplemented by two horse mills and a 'hand mill' for fulling cloth and grinding flour and malt.[43] But the foundries were the most important industries of the borough: indeed, the location of industries serving wide areas of England and Wales with bells, brass and copper domestic utensils and pewter in such a small and remote town as Wigan in the seventeenth century is a remarkable testimony to the complexity of the industrial geography of 'pre-industrial' England. A very large proportion of the church towers in the tract west of the Pennines from Cumberland in the north to Merioneth and Cheshire in the south contained bells cast in the Wigan foundry of the Scotts.[44] The existence of four separate Companies of non-ferrous metalworkers by the 1670s illustrates their numerical importance and the degree of specialisation which had by then occurred. There were two braziers' gilds in the early 1660s, the 'braziers and potters' and the 'braziers and panmakers'. Those who founded the heavy items such as bells and pewter moulds probably belonged to the former and the makers of household utensils to the latter. Whether the 'potters' whose excavations on the waste for clay caused considerable opposition[45] actually made earthenware, or whether their 'potting' was limited to the making of clay moulds into which their brass and gunmetal castings were poured, or to the casting of brass pots, is not known. In 1670 the braziers' Companies were named as the 'braziers and panmakers' and the 'braziers and founders' and in 1672 they multiplied into three, the 'founders' becoming separate from the 'braziers or pottmakers'.[46] The distinctions, perhaps, were that the founders cast heavy goods, the potters lighter wares and the panmakers made utensils that were composed of a number of soldered pieces and wrought by battery work. Whatever the case, some of these craftsmen, besides the bell-founders, obviously operated on a large scale notwithstanding their medieval style of organisation: the stock in trade, raw materials, semi-finished goods and equipment of William Marsden, a brazier, were valued at £129. 4. 8 in 1673.[47]

Despite the numerousness of the braziers' Companies they were probably not so important as the pewterers, for whom they provided casting moulds. By the fourth quarter of the seventeenth century Wigan was second only to London as a pewter manufacturing centre and its workshops served a market which stretched across the Pennines and down into the south midlands.[48] Two-hundred-and-fifty pewterers are mentioned in the borough records of the seventeenth century and in 1683 it was claimed that there were between seventy and eighty at work in the town. This number might well have contained apprentices and journeymen, as well as some craftsmen whose interests were in brass as well as pewter, and it can hardly have justified the claim that the town consisted 'chiefly of pewterers'.[49] Nonetheless, the pewterers did dominate the higher offices of administration between 1660 and 1690,[50] and the pewterers and braziers together accounted for nearly a quarter of the 419 separate persons whose occupations were recorded in the town records between 1660 and 1679. It was probably their numerous industrial fires which gave Wigan taxpayers such a high average number of hearths per household.

There must, therefore, have been a considerable demand for fuel in the industrial hearths and foundries of Wigan and the hamlets around it. Whether or not they burned coal is impossible to prove. Fuel stocks appear in only one of the inventories of braziers and pewterers, that of a panmaker who had about twelve tons of cannel in 1632.[51] Barker and Hatcher have suggested that coal was 'almost certainly used' before the seventeenth century in the manufacture of pewter,[52] and if this was so then this trade was in advance of brass-founding in this respect, where coal only seems to have become important during the course of the seventeenth century.[53] It is possible that this transition occurred sooner in Wigan that elsewhere and that it was in the peculiarities of two local coals that the town's early importance as a non-ferrous metalworking centre rested. The major disadvantages of a coal fire to the metalworker were the smoke and gases given off in combustion, particularly sulphur, which combined with the metal at high temperatures to give a brittle finished product, and the low temperature of a coal fire compared with one made from charcoal. But the cannel and smith's coals were largely free from these disabilities. Cannel, which was found in exploitable quantities only in Haigh and Aspull to the north east of Wigan, has a very high calorific value and a low sulphur content, and the nomenclature of the Smith seam – and, in the eighteenth century, its widespread reputation and the price of coal from it – indicate its particular suitability for metalworking. This seam comes to surface to the west and south west of Wigan, and it might be significant that one of the colliery owners who claimed that their fuel was necessary for industrial craftsmen must have mined this seam, the highest in that part of Orrell where the colliery was located. The other who made this claim worked the Cannel seam in Aspull and it is suggestive that the panmaker's fuel stocks were specified as this kind of coal. The evidence is not of course conclusive, but the metalworkers of Wigan and the Central area thrived in a place devoid of all metalliferous raw materials and remote from markets, but near to supplies of special fuels whose peculiarities were already appreciated. That it was the fuels which caused them to flourish so early and so massively is a logical, if not an incontrovertible, inference. But, whatever the direction of causal influence in the process of growth, it is clear that the Central area contained a significant market for industrial coal, augmenting that of the domestic hearths of Wigan and the thickly settled rural areas around it.

*Exports.* Before 1590, thirty years had passed during which embargoes against coal exports and fines on road-breaking coal carts were imposed in Liverpool.[54] These civic enactments were either ineffectual or waived and exports of 896 tons passed through the port in 1592. By 1605, outward shipments from the port of Chester had reached 1328 tons, but the proportions in which this quantity was split between Liverpool and Mostyn are unknown.[55] The Corporation continued to pass edicts to regulate the coal trade, but contraventions were immediate and by 1617 Liverpool was exporting 3759 tons to Ireland.[56] At this time, Liverpool probably shared the Irish market equally with Flintshire and Cumberland and its exports continued to rise, providing a lucrative source of income to those who had rights of way to the waterside and banking space there.[57] A peak of 4370 tons was reached in 1664/5, but the trend then turned and exports fell to a mere 194 tons by 1680/1.[58]

One of the reasons why the Irish market was lost to Mostyn, which shipped 12,908 tons in 1680/1, and Whitehaven, which shipped 11,348 tons,[59] was the growth of demand in Liverpool itself. The edicts against coal exports might well have been aimed primarily at preventing coal which had entered the town from leaving it, and in 1698 the Corporation ordered the expenditure of 'a sume not exceeding tenne pounds . . . at the towns charge this summer for an experiment to bore for Coales on the Comon'.[60] Because exports plummeted as the town began to grow rapidly, Liverpool's increased population did not necessarily mean an expansion in the demand for coal in the South West in the latter part of the period, but on the other hand, of course, the exports of the years before c.1670 meant that demand was much higher there than the size of the population and the lack of industry would suggest.

Liverpool was not the only point of export for Lancashire coal. In 1661 the inhabitants of Tarbock petitioned in the Quarter Sessions Court that 'Our said Towne of Torbocke lyinge betwixt Whiston where many Coales are gotten and Halebank, & Ditton whereunto many coales are led and layed the Leanes in O$^r$. sd. Towne in many places are soe Mizzie, Spungie, and unsound . . . beinge so Extreordinarilly beaten outt, & impassable.'[61] In 1646, Richard Case, who owned coal in Whiston (25 and 37), had three debtors in Halebank and in 1668 Henry Ashton (36) owned over 1447 tons of coal lying banked at Dungeon, a point on the Mersey in Halebank.[62] It may be that Dungeon Point and other places on the estuary were used as alternative shipping places to Liverpool for coal that was outward bound because of the tolls on coal carts and wharfage costs in the port. Alternatively, the coal in Halebank and Ditton may have been bound up-river for Warrington, or for the Weaver and thence to the Cheshire salt pans. 'Pitcoals' had become the usual fuel in brine evaporation by the late 1660s and Lancashire coal found its way into the Cheshire saltfield in the 1690s, although it was reported in 1669 that Staffordshire was the source of the coal used there.[63] Wherever they were bound, exports of coal from points on the upper reaches of the Mersey estuary supplemented demand in the South Western area of the coalfield in the second half of the seventeenth century.

*Alternative sources of fuel*

Many of the markets described above, and particularly the rural and urban domestic markets, could have been satisfied by other fuels than coal. Even though Lancashire coal was cheap, price was not the only criterion used by a customer for household fuel. Coal burned noisomely: its smoke 'annoyeth all thinges neere it, as fine Lynen, and mens handes that warme themselves by it' and complaints were made against its combustion in London as late as 1661.[64] In Britain generally, coal came into use virtually of necessity as the shortage of timber became acute.[65] The acreage of woodland on the Bankes' estate in Winstanley, at least, was small and there was a complaint about 'excessive gathering of wood', some of which was for fuel in Warrington, in Haydock in 1630.[66]

In Britain generally, firewood was the most effective competitor with coal – indeed, the only substitute. But in South West Lancashire there were extensive peat deposits and this coalfield was unusual, if not unique, in the existence of large reserves of another fuel than coal.[67] As late as 1784 (see figure 12), after wholesale drainage and clearance, peat mosses covered large areas of low-lying ground on and around the coalfield, and rights of turbary were normally held by manorial tenants as part of their customary privileges. For all areas of the coalfield but the North, considerable evidence of peat digging has survived. Much of the West was blanketed with mossland. In 1635 there were ten turf rooms on Scarth Moss, rented from the Earl of Derby by persons from Ormskirk, Aughton and Lathom, and it was claimed then that people from Ormskirk had 'been used and accustomed to get Sodds, Clodds and Ridgeing Turfe' there since the reign of Elizabeth I.[68] Just to the south in Bickerstaffe, turf digging had become important at least by the end of the period, when the Stanleys of Aughton held a right of way for 'Leading fetching or Carrying of Turves of and from the said Mosse'.[69] But it was neighbouring Rainford that provided most peat in the West. An acre of turbary was part of the rights attached to leasehold properties in 1642 and 1687, and in the latter year the manor court of Rainford ruled that a penalty of 2d or four hours in the stocks be imposed upon anyone caught stealing a 'burden' of turf, surely an indication of the importance of the peat of the township as a source of fuel.[70]

The tenants of Sutton, in the South, had the right to coal from their lord of the manor's pit, but they continued to exercise their right to dig turf as they had done for 'time out of mind without hurt'. In 1620 large stacks of turf and tanning bark at the doors of their homes were 'pulled down and destroyed' by Richard Bold, lord of Sutton manor, who did £200 worth of damage.[71] Just to the east, the heavy penalty of 10d was fixed in 1596 for persons from Golborne, Newton and Lowton who got turf on Golborne Moss without 'lawful cause', and the fine was levied in 1596 and 1602.[72] Heaviest exploitation seems to have occurred on Amberswood Moss, to the south east of Wigan. Tenants of the Gerards and Inces, joint lords of the manor of Ince, dug fuel in the turbaries of Amberswood Moss as early as 1587 and tenants of adjacent Hindley began to exploit the moss in at least 1636. Turf rooms were also let out commercially between 1586 and 1605, and it was doubtless this heavy use by tenants from two townships and turbary lessees which caused the drainage problems of 1605, when several turf rooms were flooded out. Nonetheless, this did not curtail exploitation, and 'severall mosse roomes' were still leased out for annual rents in 1639.[73]

Turf was brought into Ormskirk from Scarth Moss, but it was dug within Liverpool itself. In 1667 Edward Moore advised his son, 'when you get this watercourse opened it will make the turf room so dry that you may sell fifty pounds worth at least to the town in a year, for of my knowledge you have good black turf at least 4 yds. deep; if so it may be worth two hundred pounds an acre if you have ten acres of it'.[74] Moreover, the Corporation edicts against the export of coal applied equally to turf, which was obviously considered to be an important palliative for the port's fuel shortage.

### Coal markets and the colliery location pattern

How the various elements described above were compounded into a pattern of demand is, of course, difficult to discern. It is clear that the domestic market was the main consumer over most of the coalfield. Population was relatively thickly spread and there was a number of small towns, but these domestic hearths did not automatically provide a market for coal. Peat usurped a part of it, burned even by people in the countryside who did not have rights of turbary as manorial tenants and by town dwellers. But turf was not burned in preference to coal wherever it was available. The inventories of people who lived in the heart of the western mosslands contain valuations of coal as well as turf[75] and the tenants of a number of coalfield manors possessed the right to coal as well as peat, and they presumably exercised both rights to capacity. The Shuttleworths of Smithils burned both turf and coal hauled over long distances (see figure 9) between 1583 and 1605, and the parishioners of Sefton, who lived close to sources of peat and ten miles from the coalfield, complained in 1662 that coal was 'very scarce and extraordinary chargeable in fetching or Carryeing unto there'. But they obviously burned some coal, and their predilection for it was demonstrated when they agreed that four or more people from each quarter of the parish should jointly bear the expense of paying three miners hired to prospect for coal.[76]

It seems, then, that people burned both turf and coal in the same domestic hearth, presumably satisfying their needs with whatever was available at least trouble and cost when they arose. Thus, the growth of population did not necessarily increase the market for coal and if the lord of a manor, or anyone else, closed a colliery it did not necessarily mean that a profitable local coal market existed for a subsequent adventurer in mining. This was just the kind of market which would be conducive to short-lived ventures mining areas sporadically in space and time. The locational flux which was observed over most of the coalfield does, therefore, seem to have been related to a market of scattered domestic hearths which had other kinds of fuel available besides coal. It was only in areas where this domestic market was supplemented by industry or exports that continuous mining through relatively large collieries occurred. The thick cluster of workings in the South West served Ireland, Cheshire saltboilers and the growing town of Liverpool during the course of the century; the more numerous but more scattered collieries of the Centre were located amongst domestic metalworkers and around the important brass and pewter foundries of Wigan. These two areas were probably responsible for seventy per cent of the total output of the coalfield. This is not, of course, to argue that exports and industry consumed this large proportion of the coalfield's output. If permanent exploitation occurred in an area in response to industrial or export demands, then the collieries could be used by domestic consumers whose existence would not alone have brought them into being and they might even have produced the bulk of the market which these collieries served. But whatever the relative importance of local domestic consumers to these large collieries clustered in districts of continuous exploitation, the point remains that such exploitation only occurred where export or industrial markets were also available. Given this, and the small-scale intermittency of working elsewhere than around Prescot and Wigan, an explanation of the location pattern of collieries in terms of demand orientation seems to hold considerable validity.

The influence of demand seems to have been apparent, too, in the trend of output and the pattern of prices. Because of the availability of peat and the sketchiness of the evidence on markets, nothing more than tentative statements can be made about changes in the demand for coal between 1590 and 1690. However, it does seem to be overwhelmingly probable that the increase of about sixty per cent in total coalfield output, with a more rapid rise in the Centre, was related to the growth of population in the region as a whole and of industry in Wigan and the hamlets and villages around it, precursive of the more rapid growth of the following century. However, it is noteworthy that the evidence on output suggested that production at best remained steady in the South West, where the growth of Liverpool and the Cheshire saltboiling industry, supplementing the Irish market, must have meant the existence of possibilities for increased output.

Prices seem to have risen gently through the period, without the 'steep' or 'startling' increases which marked the trend of English prices generally.[77] Surviving statements of prices have been standardised for variations in measure contents and presented on figure 11. The upper illustration contains a regression line which provides an estimate of the trend of the average pithead price over the coalfield as a whole, and on this evidence it seems that South West Lancashire coal prices rose less steeply than prices in general over the century. It is clear from the graph, however, that this average is compounded of two quite different subregional price levels. This geographical variation is brought out much more clearly on the lower illustration of figure 11, where the standardised prices are plotted as

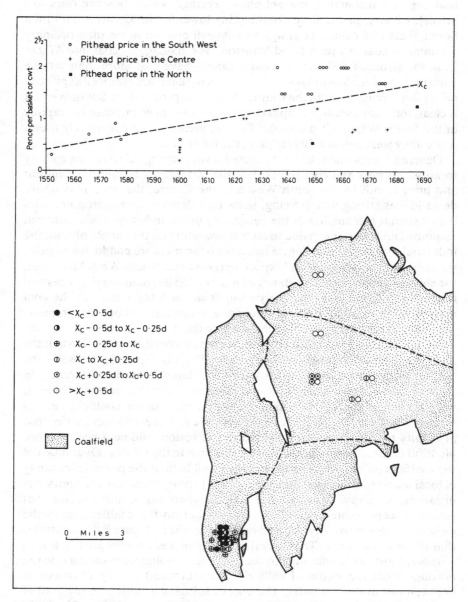

11. Pithead coal prices. 1550-1690. Mapped symbols represent residuals from the regression line $X_c$ plotted on the graph.

residuals from the regression line (these residuals are of the form $X_c$-x, so that negative residuals represent above average prices, positive ones low prices). Clearly, prices were considerably lower in the Centre than in the South West and cannel at Haigh and Aspull cost no more than ordinary bituminous coal at Prescot and Whiston. The haulage of coal into Wigan from the girdle of collieries around it cannot have cost more than cartage from the pits of the South West to Liverpool, so that market as well as pithead prices were much lower in the Centre than in the port and the South West. It is clear, too, that although output does not seem to have increased as rapidly in the South West as it did in the Centre, prices rose most quickly there, whilst they were remarkably stable in the latter area.

Demand seems, then, to have exerted a very strong influence on colliery location, outputs and coal prices. But the differences between the output and price trends in the South West and the Centre, the two areas where demand was strong and growing, show that demand orientation provides only a partial explanation of the geography of the industry. Their different responses must have been due to factors operating on the supply of coal: the industrial markets of the Centre can have been no more conducive to price stability than the domestic and export markets of the South West. Moreover, the short duration or intermittency of most collieries, their varying sizes and fluctuating outputs were certainly *compatible* with the nature of the coal market, but particular kinds of resource endowment and/or entrepreneurship were necessary, in addition, to *produce* them.

Other facets of the location pattern seem to have completely flown in the face of a demand orientation hypothesis. The discrepancies between the outputs of the five coalfield areas were much larger than the differences in their market potentials. There is no reason to doubt that domestic hearths were the major source of demand and other parts of the coalfield were as densely populated as the Centre and South West. The South was also in close proximity to Warrington and the North to Preston, and peat was no more plentiful in these areas – probably less so – than in the Centre. Over much of the coalfield output seems to have lagged well behind the potential capacity of local markets. Moreover, because of heavy transport costs over even short distances, a simple demand orientation hypothesis would require that collieries were located actually at the market on the coalfield, or in the closest possible proximity to extra-coalfield markets. Figure 8 demonstrates that this did not occur. The collieries at Wigan were conspicuously few in number, short in life and small in size, and fuel supplies were carted into the borough from the girdle of collieries that surrounded it at distances of between one and three miles. The cost of bringing a twelve-penny load of coal into the town in 1657 was 8d and the road to the cannel pits at Haigh (2) and Aspull (12) was particularly heavily used by fuel carts.[78] The pits of the South Western area that supplied Liverpool and Dungeon Point were situated on the east central edge of the coalmeasures promontory, one of the furthest parts of it from both destinations, whilst the Croxteth Park outlier, much nearer to Liverpool, seems to have been completely unexploited. Throughout the coalfield there was a singular lack of small-scale correlation between markets, as represented by heavily settled rural townships and urban areas, and colliery locations.

A further small-scale feature of the colliery location pattern which is inexplicable in terms of the nature of demand and Löschian supply conditions is the clustering of collieries. 'Twin' workings on immediately adjacent holdings were quite common and the pattern as a whole coalesced into small groups of points; it was not one of even dispersion with densities varying with the intensity of demand.

**Supply**

*Coal seams*

The coal measures of Lancashire are, geologically, the thickest and contain the greatest depth of coal of any in Britain.[79] There are (or were) thirty individual seams in the coalmeasures themselves and two in the Millstone Grit which underlies them and outcrops to the north of the coalfield proper and in the lens-like inlier between the West and Central areas. The coalmeasures contain three major stratigraphical groups: the *Anthraconaia Lenisulcata,* or lower coalmeasures; the *Communis, Modiolaris* and *Lower Similis Pulchra* zones, which are collectively known as the main productive coalmeasures, and the upper coalmeasures. The last were important in the Manchester area but occur only as a barren southern fringe in South West Lancashire (see figure 6).

The lower coalmeasures contain only five seams and although these vary in quality and thickness, generally only the Pasture and the two Mountain mines are workable. These three seams are separated by upwards of 200 feet of barren rock and rarely exceed two feet in thickness, but their coal is of excellent quality and cokes well. As their name implies, the main productive coalmeasures that lay above the *Lenisulcata* rocks contained the richest reserves. Twenty-four seams interspersed this group if the Arley mine, which marked its lower boundary, is counted. These range in thickness from over seven feet to less than two feet and in quality from the first class Arley, Smith, Bone and peculiar Cannel seams to the softer coals of the Ince group. Generally, the lower coals were the best, quality decreasing in upward succession, and every type of coal but anthracite was represented. Hard coking coal was least common, present in only the Arley and Smith mines, and the character of the first, considered the best in the coalfield in modern times, made it of particular value in glass-smelting. The Bone mine contained an unusually high proportion of *durain,* rendering it brown in colour and of very low ash content. The Cannel mine attained a workable thickness only to the north, east and south east of Wigan. It is different from ordinary coal because it was formed from plant spores deposited in stagnant water, rather than woody plant tissues. In consequence, cannel is very light in weight; it can be carved and cut like jet; has a high gas content of 13,000 cubic feet per ton, or up to one-fifth of its weight, which provides an extremely luminous blaze; a very high calorific value of 14,251 Btu per ton and a volatile content of 48.4 per cent, the highest of any coal, which means that it can be lit without kindling. It also has particularly low sulphur and ash contents. Steam and furnace coals abounded in the Wigan and Pemberton groups of seams, which were also renowned in the nineteenth century for the quality of the house-coals they provided. These numerous seams, rich in variety and quantity, studded the main productive coalmeasures in South West Lancashire, often separated by less than thirty yards of barren rock. In spatial terms, the whole series was generally of highest quality around and to the north of Wigan and the quality and thickness of many seams deteriorated to the south and south west, where the coals were less in variety and usually of a lower quality in any particular seam.

The highest of these coals originally lay 500 feet below the top of the coalmeasures series, and the lowest and best were more than 3000 feet deep. But Lancashire, like most of the western coalfields, was very heavily folded and faulted. Tectonic processes shattered the coalmeasures, completely destroying their stratigraphical sequence, and subsequent erosion planed off the higher strata in many areas, so that the Millstone Grit actually outcrops within the coalmeasures to the west of Wigan. Lower coals were brought to the surface on the margins of the large lower coalmeasures outcrops, and the higher seams occur only in the downthrow marked by the north–south trend of outcrops passing from the Northern area, east of Wigan to the Southern area, and along the southern margin of the coalfield in the South and South West. This combination of numerous seams and heavy folding and faulting produced a profusion of near-surface coals, serried ranks of outcropping seams of different qualities.

Over much of the coalfield the solid geological strata were smeared with glacial drift (see figure 12). Those collieries whose locations can be fixed with sufficient accuracy relative to the distribution of drift do seem to have tended towards locations where the mantle had been eroded away, although the correlation was by no means perfect and many workings penetrated drift cover. This was not thick except in the mosslands, and the vast abundance of coal near the surface could easily be revealed by chance encounters in valley sides, ditches and marl pits.[80] The main productive coalmeasures thus yielded evidence of their riches with ease; it is noteworthy that only one colliery (17) existed on the lower coalmeasures (see figure 6), where seam outcrops were much rarer.

The many outcrops and thin drift cover of the main productive coalmeasures also meant that workings were shallow almost everywhere,

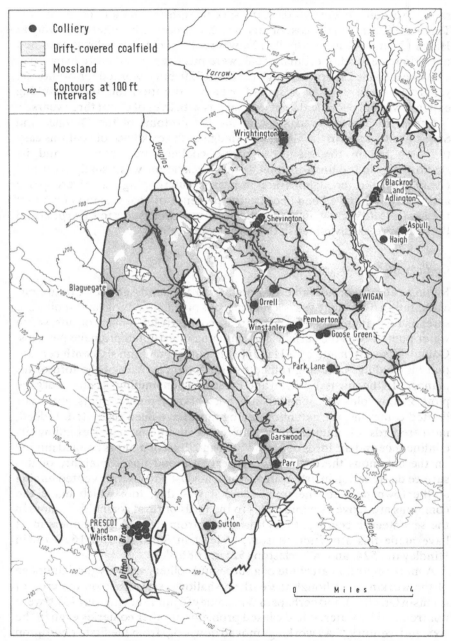

12. Relief and the distribution of collieries, 1590-1689. Only those collieries which can be located reasonably accurately in relation to relief features and drift are plotted.

rarely exceeding twenty yards in the seventeenth century and never forty.[81] This was reflected in the cost of pits, which was normally under £20. Only at Haigh (2) and possibly Aspull (12), where the choice cannel was searched out below the level of ordinary coal, were pits deeper and more expensive. The Great Stone Low pit at Haigh (2) cost £320, but it worked for twenty-six years and when the cost is spread over this duration the annual sinking expense was only double that of normal pits with lives of about three years.[82]

Except in peat-covered areas in the flat bottoms of the Douglas and Sankey valleys and in the West, knowledge of the existence of coal was easy to come by on the main productive coalmeasure outcrops, and its shallowness and abundance meant that excavation was usually easy and cheap. On the lower coalmeasures, however, outcropping seams were sparse and it may well have been because the presence of coal was unknown that the *Lenisulcata* rocks which were the nearest coalmeasures to Liverpool and Preston were hardly mined in the seventeenth century.

*Colliery sites*

Mining caused considerable destruction to the land surface. Continuous exploitation over many years left numerous pits; banked coal and spoil were spread indiscriminately over large areas, coal carts left a maze of ruts, waste water poured over the surface and slumping interrupted drainage patterns. Complaints were made by a Haigh (2) tenant even in the sixteenth century that his holding was being ruined by the mining activity of the lord of his manor.[83] In the towns of Wigan and Prescot, where mining on small burgage plots or copyholds was bound to affect neighbours, public streets, gutters or limited commons, the extent of mining must have been restricted. Indeed, most records of mining in Wigan concern injunctions forbidding its continuance and the lords of both Wigan and Prescot manors banned mining on the commons there.[84] It was the expense and unacceptability of this surface depredation in urban areas which at least partially prevented mining at the town markets themselves and it was this, too, which made the commons attractive as mining sites to the lords of rural coalfields manors in the seventeenth century, though law suits from injured commoners must have made this attraction illusory, at least in Sutton (3) in 1620-74, in Hindley in 1524, and in Eccleston (52) in 1686.[85]

A more important attribute of a successful colliery site was good drainage of the workings. Although the earliest shallow excavations on the crop of seams in Orrell (7), and perhaps in Winstanley (6) and Shevington (48),[86] were not troubled by water, it had caused problems at other collieries even in the sixteenth century.[87] As workings moved down-dip from the first penetration on the crop, drainage problems increased inexorably even in Lancashire's shallow pits. The most primitive way of raising water from a pit was by ladling it out of a pit-bottom sump in buckets carried to the surface by workmen. This method may have been used at first in very shallow pits, but it was laborious and inefficient and it was a simple improvement to attach the buckets to an endless chain or rope moved by a windlass or 'turn'.[88] The continuous operation of such a device was expensive of labour and the substitution of horses to provide more power greatly increased the expense.[89] Water powered drainage engines were much cheaper to run, and they were used by at least five, possibly seven, collieries during the century.[90] In most cases the rotary motion of the wheel was probably transmitted to a horizontal windlass at the pit eye, to which an endless chain of buckets was attached, scooping the water from a pit-bottom sump and pouring it out onto the pit bank.[91] It is possible that a more complex rag and chain or suction pump was used at Hindley (13) between 1607 and 1624.[92] Without exception, power was derived from small streams, either directly or through weirs and sluices. The rivers of the coalfield, though not large, were more difficult and expensive to control and their greater power was not yet needed in the shallow workings.

Water wheel drainage engines had a number of disadvantages. They were most useful where the seam sloped downwards from the working face to the shaft, otherwise a tunnel or 'sough' must be dug to a deep sump in the pit bottom to provide slope from the workings to the buckets. When most workings moved down-dip, often quite steeply in Lancashire's fractured seams, conditions were not most propitious for engine drainage. Moreover, the endless chain of buckets spilled a continuous cascade of water down the shaft, and wheels were dependent on the vagaries of the weather, especially when they derived power from small streams.

Sough drainage was most popular. Soughs were tunnels dug from the workings to outlets below their lowest level and the shallowness of the region's pits made soughing a relatively easy proposition. It seems likely that water wheels were only used when soughing was not possible, either because sufficient fall was absent or because permission could not be gained to sough beneath a neighbour's property. This was definitely the case at Sutton (3), where a neighbour refused to allow a sough outlet on his land,[93] and the fragmentation of holdings in Whiston and Prescot was completely unconducive to the assembly of soughing rights. At both Shevington (49) and Wrightington (55) towards the end of the century it was expressly stated that pumping was only to be resorted to if soughing failed to keep the seam dry.[94]

Although it could be relatively easily effected in shallow coal, soughing seems not to have been very much cheaper than engine drainage. Cost figures are, of course, few and variations from pit to pit were enormous according to depth, the hardness of the rocks, the wetness of the seams and the length of the soughs. All the surface and subterranean equipment and installations, including a 'water Engine', were valued at only £10 at Henry Ashton's Whiston colliery (36) in 1668, whilst the Southworths claimed to

have spent £200 on their 'Timber works; pumpes; to draw and exhaust water
. . . and other buildings and necessaries' at Hindley (13) in c.1605.[95] These
are almost certainly extreme figures for engine drained collieries. The three
sections of Sir William Gerard's sough at Park Lane (40) cost £43. 5. 1d in
the 1660s and 1670s and the Winstanley (6) sough of 1678 cost only
£16. 9. 1.[96] Soughs were probably preferable to engines because of their low
running costs. Whereas engines needed continuous attention and horses,
stabling and fodder if they were not powered by water wheels, sough
maintenance seems to have involved negligible expense: even the
'dangerous faults' in the Haigh (2) sough were only 'to bee lookit into onse a
quarter', and the collapse of the quicksand-plagued Aspull (12) sough cost
only five shillings to repair in the 1620s.[97]

It is clear that sough drainage was preferred, with water powered engines
second best. Both required one essential condition at the colliery site –
sloping ground. It is abundantly clear from figure 12 that collieries clustered
in the small valleys which tumbled into the major rivers across the edges of
the main erosion surfaces. Many of these little notches are over fifty feet
deep, and with the depth of collieries never yet surpassing them they
provided ideal sites for soughing, or for harnessing water power. The strong
locational attraction exerted by these valleys explains a number of the
distributional trends that are anomalous to demand orientation. The
collieries of the Centre may have been a few miles from Wigan, but their
locations were geomorphologically optimal. Those of the South West were
relatively distant from Liverpool and the Mersey, but they were tightly
packed on the side of the Ditton Brook valley on sites that were conducive to
drainage. Furthermore, the paucity of mining in the mosslands was probably
due as much to flat site conditions as it was to the thick peat cover; no one
would have thought of looking for coal at places where it could not be
efficiently drained and where mining was therefore impossible.

There were two very marked exceptions to this rule of orientation towards
small valleys. These were the cannel collieries of Haigh (2) and Aspull (12),
which were stuck up on the crest of a spur between the river Douglas and the
Borsdane Brook in Hindley to the east (see figure 12). They were exceptions
which prove the strength of the rule. Haigh colliery (2) was the largest and
longest lived on the coalfield and Aspull (12) worked for far longer than the
average. The only place where the Cannel seam outcrops and has a
considerable extent at a shallow depth is on the crest of this blunt spur, 300
feet above the bottom of the Douglas valley which lay a mile away. To open
up the cannel there required enormous soughs and deep pits. The Great
Sough at Haigh (2) was six feet high and four feet wide, running for
two-thirds of a mile. It took seventeen years to complete.[98] The Gerards
claimed to have spent £3000 on searching for, sinking to and draining their
Aspull (12) cannel in the early part of the century: even allowing for the
degree of exaggeration necessary to impress their creditors, the scale of their

spending was many, many times that of the owners of ordinary collieries.[99] Only the existence of an exceptionally choice fuel within reach of a town at manageable – though for this coalfield, unusual – depth could prompt such expenditure to open up unpropitious sites. Elsewhere, the little obsequent valleys obviated the need for heavy spending. Short soughs or water powered engines sufficed the small shallow works there; because of the abundance of seam outcrops and valley sites they could be abandoned for fresh ones when the inevitable happened and their crude drainage installations were overwhelmed. Flooding caused many closures. It was the reason for the ephemerality of most of the collieries. The long-lived workings at Haigh (2), Aspull (12) and Winstanley (6) tapped the Cannel or the Smith seams, but to plug away at the same site at ever increasing depths to get these special coals involved completely exceptional expenditure.

*Labour supplies*

The terms collier, coaler, coalman, coalminer and miner, the last much less common than the first in this coalfield, were already used to describe mineworkers at the start of the seventeenth century. But the existence of specialist terms and the frequency of such appelations as 'John Greenough of Parr, collier' can be misleading. Many who were so called probably spent only a part of their working life, or even of any working year, in the pits except in the most developed parts of the coalfield. The terms were usually used in accounts, leases, bonds or law suits and probably merely describe what the man was doing when the record was made or signify that he had experience which was relevant to the case being tried. Raphe Laithwaite described himself as a skinner when he was hired as auditor at Nicholas Pennington's Wigan colliery;[100] the large working partnership which leased some Whiston mines in 1647 included yeomen and a skinner as well as colliers,[101] and away from these intensively worked areas the labour situation was probably even more fluid. One of the Rainford men hired to sink for coal in Sefton was a husbandman, and Peter Martlu of Lathom was styled shoemaker in his will and 'coler' in his inventory, where the main items valued were farm stock and equipment.[102]

This kind of situation is, of course, to be expected over most of the coalfield where collieries were short-lived or intermittent. In the South West and Centre there were almost certainly already men who spent most of their working lives as coalminers. William Browne of Whiston first leased coal, which he worked himself, in 1627; he styled himself collier in a law suit concerning this lease in 1636 and again in 1650 in a case concerning another lease which he had taken out in 1647 with his son, who styled himself yeoman, and five others.[103] Peter Lowe's name occurs in lists of Haigh miners in 1635, 1636, 1637, 1645-7 and 1664 and those of Richard Lowe, James Lowe and William Birchall in 1635, 1636, 1637, and 1645-7.[104]

As Nef has demonstrated, this was the period when the separation of capital and labour began[105] and all stages of this separation can be observed in the South West Lancashire coalfield in the seventeenth century. It is possible that the colliery in Lathom (56) in the West was operated by the men who owned the coal, yeomen who had other trades and farms and dug coal when they or their neighbours needed it.[106] Other miners, too, were independent of capitalist employers. In the South West it was common for groups of colliers to band together to lease and work coal jointly, sharing the profits between themselves and the owner of the seams. One such partnership had seven members and another was split into twelve shares.[107] Other colliers hired themselves out in bands to work the coal of particular landowners, finding all their own equipment and operating as independently of the owner as the lessees of the South West.[108] Nonetheless, it was most common throughout the coalfield for colliers to be wage earners employed by the owner or lessee of a colliery. Some collieries in the South West were run this way as well as by independent colliers, as were all but one of those in the Centre for which sufficiently detailed records have survived, right from the beginning of the period. At Haigh (2) and probably at Winstanley (6) the colliers signed annual bonds in which they contracted to work for the one coalmaster.[109] At Haigh (2) they also signed agreements in which contractual penalties were fixed for numerous specified misdemeanours.[110] Here the subservience of the collier to the coalmaster was as complete as it could be.

The size and composition of the workforce of collieries varied considerably. At the smallest, one or two men dug the coal, hauled it to the pit bottom, wound it up the shaft and sold it. At the other extreme at Haigh (2) in 1636, twenty-five auditors, hewers and drawers signed the 'Orders for the Oditors, Hewers, Drawers, Winders, Treaders, Takers of Cannell at the pit Eye & all other officers and workfolks belonging to the pit & pits . . . belonging to Roger Bradshaghe of Haighe'.[111] The fact that only auditors, hewers and drawers signed perhaps suggests that many of the jobs listed were done on an *ad hoc* basis rather than by specialists, but the hewers at least seem to have been kept hard at their specialist task, for it was stated in a re-issue of the Orders in 1687 that 'without great and absolute necessity noe work besides getting Cannell shall be allow'd below ground'. If the hewers considered it necessary to do other work they had to give notice to their auditor and a winder and these two plus 'ye Overseer of ye next Pitt shall goe down and view ye same and shall allow or disallow ye same to be don as they shall see cause and shall onely allow soe much time, and such person or Persons to doe it as is absolutely necessary'.[112]

This degree of continuous activity at one specialist task was probably unusual for underground workers at other collieries. Richard Wareing sometimes hewed, sometimes drove soughs and water lanes and sometimes did odd jobs on the surface at Winstanley (6) in the 1670s and 1680s.[113]

Other specialist workers at Haigh (2) and in Parr (43) were the women drawers, who hauled the coal on sleds from the place where it was hewed to the pit bottom.[114] Not all the drawers at Haigh (2) were women; in 1636 four drawers out of the six in Mowsecroppie pit were and three out of five in the Halle Croft pit. Usually, but not always, the surnames of the women drawers were the same as those of hewers in the same pit, and a man and wife (or father and daughter or mother and son) team, Helen and John Lowe, also held the auditorship of Halle Croft pit in 1636.[115] This degree of specialisation at Haigh (2) was a reflection of the tight organisation of the colliery, which in turn reflected the ability of its owner and the size, efficient drainage and remarkable forward planning that ensured its smooth continuous operation. Elsewhere, men's jobs were changed as water levels rose, as unexpected faults were hit, as pits became exhausted, requiring very long draws and quick new sinkings, and as sales slackened in winter or became almost overwhelming in summer.

Occasionally the hewers sold the coal they mined, but it was much more common for an auditor – banksman, overseer or lord's officer – to handle the business side of the operations. This specialist existed everywhere but his duties varied from colliery to colliery. At Winstanley (6) the 'Auditor and Banksman's' jobs were to 'wind Rook sell & secure ye money' and 'to clear Accompts every Newyears day'.[116] At Haigh (2) each pit had its own auditor and came to be named after him. There, the tallying of each hewer's get during the week, ensuring that full and proper measures were sent up, tallying sales from the bank and periodically inspecting the conditions underground were the auditors' jobs. Every Saturday between four and five o'clock they took their tallies to Haigh Hall, where the money they owed, comprising receipts plus fines levied on workmen minus wages, was paid over. The complex tallying procedures and the frequency of settling with the steward were probably because the auditors were illiterate.[117] At nearby Aspull (12) in 1626 there were two overseers. They generally supervised the colliery, hired labour, negotiated wayleaves, received money from the auditors and issued instructions to them as agents of the coalowner, who lived a few miles away at Ince. As at Haigh (2), each pit had an auditor as well, and here they were hired on annual contracts like the Haigh (2) colliers. Their duties were to wind and bank coal, receive customers' money and pay the workmen's weekly wages, giving a weekly account to the overseer.[118]

Little evidence of wages has survived. At Aspull (12) the auditors got 4/- and 1/6d worth of cannel per week in 1627; at Shevington (48) in 1666, 7d per day and 'soe many baskets of Coales . . . as hath beene accustomed to allow ye Oditr of Shevington Cole Pitts', and at Winstanley (6) in 1670–90, £3 plus £1 worth of coal a year plus 1d a load mined until 1684, ½d a load thereafter. Thus, auditors' wages were expressed differently in different places and varied quite considerably: they were 5/6d a week at

Aspull (12), about 5/- a week at Shevington (48) and about 7/2d per week, falling to about 4/5d after 1684, at Winstanley (6). At the last colliery the auditor's wages would vary as weekly output varied, but in 1684 the practice was begun of paying him for attendance on the bank when no coal was got at the rate of 3/3d per week and this must have compensated his cut in load payment. Other perquisites were also available. At Haigh (2) the auditors bought cheaply the colliers' concessionary coal, raised in 'extravagant great baskets'; at Aspull (12) they paid wages in truck, and at Winstanley they received a bonus of 2/6d when accounts were settled on New Year's Day. Dishonesty was also profitable. At Haigh (2) there were complex orders to prevent wrong reckoning, which incurred swingeing penalties, and one of the Aspull (12) auditors was sued for embezzlement in 1626, although he claimed that it was the overseer who had cheated 'both Mr. Gerrard & the poore auditors' on that and many other occasions, 'enriching . . . himself, who was of late times very poore, but is now comen to great welth and that within fewe years'. At Winstanley (6) the frequency with which people were 'remov'd from being auditor' suggests that the coalowner there had similar trouble.[119]

The wages of colliers were reckoned in equally complicated ways. Hewing was paid on piece rates at Winstanley (6) and Haigh (2), but at Orrell (7) in 1601 it was paid by the day, probably because there was no auditor to tally output. Coal cut when opening tunnels was paid by the yard at Winstanley (6) until 1680, when it was decided to pay by the load, as for ordinary hewing, 'because bi' th' yard they are apt to waste coal by burying them in hollows to spare labour of wheeling'. Other jobs at Winstanley (6) were paid by the day: 10d for sinking and soughing and 8d for winding and labouring for a sougher. These rates were also paid for any coal got in the course of such work, compared with a normal 4d for getting and 1d for drawing a load.[120] In such circumstances it is obviously impossible to generalise about colliers' wages; indeed, it is impossible even to estimate them reliably from the kinds of records which have survived. If, as is likely, three hewers worked in the Winstanley (6) pits, they would have cut on average about twenty loads per week, yielding a wage of about 6/8d for the hewers and 1/8d for the drawers. Employed full-time on sinking or soughing a collier would have made 5/- a week, 4/- at winding or labouring. At Orrell (7) in 1601 day wage men earned about 3/4d a week.[121] Besides wages the colliers received drink and tobacco and sometimes an earnest at the annual hirings at Winstanley (6) and Haigh (2) and they also got concessionary coal. At Haigh (2) this was worth 2d a week and some colliers took cash instead. In addition to normal wages, compensation was paid in either coal or cash when seams were bad for hewing or if drawing was long.

On the debit side, fines had to be paid for misdemeanours at Haigh (2). In the Orders, penalties were imposed for numerous different offences, ranging from a few pence to £2 for cutting into pillars. However, in 1636 and

1637, the first two years the Orders were in operation, the fines were all 4/-, levied in late June.[122] In these complicated circumstances it is just not possible to find what a collier earned; that they could get as much as 6/8d at Winstanley (6) in the 1670s and 1680s, more than auditors got earlier elsewhere, is perhaps surprising. This, and the earliest Orrell (7) rates, might well suggest that money wages rose quite steeply in this coalfield during the course of the century, although whether real wages rose too is, of course, another matter.

Working conditions were bad and colliers were killed and maimed in all parts of the coalfield, women as well as men.[123] If the disbursement entry in a Haigh (2) account book, 'Lowe wiffe in Whelley Lane whose husband was lost in ye pit . . . 2/-', reflects the value put on colliers' lives, then they were not held in high esteem.[124] Injuries were occasionally severe. Two Whiston colliers, forced by an overbearing auditor into a pit (58) from which fire-damp had not been cleared, ignited gas with their candles and were thrown 'from one side of the drift to another & left . . . at the coale pit eye with or flesh torne off or armes, and armes & bodies so ill burnt that to looke upon we were thought to be mortally wounded insomuch that the pains made us madd'. They were left unable to work, owing bills to a doctor and an apothecary, and threw themselves and their families onto the mercy of the parish poor relief.[125] Some did manage to work until they were 'so old and weak' that they had to seek relief,[126] and others could find nowhere to live when in employment: even in Haigh (2) it was not customary, as it apparently was elsewhere, to provide workers with accommodation.[127] On top of all this, bad weather could cause complete loss of earnings for long periods in winter.

The miner's job might have been well paid, then, but it was also dangerous and a number of quite specialist skills were required both underground and from the auditor on the bank. Moreover, in Haigh and Aspull and in the South West, at least forty or fifty miners were required in the clusters of collieries. The annual bonds, an agreement between the owners of Aspull (12) and Haigh (2) not to employ each other's workers, and the prevention of one of his tenants from working at Haigh (2) by the lord of Aspull, all suggest that labour of the required kind was scarce around the Cannel pits.[128] This might, too, have been the reason for employing female relatives as drawers at Haigh (2), although it is more likely to have been so that the hewer could be confident that there would be no collusion between the drawer and auditor falsely to reckon his get, or spuriously to claim that his baskets were of insufficient fullness or quality.

There was some labour mobility,[129] despite the Poor Laws and accommodation problems, but it seems to have been more usual for colliers to travel miles to work. Ashton colliers worked in the Prescot area and at Winstanley (6); Ashton and Eccleston colliers worked in Parr (43) and many of the Haigh (2) colliers were from the Whelley area of Wigan.[130]

The short or intermittent lives of many collieries and the possession of skills would, of course, encourage this.

One of the most striking features of accounts and law suits is the recurrence of particular surnames: mining was undoubtedly an occupation that ran strongly in families. There were Lowes in every record of names at Haigh (2) in the seventeenth century. In 1636 there were four men with this surname, three women, two women named as 'Rothwell alias Lowe' and three male Rothwells – nearly half the labour force.[131] Other names which were common at Haigh (2), occurring more than once at a time at one pit or over widely spread intervals, were Bradley, Birchall, Leatherbarrow and Bibby. There were Birchalls in Winstanley (6), Orrell and Prescot (1) as well and Bibbys in Adlington (54). There are no accounts for South Western collieries, so names are much scarcer, but Forber, Tickle, Alcock and Browne occur with unusual frequency in leases and law suits and Greenhough was a common miners' name in the South.[132]

It seems, then, that labour shortages did not limit the extent of mining. In the less developed parts of the coalfield men probably moved into and out of the pits, travelled to work elsewhere or migrated as collieries opened and closed, so that a labour force of the required small size was latent everywhere. Whether or not wage differences gave advantages to some parts of the coalfield over others cannot be discovered from the evidence that has survived. It is quite obvious, however, that the quality of labour was important. Bad workmanship underground caused collapses and flooding at Haigh (2) and Aspull (12) in the first three decades of the seventeenth century and the litigation which ensued between the owners of the two adjacent cannel collieries must have been extremely expensive.[133] Unreliable auditors could also prove costly and the Colliery Orders, first issued at Haigh (2) in 1636 and in continuous operation afterwards, were an attempt to overcome these labour problems. Gradually, a skilled and effective labour force must have emerged amongst the groups of mining families in the cannel mining area and probably, too, in Winstanley and neighbouring Orrell and in the South West, a workforce which could easily be expanded through the recruitment of children and in-laws. It must have represented a significant external economy to those coalmasters who could take advantage of its skills and discipline.

## Capital and entrepreneurship

The statements of profits which have survived are given in table 3, 'profits' comprising the sums which remained in hand when annual accounts were closed minus any balance which had been carried over at the beginning of the year. Profitability would obviously vary with a host of factors ranging from output to depth and miners' skill, but an 'average' colliery raising about 1500 tons a year, such as Winstanley (6), could yield a steady and substantial

TABLE 3. *Colliery profits, 1590-1689*

| Colliery | Area | Year | Annual profit £.s.d | Annual output tons | Profit as percentage of receipts |
|---|---|---|---|---|---|
| Sutton Heath (3) | South | 1614 | 20. 0. 0 | | |
| Hindley (13) | Centre | 1624 | 200. 0. 0 | | |
| Aspull (12)* | Centre | 1624-5 | 200. 0. 0 | 2080 | 79 |
| | | | 100. 0. 0 | 1600 | 50 |
| Shevington (47) | Centre | 1669 | 15. 0. 0 | 672 | 36 |
| Winstanley (6) | Centre | 1677 | 95. 6. 0½ | 1070 | 72 |
| | | 1678† | 70. 1. 0½ | 1292 | 43 |
| | | 1679 | 79. 5. 4 | 1531 | 42 |
| | | 1680 | 130. 5. 4½ | 1854 | 56 |
| | | 1681 | 126. 6. 5 | 2352 | 43 |
| | | 1682 | 97. 7. 8 | 1062 | 73 |
| | | 1683 | 99. 4. 7½ | 1229 | 66 |
| | | 1684 | 93.14. 2 | 1483 | 50 |
| | | 1685 | 70. 0. 4½ | 1194 | 46 |
| | | 1686† | 62. 2. 9 | 1379 | 40 |
| | | 1687 | 104. 0. 7½ | 1475 | 57 |
| | | 1688 | 110. 5. 3½ | 1689 | 53 |
| | | 1689 | 60. 7. 4 | 1490 | 33 |

\* Two workers made estimates for Aspull colliery as evidence in a court case.
† Years in which new pits were definitely sunk at Winstanley.

*Sources:* PRO, DL4/62/5, 74/24, and 75/13; PRO, PL6/30/108; LRO, DDBa/Coal Pitt Accompts, 1676-96.

income, and the proportion of the receipts which went into the proprietor's pocket was encouragingly large, ranging from thirty-three per cent to nearly eighty per cent.

It is evident from the sections on seams and sites that to begin mining in the shallow coals of this coalfield was not inordinately expensive. Generally the cost of pits and the initial expense of draining them with an engine or small sough, plus the purchase of tools, a hovel at the pithead and so on, cannot have cost much more than £30 or £40, although at Hindley (13) the opening of a colliery cost five times this and at Aspull (12), in the deep cannel, the owner claimed it cost him nearly ten times this normal amount to re-open his colliery in the early seventeenth century.[134] The heaviest costs were usually incurred after mining had been begun as the workings moved down-dip from the crop of the seam. As successive pits were dug about every three years they became deeper, and as pits became deeper the cost of soughing them dry, as well as of sinking them, increased. These recurrent costs were paid out of receipts accumulated in the current account, where up to half a year's profit was carried over from one year to the next to meet such needs. It is clear from table 3 that profits dropped substantially at Winstanley (6) in years when new pits had to be sunk and equipped. At Haigh (2) the inordinately expensive Great Sough was only begun thirty years or more after the colliery, and even then its cost was spread over a further seventeen years.[135]

Prudent financial management could ensure that there was normally enough cash in hand to cover these recurring charges, but even so there must have been occasions at all collieries when the balance in hand was insufficient to meet a particular expense. Then the coalmaster must dip into reserves accumulated from the sums 'paid to my master for his own use' by the auditor. If these had been spent, as must have been usual, he would have to scrape together money from elsewhere until profits accumulated to wipe out the debit. It is clear, then, that it was not beginning mining which was most taxing of recourses of money and ingenuity, but continuing it once it had been begun. Moreover, surveying for fresh reserves, overcoming faults, organising work in deeper pits more prone to collapse and flooding, calculating drainage levels and so on would all become progressively more difficult as time passed. The sough and the Orders at Haigh (2), the expense at Aspull (12) and the careful accounting procedures at Winstanley (6) all testify to this, as does the fact that it took the vast experience of William Browne to make the Whiston mines workable.[136] Because the seams sloped steeply from their crops in a wet coalfield, problems of financial management, labour organisation and engineering would crowd in on the coalmaster very quickly in South West Lancashire. Where did the men who provided the money and skills come from?

Figure 13 shows that the landed gentry provided most of the investment and ability, their relative importance sinking only in the 1620s and 1630s when an exceptional crop of small non-gentry workings was recorded.[137] They were overwhelmingly dominant in all areas except the South West. Outside that area the only lords of manors who consistently leased out coal were the absentee owners of Shevington and Aspull in the Centre and Wrightington in the North. The infrequency of leaseholds outside the South West was reflected – and perhaps caused – by high rents, which were crudely expressed as so much per year or score of coal, irrespective of output and usually irrespective or prices.[138] Rents ranged between forty per cent and seventy-nine per cent of receipts in the Central area; the one lease granted in the North cost twenty-four per cent of receipts and two leases were granted in the South at twenty-six per cent and fifty per cent (figure 14).

These rents suggest that the landed gentry were not keen to divest themselves of a direct interest in mining, especially in the important Central area. Indeed, the Bradshaighs and the Bankeses probably operated their collieries at Haigh (2) and Winstanley (6) throughout the century, consistently ploughing back a proportion of receipts to yield substantial continuous incomes. But most of the gentry families, unlike these two, were Papist Royalists, sorely troubled by debt and after the 1640s recusancy and sequestration; 'most men's estates being much drained by the wars and now almost exhausted by the present scarcity and many other burdens incumbent

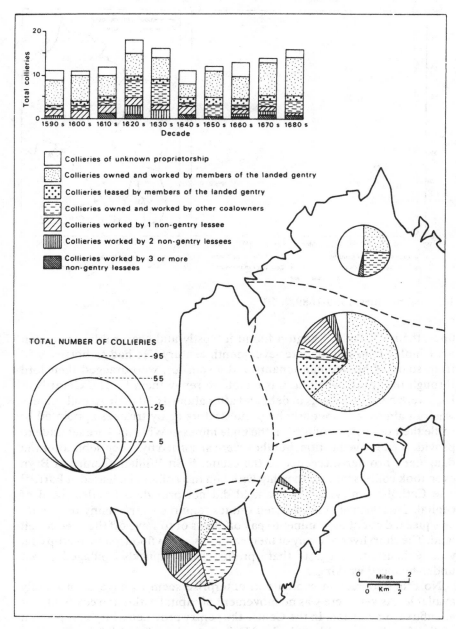

13. Sources of capital and entrepreneurship, 1590-1689. The volumetric symbols on the map include a 'colliery' for each colliery recorded in each decade. See Langton, 'Development of coalmining', for a detailed discussion of the categorisations used.

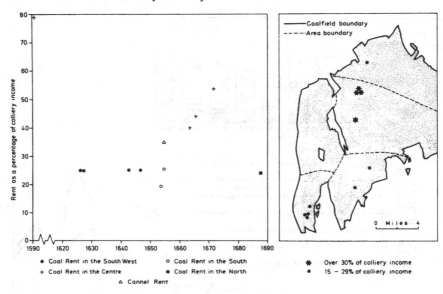

14. Coal royalties, 1590–1689.

upon them'.[139] Philip Langton found it 'costly and chargeable' to reopen his Hindley mines in the late seventeenth century. He had to borrow £50 from some of his yeoman tenants and a spinster, who dragged their lord through the courts when he was unable to repay them.[140] The Gerards of Ince were head over heels in debt and after about 1650 their Aspull colliery was run almost continuously by sequestrators or by trustees appointed to settle their debts.[141] Profits from the Ogle mines in Whiston were set aside to provide a marriage portion, but they were absorbed by other debts after the daughter's brother succeeded to the estate.[142] Sir William Gerard of Bryn even took colliers into partnership with him in a colliery he leased at Parr.[143] The Catholic gentry of this coalfield did not provide a limitless fund of capital. Their heavily encumbered estates required every penny that could be squeezed out of their mines to pay off debts or to stave off the need to sell land. The short lives of many of their collieries, often flooded out as crop coal was exhausted, suggest that profits were quickly pillaged from under-financed workings.

No other sources of capital and enterprise seem to have been readily available. As yet, there was no movement of capital within the coalfield, nor an influx from outside. In the Centre the gentry provided a large proportion of the few lessors, although the Heskeths' mines in Wrightington were opened by a partnership of four local 'gentlemen' and yeomen.[144] John Rigby, 'gentleman', had interests in a number of collieries in Shevington. He usually shared costs with the lessor and took partners, who sometimes changed quickly, into each of his ventures, which were all short-lived: the last was flooded out in 1679 after he had 'of late received great loses'.[145] One cannot but conclude that the function of such partnerships and arrangements with lessors was to scrape together capital and that the scrapings were not always sufficient to keep collieries going as expenses and difficulties inevitably increased.

In the South West, collieries worked by members of the landed gentry were in a small minority (see Figure 13). The 'middling gentry' who owned the manors and ran the collieries elsewhere were absent from this area. Prescot manor was owned by King's College Cambridge; the Earl of Derby and Viscount Molyneux, who did not as yet bother themselves with mining, each owned a number of others, and most of the remainder in Whiston, Eccleston and Huyton were owned by Papist Royalists who were forced to sell many small freeholds from which coal rights were not reserved.[146] Of these, only the Ogles and Ecclestons mined spasmodically, although the former more usually leased their coal out.[147] The apparent absence of direct gentry interest in this area, then, was partially a result of absenteeism, the concentration of manors in noble hands, and of freehold sales which not only made the arrangement of sough drainage difficult, but put coal rights over large areas into non-gentry hands. It was not a result of a lack of profitability in mining. Coal prices in the South West were double those of other areas (see figure 11), and one local freeholder who had few interests outside mining styled himself 'gent' in his will, which disposed of over £1500 in cash, banked coal and debts owed to him.[148] But there was no superfluity of capital outside gentry hands here, either. Rents were lower and almost always expressed as twenty-five per cent of receipts in this area, and lessees were much more numerous. Generally they were yeomen, husbandmen, colliers, skinners, and so on, who banded together in partnerships larger than elsewhere (see figure 13). As we saw, one contained seven members and another had twelve shares.[149]

The whole situation was, therefore, evocative of a shortage of capital and the experience and ability to run even such small concerns as the collieries of the time. This must have been one of the major reasons for the short lives of most of the workings, whose problems and expenses increased quickly after a short period of windfall gains. The middling gentry and their purses did not usually have sufficient capacity to run collieries successfully over long periods: those few who did were conspicuously able in other fields besides mining.[150] Amongst the coal lessees of the South West and within the ranks of the coal-working gentry of the Centre, capital and expertise were slowly being accumulated. It was these accumulations which ensured the longevity of the collieries in Haigh (2) and Winstanley (6) and the continuity of operations over various sites in Whiston and Prescot, where lessees did not have the incentives of landowners to keep one particular colliery running. Where market opportunities and geological and site conditions were favourable, then, sometimes commensurate capital reserves and ability also developed. But this was by no means always the case, and although the gentry who succeeded in deriving a steady and continuous income from their mines in the Centre worked cannel or smith's coal, others with similar geological resources failed to do so.[151]

### Competition between collieries

Variations in colliery size must have owed much to an interplay between the pattern of demand and the ability of individual coalmasters to marshal capital and expertise. They might also have been due to certain kinds of competition between the owners of collieries. In Lösch's scheme perfect monopolistic competition prevails, each colliery having the exclusive custom of the consumers in an area surrounding it. Such areas might extend much further off the coalfield than on it, and this might well have been the reason for the numerousness and size of the collieries in the South West which served Liverpool and Cheshire. The cost of transport and the abundance of near-surface coal must have limited the distances over which coal was hauled on the coalfield, and therefore the size to which non-peripheral collieries could grow. Surprisingly long hauls occurred on the coalfield, from Stanley Gate to Smithils and from Winstanley to Liverpool, but they probably represented back-carriage in carts whose main purpose was to take something else in the opposite direction.[152] Such fortuitous hauls as these could not have disturbed the pattern of local demand and supply.

The stability of colliery numbers and output in the Centre and South West was attributed to the existence of large markets in Wigan and on the Mersey, at some distance from the collieries themselves. The differences in prices in these two areas must have meant that there was little overlap between their markets: the similarity in prices within them must have meant that there was some kind of internal competitive integration (see figure 11). In the Centre, collieries were scattered around Wigan and relatively distant from each other and it was probably competition between them in the equidistant Wigan market which kept prices low there. In the South West the collieries which served Liverpool and the Mersey were clustered together and this, and their isolation from other coal-producing areas, allowed greater opportunities for collusion and the command of inflated prices. Certainly, the only evidence of price cutting which has survived concerns the Centre; the strategy followed in the South West was to try to get competitors closed down.[153] The low prices of the Centre might, too, have been responsible for lack of mining in adjacent parts of the South and North; the infiltration of coal cheapened by sharp competition elsewhere could have combined with the presence of peat to make mining unworthwhile in these rural areas. Thus, competitive strategies and the way they differed over the coalfield may have been an important cause of discrepancies between markets and mining and of variations in price.

These competitive effects could occur in a situation where the products of collieries were undifferentiated. But differences in the quality of coals from different collieries may have been equally significant determinants of the pattern. The cannel of Haigh (2) and Aspull (12) and the smith's coal of Winstanley (6) and Orrell (1 and 7) certainly allowed those collieries to wax fat at the expense of nearby sites where these choice fuels were not available. The peculiarities of cannel were fully appreciated at the time, not only as an industrial fuel but also as a house coal: in 1634 Thomas Gerard claimed that his Aspull (12) cannel was better than that from Haigh (2) for 'giving light in the fyer', and in 1676 Roger North observed that light from the grate was the only illumination of rooms in the poorer households of the area.[154] Cannel was 'the most pleasant and agreeable fuel that can be found',[155] its lightness in weight, freedom from dust and easy kindling compounding the advantage it gained from luminosity and lack of sulphur and therefore noxious smoke. It was twice as expensive as the ordinary coal of the Centre but as cheap as that of the South West.[156] Its price was kept down by competition between the Haigh (2) and Aspull (12) collieries, where price cutting and the changing of basket sizes manifested rivalry on different occasions,[157] and by competition with smith's coal from Orrell (1 and 7) and Winstanley (6) in Wigan. At its relatively low price it was a popular fuel; people were willing to queue for two days at Haigh (2) and large quantities were carted across the Douglas to the West, to the great detriment of the road in the valley bottom, some as far as Heskin, eight miles away.[158] The size of the colliery at Haigh (2) in particular, but also of those at Aspull (12), Orrell (1 and 7) and Winstanley (6), must have been due to the excellence of the fuels they provided, which penetrated relatively great distances from their pitheads and cast a blight on the mining of ordinary coal over considerable areas around them.

## The geography of the South West Lancashire mining industry, 1590-1689

The South West Lancashire mining industry was an amalgam of small-scale affairs in the seventeenth century. Although total output increased slowly and reasonably continuously, maximum colliery size did not and most collieries were small and short-lived or intermittent, scattered over the coalfield in a pattern which shifted constantly around the stable dominance of the Centre and South West. This kaleidoscopic picture was produced by many variables interacting in ways which would be impossible to distinguish in detail even if the surviving records allowed more than a glimpse of their operation. It seems reasonably clear that markets were the major control: collieries were only large, long-lived and numerous where thickly spread domestic hearths were accompanied by export or industrial markets;

moreover, the fitfulness of mining elsewhere is compatible with the way that rural domestic demand would be expected to influence mining. Ubiquitous shallow coal on the main productive coalmeasures, sloping steeply from outcrops and prone to flooding, and the thin purses of the coalmasters were also conducive to small-scale short-lived works and they made the little obsequent valleys attractive sites. Only where especially desirable coal came near the surface in proximity to an urban industrial market did the investment necessary to ensure permanent operation occur, and the collieries which tapped the cannel and smith's coal were also amongst the largest and best run. Unskilled labour might, too, have contributed to the fitfulness of working in most areas, just as the force of trained, tightly controlled and long-inured workers at Haigh and Winstanley must have contributed to the long lives and profitability of the collieries there.

Throughout this examination the coalfield has been treated as a whole, but it has become quite clear that it was not a coherent economic unit in the seventeenth century. The large differences in the levels and trends of prices in the Centre and South West demonstrate that their markets must have been quite separate, and no labour or capital moved between them. Rented collieries were rare in the Centre but common in the South West, where the collier-coalmaster was as normal as the gentleman-entrepreneur of the other area. In the Centre rents were expressed as so much money per unit of coal mined and they were high; in the South West they were levied at a percentage of the value of coal mined, usually twenty-five per cent. The measure systems were so different that they are difficult to equate (see Appendix 1); basket sizes and the aggregative units into which baskets were combined were dissimilar. These deep differences between the mining economies of different parts of the coalfield are a forceful testimony to the low level of economic development of the region; there was no articulation or cohesion amongst the small-scale economic fragments into an 'economic region', no coalfield-wide economic system through which interactions and growth could pulse. But within the Centre and South West prices, rents and measures were reasonably similar and workers and entrepreneurs moved with some frequency over quite long distances. This integration occurred around Wigan and in the vicinity of the Mersey where industry and export points quickened demand, at points where the economy of the coalfield was integrated into wider and richer economic systems. The mining industry was at its most developed where it was touched by contacts which linked it, directly or indirectly through the non-ferrous metal industries, with stimuli propelled from outside the region. The two sets of stimuli were different in origin and kind and they produced two quite distinctive and separate mining regions on the coalfield.

# PART TWO:
# THE BEGINNINGS OF GROWTH

CHAPTER 4

# THE COLLIERY LOCATION PATTERN, 1690-1739

## The numbers and distribution of collieries

The number of collieries working the coalfield increased in each decade after the 1680s, except for an interruption in the 1700s which was so slight as to be well within the margin of error of the sketchy sample data used to reconstruct the trend (figure 15). The increase from sixteen collieries in the 1680s to twenty-six in the 1730s was greater than that of the whole preceding century.

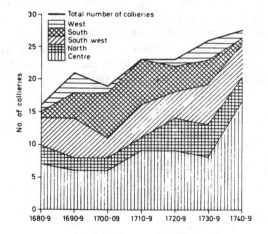

15. Numbers of collieries, 1680-1749.

The pattern of distribution changed considerably as the number of collieries increased and the joint dominance of the Centre and South West, a continuous feature of the preceding century, was broken. Numbers were stable in the Centre; the West remained of small importance, and there was a noticeable quickening in the North in the 1720s and 1730s. The greatest changes occurred in the two areas on the southern fringes of the coalfield, where numerical trends moved in complementary negative correlation. The total colliery population of the South and South West expanded considerably in the 1690s and remained large until the 1730s. Until 1720 it was the previously unexploited South which contributed the lion's share of the total, whilst the South West declined: in the 1720s and 1730s there was a resurgence in the South West and a decline back to the numerical level of the 1680s in the South. In the 1740s another sharp subregional spurt occurred, this time in the Centre. This, and the waning of the South, marked the end of a period of reasonably constant trends in the changing subregional patterns. The period 1689-1739 seems to have been one of considerable geographical continuity.

The movement of the centroids calculated from decennial distribution maps give a further illustration of the constancy of the trend of spatial change between 1699 and 1739 (figure 16). After c.1710 colliery sizes varied much more than previously and output data are reasonably plentiful, so that it is both desirable and possible to incorporate colliery outputs into the centroid calculations: after 1710 the centroids depict the geographical centre of coal production and are differentiated on figure 16 as solid symbols. The southward lurches of the centroid in the 1690s and 1730s demonstrate the growth of the southern parts of the coalfield and the movement northwards in the 1740s reflects the spurt in the Centre. Although apparently aimless shifts occurred around the clearly directed dominant trend of the centroid, they were less random in direction and long in distance than those of the previous century. The movement east, then west, from the 1690s to the 1720s reflects the changing fortunes of the Southern and South Western areas revealed by the graph of figure 15.

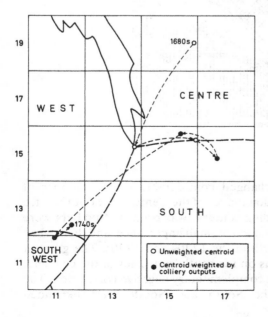

16. Shifts in the colliery centroid, 1680-1749. See figure 5 for location of grid. Output data and estimates were incorporated into the calculation of the centroids from the 1710s onwards, so that the shaded centroids represent approximations of the centre of gravity of coal *production*.

**The duration of collieries**

The relative unimportance of random shifts in the movement of the centroid indicates a slackening in the volatility of the location pattern which resulted from the emergence and extinction of numerous short-lived collieries, a pervasive tendency before 1690. Although a majority of collieries was still short-lived after 1690, the size of this majority seems to have been much smaller than earlier. Individual collieries were worked for over twenty-five years in each area but the West, five in the Centre and three in the South West, and in the Centre Haigh (2) and Winstanley (6) continued inexorably on beyond a century of continuous operation. It is clear, too, from figure 17, that short-lived workings clustered around collieries of longer duration in mining districts in each area. The ephemeral works operated at different times, so that coal extraction now stretched over much longer periods of time in particular localities, even if the majority of collieries was still short-lived. The spatial pattern of mining was thus much more stable on the small scale after 1689 than it had been during the previous century.

Of course, it would be foolish not to acknowledge that much of this impression of massive short-term change could be due to the sporadic and fragmented nature of the evidence that yields it. Moreover, perhaps one should expect more abundant evidence to survive from a more recent period: it may therefore be that the tendency towards greater locational stability is to some extent spurious. But this cannot be completely so. On the one hand, reliable evidence of short-term working, which exists for eight collieries, is more abundant than for the previous century, whilst on the other hand there is also much more trustworthy evidence in account books and court cases of the continuous operation of particular collieries for long periods of time.[1] Furthermore, the long spate of litigation which was instrumental in the closure of many seventeenth-century collieries abated after 1690, affecting only nine out of fifty-seven, compared with twenty-two out of fifty-seven collieries before 1690.[2] Thus, all the evidence which has survived, scrappy though it is, points to the conclusion that although most collieries still had short lives, many more than had done so previously lasted for over ten years. At least ten per cent of the colliery population worked for more than twenty-five years between 1690 and 1739 and the long-lived colliery became a feature of each coalfield area except the West.

**The size of collieries**

Many more statements of colliery output, and indexes from which output can be estimated, have survived for this period than for the century before it. As a result, the plot of colliery sizes on figure 18 is much denser than on figure 7. The range of accounted outputs expanded at both ends in the early eighteenth century, particularly upwards. Before 1689, 2500 tons a year was

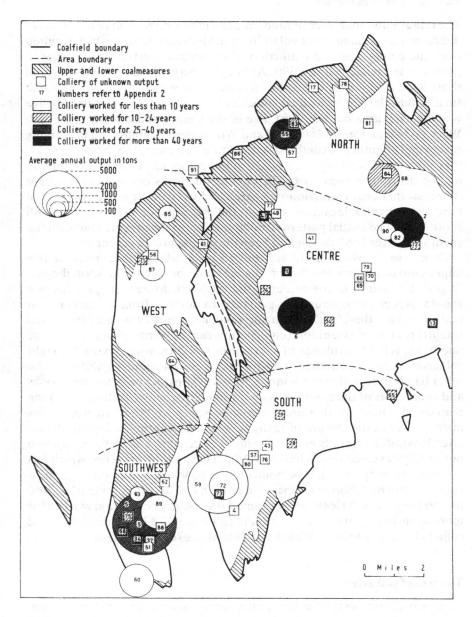

17. Distribution, duration and size of collieries, 1690-1739. See figure 6 for explanation of volumetric symbols.

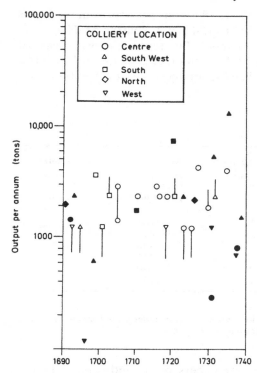

18. Recorded and estimated colliery outputs, 1690-1739. Shaded symbols represent recorded outputs, open symbols estimates. Vertical lines show possible ranges of error of estimates.

the largest recorded output, 4000 tons a year was the highest estimate and the vast majority of statements and estimates fell below 2000 tons a year. After 1690, one colliery averaged 10,000 tons a year in the 1730s and three outputs of over 4000 tons a year were recorded.[3] Clearly, the production of particular collieries increased considerably as the number of collieries increased. This increase in colliery size, like that of numbers, was particularly marked only in the southern parts of the coalfield.

Although there was a small expansion of the colliery population in the West, and although the Blaguegate area of Lathom was almost continuously exploited after 1689, all the outputs recorded in that area were small, even by the standards of a century earlier. No colliery there raised as much as 1500 tons a year. Of the three for which records have survived, Rainford (64) produced only 365 tons in 1696 and 1697; Newburgh (85) averaged only 614 tons a year 1736-9, and Westhead (87) raised 1279 tons in 1731.[4] There is therefore no evidence of expansion in the size of collieries beyond the average of the previous century in the West. In the North, two collieries produced over 2000 tons a year, one at the very beginning of the period and one in the 1720s, and although numbers grew accounted outputs did not rise beyond the size of the normal colliery of the seventeenth century here, either.[5] Neither did they do so by much in the Central area, where

Winstanley (6) continued to raise about 1500 tons per annum in the 1690s and, probably, in the 1710s.[6] The Bradshaigh's colliery at Haigh (2) perhaps grew to an average output of about 4000 tons a year, with peaks of over 6000 tons a year in the early 1730s,[7] but other collieries at Haigh (82 and 90), leased out by the Bradshaighs, produced less than 1000 tons a year.[8]

TABLE 4. *Productivity per pit, 1690-1739*

| Colliery | Years | Av. no. of pits p.a. | Av. annual output per pit in tons (o) | Coefficient of annual variation of (o) |
|---|---|---|---|---|
| Thatto Heath (59) | 1718-22 | 7 | 1166 | |
| Haigh (2) | 1725-55 | 3.46 | * (i) 1441 | 32% |
| | | | †(ii) 1490 | 14% |
| Haigh (82) | 1728-30 | 1 | 413 | |
| Westhead (87) | 1731 | 2 | 640 | |
| Haigh (90) | 1737-56 | 1 | 1132 | |

* See Appendix 4 for method of calculation
† As (i) but ignoring years of exceptionally low output due to an underground fire and two years when output was exceptionally high as new pits were opened.

Estimates made from pit and hewer numbers, profits and so on, reinforce this impression that collieries grew little in size in the three areas away from the southern margins of the coalfield (see figure 18). Information from which estimates of output per pit can be made has survived for only six collieries between 1690 and 1739. Table 4 shows that the annual productivity of individual pits remained at the seventeenth-century level, running up to c.1500 tons at pits that were operated continuously at full capacity and down to minuscule amounts at pits worked intermittently by only one or two hewers. Even at the fully exploited pits, productivity still varied considerably from year to year according to underground and market conditions. If it is concluded that the same pit and hewer productivity levels as before 1690 continued to hold, then most collieries in the North, West and Centre grew little from the normal pre-1690 sizes, although the greater relative frequency of collieries with more than one pit suggests a general unspectacular increase in colliery capacities and it is possible that a Shevington colliery (49) raised nearly as much as Haigh (2) in c.1730.[9] It seems then, that the slight rise in colliery numbers which occurred for most of this period over most of these areas was paralleled by an increase in colliery sizes, but that the burst in numbers in the North in the 1720s and 1730s was not matched by a commensurate expansion in the capacities of the collieries there.[10]

The size of collieries in the South and South West grew and shrank in unison with numerical trends. In the South, Edge Green colliery (65) was probably working through three pits by about 1700 and producing, therefore, somewhere between 4000 and 5000 tons a year then.[11] A colliery larger than any other on the coalfield had appeared in a previously almost unexploited area. In 1701 Parr colliery (43) had more than one pit and so did nearby Garswood colliery (29) in 1721,[12] whilst two Sutton collieries (72 and 73) raised about 4000 tons between them in 1710.[13] These were quite large collieries by the standard of the rest of the coalfield. By that same standard, Thatto Heath colliery (59) was enormous, raising 900 tons in 'a few summer months' of 1717 and averaging about 8000 tons a year – twice as much as any other past or current works – through seven shafts between 1718 and 1722.[14] No later outputs were recorded for any of these collieries of the South and most of them seem to have ceased production during the late 1720s and 1730s.[15]

Whilst these collieries thrived in the South none in the South West surpassed them. The new Tarbock works (60) raised about 2500 tons during each of the two years it functioned in the 1690s and so did Prescot Hall (5) in 1721.[16] Knowsley colliery (63) produced a minuscule 500 tons in 1697.[17] After the 1720s, when the Southern collieries waned in size and numbers, those of the South West waxed. Knowsley (63) expanded to two pits[18] and at Prescot Hall (5) output more than doubled between 1721 and 1731, when 6000 tons were raised.[19] By 1735 output had more than doubled again to 15,000 tons a year.[20] By that time, too, two other South Western collieries may well have been of the same size. In 1730 Whiston (34) was described by the lessee of Prescot Hall (5) as a 'vast colliery which must be a prejudice to that at Prescot',[21] and in 1735 an agent of Prescot Hall (5) reported that 'there is . . . a large colliery . . . about a mile and a half nearer to Liverpool, which is a disadvantage to ye sale of ye coals at Prescot: and tis believed that M[r] Case may (if he please) increase his quantity there'.[22] It seems reasonable to conclude that these collieries were of the same order of magnitude as the leviathan at Prescot Hall.

## Supplementary evidence

It is clear that large-scale changes began to occur in the South West Lancashire mining industry after 1690 and that the chronology of change was complex in the southern parts of the coalfield. But the improved evidence of colliery numbers and sizes is not commensurate with the increased geographical and temporal variety, and many of the inferences about the course of change drawn in the previous section are very tentative. Large collieries were few and badly documented: even one more piece of evidence of large-scale mining would throw awry the spatial and chronological patterns reconstructed in the previous section. It seems prudent, therefore, to attempt some kind of corroboration from a completely independent body of evidence.

If they were as detailed as they can sometimes be, parish registers would provide such data in the occupations of the fathers of baptised children.[23] Eleven parishes lay within or impinged upon the coalfield. Some of them were large and many chapels had been established, so that there were nineteen places at which vital events were registered by residents of the coalfield in the early eighteenth century. People from particular townships often used different places of registration, some going to their parish church, others to one or more chapels in their parish. To complicate matters further, parts of some townships were much nearer to a church or chapel in a neighbouring parish than to any place of worship in their own and it is quite clear that in some parts of the coalfield people registered baptisms in parishes in which they did not live. A reconstruction of the distribution of colliers thus requires the registration of places of residence as well as occupations. Such detail is never simultaneously available in all the church and chapel registers – indeed, some registers never contained this amount of detail at any time in the eighteenth century. However, in the 1720s enough come together to allow a reasonable reconstruction of the frequency with which colliers resident in most of the townships on the coalfield had children baptised.

Figure 19 shows those townships in which the registers of the local church or chapel did not contain occupations and the numbers of colliers who were ascribed to each township in the registers which do contain places of residence and occupations in the 1720s. These numbers have been converted to annual frequencies because many of the registers do not contain the information for the full ten years. The distribution of colliers over those parts of the coalfield for which the evidence exists is very similar to the reconstructed pattern of collieries (compare figures 19 and 17). The South and especially the South West had by far the largest number of collier-fathers and there was also a significant concentration in and around Wigan in the Centre. There were five collieries within or adjacent to Wrightington in the North, and that township, too, appears as one with relatively large numbers of collier-fathers on figure 19. Thus, despite the incompleteness of the evidence, the parish register data for the 1720s do seem to confirm the broad lineaments of the distribution pattern of collieries reconstructed from the mining records. There is perhaps one area where the degree of similarity is markedly less than elsewhere: there seem to have been relatively few colliers in Haigh and Aspull to the north east of Wigan, whilst this was undoubtedly an important mining district (see figure 17). This is probably because many of the workers at the Haigh and Aspull collieries lived in Wigan itself, because of an under-registration of miners in these townships due to the Wigan registers having an incomplete listing of occupations, and because of the use of Standish and Blackrod churches,

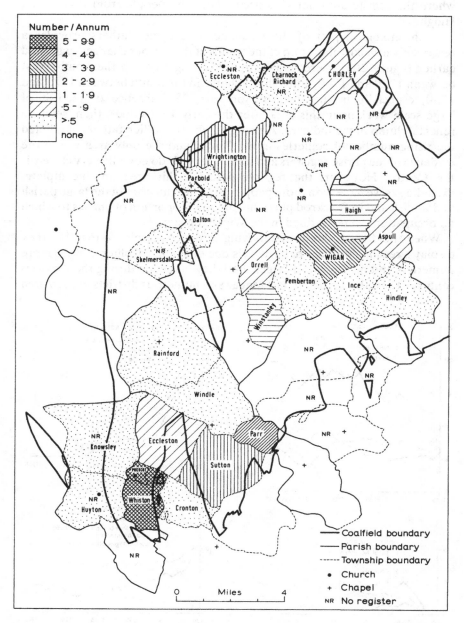

19. Distribution by township of fathers recorded as colliers in parish baptism registers of the 1720s. Multiple mentions of particular individuals are included. Entries from different registers have been accumulated for each township.

where the registers do not give occupations, by people from Aspull and Haigh.

A chronological plot of the frequencies of collier-fathers provides a reasonably firm corroboration of the trends of the various changes suggested earlier (figure 20). There was a strongly rising trend in the South West between 1720 and 1740, and a sharp plunge in the South between 1725 and 1730, especially in the township of Parr. The absence of records of large-scale working in this area after the early 1720s does, then, seem to reflect a slump in the industry. A rising number of collier-fathers paralleled the rising number of collieries in the North, and the only area where the evidence of the parish registers and mining records do not tally very closely is the Centre. However, the register data for this area are incomplete, excluding as we have seen those people using the churches in the large parish of Standish and in Blackrod plus a fluctuating proportion of entries to which no occupations were ascribed in the Wigan register.

When the diverse shreds of surviving evidence are pieced together they display a reasonably clear pattern. It is one of change. Collieries were more numerous and on the whole larger than they had been during the previous century. They also tended to last longer and numerous districts in each area

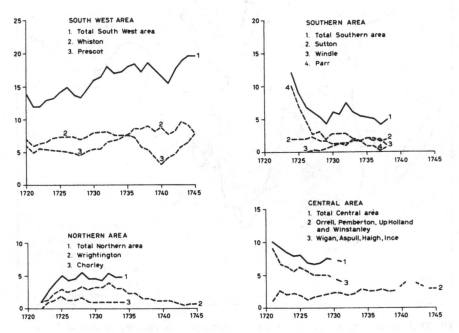

20. Numbers of fathers recorded as colliers in parish baptism registers of certain townships, 1720-45. Multiple mentions of particular individuals are included. Entries from different registers have been accumulated for each township.

of the coalfield were worked more or less continuously through this period by one long-lived or a few sequent short-lived collieries. Whereas openings and closures had caused massive change in a context of very slow overall growth in the seventeenth century, in the early eighteenth century a steadier industry cumulatively expanded. To be sure, many collieries were still short lived, but a changing tendency was clearly evident in this respect.

There was expansion everywhere. An attempt has been made to depict it on table 5.[24] In the West growth was very slow, bringing the industry up to a level reached over a century earlier in other parts of the coalfield. In the Centre, which lost its pre-eminence, the colliery population expanded slowly and the maximum and median size of collieries moved in the same way, quickening the trend of the previous century slightly but perceptibly. There was a spurt of growth in the North in the 1720s when the colliery population doubled, the largest works were as big as most of those in the Centre, and the collier population of Wrightington had become as numerous as that of any other rural area of the coalfield outside the South West. But no colliery which regularly produced over 4000 tons a year was recorded in any of these three areas, although 6000 tons was occasionally exceeded at Haigh (2) in the 1730s. Even this was dwarfed by the tonnages raised at collieries in the South and South West. The industry grew strongly in the former before the mid-1720s, with one colliery raising over 8000 tons a year, but this expansion seems to have been stifled in the last fifteen years of the period. At the same time, the industry suddenly expanded in the South West, which probably produced more per year in the late 1730s than the whole coalfield had twenty or so years earlier. This massive upsurge was marked by the development of collieries which were nearly four times as large as any which had previously worked on the coalfield except Thatto Heath (59).[25]

TABLE 5. *Estimates of the annual outputs of coalfield areas, 1690-1739*

|  | West | North | Centre | South | South West | Total |
|---|---|---|---|---|---|---|
| 1690s early | 1,900 | 3,000 | 10,500 | 5,700 | 6,900 | 28,000 |
| 1720s late | 1,500 | 6,500 | 13,400 | 16,600 | 7,500 | 45,500 |
| 1730s | 2,700 | 7,400 | 13,200 | 5,800 | 39,800 | 68,900 |

CHAPTER 5

# FACTORS AFFECTING THE LOCATION PATTERN OF COLLIERIES, 1690-1739

## Demand

In the seventeenth century, overland transport costs were high and about equal everywhere so that collieries could not be located far from the markets they served. The all-pervasiveness of this close link between collieries and markets was broken by transport innovations in the early eighteenth century. In this chapter it is appropriate to consider transportation as a variable which could skew demand away from a simple and complete reflection of market locations, which will be described first.

### The distribution of coal markets

*Households.* Information about how the coal consumption of individual households changed in the early eighteenth century does not exist. It seems reasonable to assume that it remained at its relatively high seventeenth-century level on a coalfield where output was increasing. If this was so, domestic demand must have expanded at the same rate as population, which seems to have grown very quickly. There is a number of estimates of Lancashire's population in the eighteenth century: those of Toynbee and Gonner, reproduced on table 6, lie at the two extremes of the range of calculations.[1] Despite detailed differences, they agree that the population of the county as a whole increased at an unprecedented and unparalleled rate. Quite clearly, if household consumption levels remained stable the increase in the domestic coal market of the county would have been greater than anywhere else in the country.

It is impossible to discover how far the south western part of the county

TABLE 6. *Population estimates for Lancashire in 1700 and 1750*

|  | Total | | Per cent increase | Density per sq. mile | | County rank in density | |
|  | 1700 | 1750 | 1700-50 | 1700 | 1750 | 1700 | 1750 |
|---|---|---|---|---|---|---|---|
| Gonner | 242,000 | 341,451 | 41 | 127 | 179 | 6 | 1 |
| Toynbee | 166,200 | 297,400 | 78 | 107 | 156 | 14 | 5 |

shared in this general increase or how its share was apportioned within the region, but it is commonly agreed that population increased most rapidly in manufacturing areas at that time and not until Preston was reached on a northward journey did 'we . . . come . . . beyond the trading part of the county'.[2] There is no reason to doubt that the rate of population growth in the south west of the county matched that in the south east, nor that the increase in the coalfield's domestic market was, therefore, much more than the county figures suggest.

Population totals derived from counts of registered live births[3] and interpolation between the distributions of 1667 and 1801 suggest that this large increment was unevenly spread over the region (see figures 21, 8 and 44). It seems to have been concentrated on the coalfield and its southern, south western and south eastern margins in the hundreds of West Derby and Leyland and the boroughs of Liverpool and Wigan. The Lancashire plain and the Western area of the coalfield were only slightly affected and the belt of land between Liverpool and Hindley along the southern flank of the coalfield, with a southward projection to Warrington, intensified its demographic pre-eminence. In 1667 there was a small northward salient of well-settled land from Wigan to Standish: by 1801 it extended through

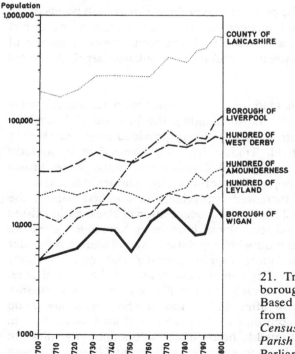

21. Trend of baptisms in selected boroughs and hundreds, 1700-99. Based on information abstracted from parish registers given in *Census of Population, 1841: Parish Register Abstracts*, British Parliamentary Papers, 1841.

Chorley and Whittle-le-Woods to Preston, creating a continuous belt of high densities from Warrington to Preston along the main north road. One can only speculate about when this extension began, but there is some evidence to suggest that it was before 1740. Standish parish, which comprised an extensive area between Wigan and Chorley, certainly seems to have experienced population growth before then. In 1664 there were 459 households in the parish according to the Hearth Tax Assessments, whilst a census of 1754 enumerated 643 families.[4] Of course, these figures can be no more than vaguely accurate and their span overlaps both ends of the period now under review, but if any inference at all can be drawn from them it is that the thickening of population evident on the 1801 map had already begun by 1740.

Most of the population increase probably occurred in the towns and villages. Moffitt estimated that Croston contained 3000 people in 1750, when Standish village had 2000 or so,[5] a large proportion of the occupants of the households enumerated in its extensive parish in 1754. The towns grew at varying rates. Wigan probably had a population of between 5000 and 8000 by 1750, Preston about the same and Ormskirk, Prescot and Warrington between 3000 and 4000. Liverpool outstripped them all as it climbed rapidly into a much higher stratum of the urban hierarchy. In 1700 its population was about 5000; by 1720 it was 10,000, and by 1750 it had reached about 22,000.[6] The port was one of the major growth points of the English economy and must have been the major stimulant to the increase of population in the towns and villages of the south western corner of Lancashire and to the industry upon which that population largely depended for its livelihood.

*Industries.* It is probable that industrial coal consumption became relatively more important after 1690, notwithstanding the increase of household demand, and that concentrations of industry provided major markets in some parts of the coalfield. Trades that were dependent upon local customers, such as smithing and limeburning, must have grown commensurately with the population: certainly, the number of blacksmiths' wills proved at Chester increased appreciably in the first half of the eighteenth century (figure 22[7]). But no concentration of blacksmiths existed outside the major towns and generally such trades as these must have been distributed in close correlation with population. Manufactures for wider markets provided bigger, more rapidly growing and more spatially concentrated fuel demands. The varied metal trades of the Central area, which had provided a considerable market for coal in the seventeenth century, continued to grow after 1690, most reaching their maximum importance in the 1720s according to the evidence of wills reproduced in figure 22. Chowbent continued to house more nailors than any village in South West Lancashire, but the scattered settlements south of Wigan

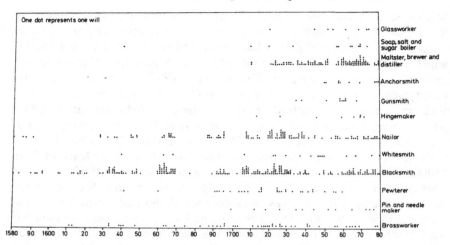

22. Growth of coal-consuming trades during the seventeenth and eighteenth centuries according to the evidence of wills.

increased considerably in importance in the early eighteenth century as the industry spread through the Central area of the coalfield towards the port which provided its major outlet to colonial markets (see figure 23).[8] The Warrington ironmongers who funnelled metal into the region were by far the largest customers for bar iron at Bodfari, Cranage and Warmingham and for rod iron at Vale Royal.[9] These supplies were supplemented by ironworks opened on the coalfield. Those at Haigh, which had operated briefly fifty years before, were reopened in 1716 by John Russell of Portway in

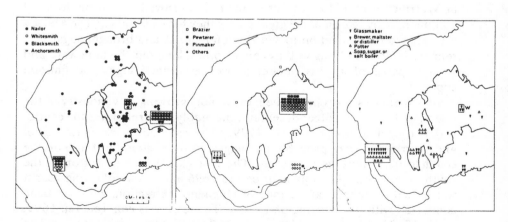

23. Distribution of coal-using tradesmen who left wills, 1690-1755. Tradesmen who died 1740-55 are included to represent those working before 1740 who died subsequently. Based on 'Indexes to wills'.

Staffordshire. During his tenancy the forge was converted into 'a Working Smithy or Shop . . . with . . . convenient wheels or other engines for that purpose and . . . several Anvils Hammers Bellows and other Implements' which were joined by a slitting mill before 1739. Another slitting mill was opened at Great Sankey near Warrington in 1727, operating for at least twenty years, and in the same year John Chadwick acquired the forge at Birkacre, a few miles south of Chorley. These ironworks were attracted by water power, coal and markets: in turn, they stimulated the development of local smelting, which began in 1720 when Edward Hall, who also had interests in Staffordshire and Furness, took out a thirty year lease of Carr Mill on the Sankey Brook. Although there were small deposits of iron ore in nearby Parr, most supplies were shipped down the Weaver from Staffordshire to Hall's furnaces and forges, which derived power from a new dam and three new mill pools. Thus, the nailors, hingemakers, locksmiths, whitesmiths and other makers of small iron wares drew the earlier processes of iron-making onto the coalfield, and all of these except smelting required coal.

Unlike the iron trades, the non-ferrous metal manufactures did not spread across the coalfield as they grew but remained tightly concentrated in Wigan, where the largest industrial concentration on the coalfield remained under firm guild control. The non-ferrous metals manufactures apparently declined in proportional importance as textiles, retailing and other services developed in the town, and the pewterers lost their dominance of borough offices, but this was still, in the 1720s, the single most important group of industries in Wigan.[10] The brass manufacturers of Wigan, Holywell and Macclesfield as well as local pinmakers must have benefited from, and perhaps even encouraged, the establishment of the Patten's copper smelter at Warrington in 1719: 'ore brought from Cornwall . . . turns to good account here by reason of the great plenty they have of coals'.[11] The Mersey was opened for navigation to Warrington in 1701 and brought sugar to a refinery built in 1717 as well as copper ore, taking out the tableware and window panes of a glass smelter as well as refined sugar and finished copper.[12]

Glassmaking was probably the most rapidly growing new industry of the region in the early eighteenth century. Inevitably, Liverpool was the centre, with two smelters in operation by 1729.[13] The waterside works in Liverpool and Warrington were probably the biggest in the region, but those on the coalfield itself were as numerous. The first glass-house in Sutton in the southern area was probably built before 1696 and in the following year it was leased by John Leafe, a man of Huguenot extraction, who had been connected with the Warrington factory. The lease was confirmed in 1699, and in 1713 Leafe purchased the works and the land on which it stood.[14] The glass bottles made at 'Thatway Heath' in 1721 probably came from this works,[15] which was joined before 1727 by that of Samuel Bowers, erected by the side of Thatto Heath colliery.[16] Thomas Cobham began glassmaking at nearby Prescot in c.1719 and his works, like those in Sutton, Liverpool and Warrington, were active until after 1740.[17]

All three coalfield glass-houses were intimately connected with the local coal industry. Bowers purchased Thatto Heath colliery (59) in the 1720s and Thomas Cobham was the partner of Thomas Makin in Prescot Hall (5) colliery. Makin was also the father-in-law of John Leafe's son Zachariah.[18] Josiah Poole, who contracted to build a glassworks in Liverpool in 1715, leased Knowsley colliery (63) until 1730. The reason for these entrepreneurial links was that glassmaking consumed large amounts of a particular kind of coal. Bowers purchased Thatto Heath colliery (59) because his glass-house 'near and adjacent unto the said Coale works required a great quantity of Coales for carrying on the same',[19] and Leafe made an agreement in 1697 which provided a monopoly of his coal requirements to Thomas Roughley, provided that the fuel 'was sufficient and good wherewith to suffice . . . the said Glasse house'.[20] In 1735 the demands of the Cobham works at Prescot kept two of the pits of Prescot Hall colliery (5) continually active. These pits produced only glass-coal from a seam of four feet which was probably the Sir John mine, a vein which outcropped near Prescot and definitely served the needs of glass smelters in 1750. Its coal was 'not fit for any other sale', presumably because its hardness made it difficult to kindle in a domestic grate.[21] The Arley mine, renowned for a hardness and lack of sulphur which made it of particular value to the St Helens' glass industry of the nineteenth and twentieth centuries, outcropped at Thatto Heath. Thus, a relationship between special coals and glassmaking developed in Prescot and Sutton similar to that which seems to have existed between cannel and smith's coal and metal manufacture in Wigan in the previous century. By 1740 the pits of Prescot (5) and Thatto Heath (59) must have been supplying at least 6000 tons of hard coal a year to glassworks in their vicinity, in Liverpool and in Warrington.

Glassmaking was located at points of raw materials extraction and at waterside markets. This split response to both supply and demand was common to the other newly emerging industries of pottery and brewing (see figure 23). Prescot, Rainford and Sutton, each with appropriate raw materials as well as coal,[22] were the centres of both industries on the coalfield, but these advantages seem to have been outweighed by those of a market location because Liverpool was overwhelmingly dominant in both industries.

The saltboiling industry also developed rapidly on the Mersey. After the discovery of rock salt in 1670 at Marbury near Northwich, the cheapest arrangement of production was to haul the raw salt to the Mersey where it met coal from Prescot and Whiston and whence the refined salt could be shipped directly for export. The first works to take advantage of the portability of rock salt were built at Frodsham on the lower Weaver some time before 1694. This joint-stock enterprise was immediately successful and declared a dividend in 1695, but it is not known how long it operated.[23] In 1696 a saltworks was built at Liverpool, probably by the Blackburnes, who kept the works going until the late eighteenth century, and in 1697 a third rock salt refinery was erected by Sir Thomas Johnson at Dungeon, the shipping point on the Mersey for Prescot coal bound for the Cheshire brine works.[24] This rapid expansion in the refining capacity of the Mersey estuary was halted in 1702 by an Act which prohibited the erection of any new rock salt works beyond ten miles from the places of extraction.[25] The links between manufacturing and coal supplies which developed in the glass industry were duplicated in salt production when Jonathan Case of Huyton, who owned two large collieries at Whiston and Huyton near Prescot (34 and 75), acquired the Dungeon works at some time before 1740.[26]

The industrial geography which resulted from these developments differed quite markedly from that of the seventeenth century. In the Central area the metal manufactures expanded and attracted iron production, but the influx of new industries into this landlocked part of the coalfield was small. In contrast, there was a considerable spawning of new manufactures on and near the Mersey. Salt, sugar and glass refining, potting and brewing were all carried on in works which must each have consumed relatively large amounts of coal, a speculation which is lent some support by the entrepreneurial links which developed between smelting and mining. Liverpool and Warrington were the fastest developing industrial locations, on the horns of the crescent of the Mersey estuary where navigable water approached closest to the coalfield. Liverpool took over from Wigan as the major industrial centre of the region and the axis of industrial coal consumption swung dramatically to the south during the course of the early eighteenth century.

*Exports.* The nadir of coal exports through Liverpool was reached at 194 tons in 1680/1. Although the quantity rose to 2606 tons in 1719/20 it slumped again thereafter and only 1290 tons were shipped out of the port of Chester, which included Mostyn as well as Liverpool, in 1749/50.[27] On the other hand, overland exports to Cheshire grew apace. The locus of the salt industry moved northwards after the discovery of rock salt. Even though a wholesale transference of the refineries to the Mersey was prevented by the Act of 1702, Northwich itself was far enough north to enable Lancashire coal to undercut the price of that from Staffordshire in the new centre of the industry, even though supplies from the latter coalfield were still cheaper in Middlewich. The passage of Lancashire coal to the main refineries was eased by the opening of the Weaver Navigation in 1732 and the quantity of coal shipped up-river reached a parity with the weight of white salt shipped down by the early 1740s: in 1732/3, 2635 tons of coal passed along the Weaver and in 1742/3, 8457 tons were shipped.[28] By 1740 Lancashire coal had a virtual monopoly in the refining of both rock salt and brine.

*Alternative sources of fuel*

The haphazard gathering of brushwood and turf by manorial tenants must have eroded the domestic coal market to some extent, as did the widespread illegal excavation of coal in roads and gardens. Charcoal supplied the iron furnaces, but the distances over which it was collected suggest that charcoal cannot have provided a reasonable substitute for coal in the forges, chaferies and slitting mills of the region.[29]

Coal's main competitor was peat, which continued to be dug on the mosses of the Centre, West and South and on the plains to the west of the coalfield. All the leases granted by the Gerards of Ince between 1690 and 1710 included the right to turbary; one lease of 1700 carried with it a specific 'Turfe-Roome in and upon Ince Moss', and Amberswood Moss in Ince was still being dug as late as 1778.[30] At Sutton in the South, Moss Reeves were appointed until at least 1750, and at nearby Knowsley in the South West the coal-owning Earl of Derby paid for the digging and dressing of large quantities of peat in the 1720s.[31]

The largest reserves of peat lay in the Western area of the coalfield in Rainford, Bickerstaffe and Lathom. By the time a complete series of Manor Court Books begins in 1728, the Moss Reeves of Rainford already had complete discretion over the organisation of peat cuttings and infringements of their directions were liable to a fine of £1. Exploitation must have been heavy, giving rise to numerous injunctions about ditching and the construction of 'Moss Gutters' to drain away the water that would otherwise accumulate in the diggings. In 1735 a communal levy was raised to maintain access roads and it was ordered that diggers 'shall not get nearer the sand or botham of the moss than two foot (viz) to have two foot of solet moss in the botham . . . also every one whoever hath great depth of Moss shall not get turfe to drown in his room nor get it no deeper than the fall will take dry'. More than half of the items in the Rainford Manor Court Book of 1735 concern peat cutting; cultivators had to regulate their activities for the convenience of users of the moss rooms in 1736, and in 1742 Rainford people were ordered by the Moss Reeves to scour their ditches 'for the Conveniency of Billinge People Carting and Carrying of Turves'.[32] All this is symptomatic of the heavy exploitation of peat, some of it for sale outside Rainford.

Despite the continued availability of peat in the West, coal was imported at what must have been considerable expense. The Blundells, who dug peat on their own estate at Little Crosby, carted coal over the whole breadth of the mosslands from Haigh (2), Blaguegate (71) and Huyton (75) and the mugworks at Rainford were supplied with coal from Parr (67) and

Garswood (29) collieries situated five and seven miles away.[33] This demand was not restricted to the gentry and industry of the West either, because 'yᵉ Poor' of the Lathom area purchased coal at local collieries (71 and 87) in 1730 despite its high price there.[34] Thus, the reason for the underdeveloped state of the mining industry in the West cannot have been simply that a substitute fuel was available in the mosslands. Perhaps this combined with a scattered rural population and cheaply mined coal in nearby townships to reduce the pressure of demand below that which would sustain the high prices necessary to excavate pits in this flat, drift-smeared landscape.

## Coal markets and the colliery location pattern

The composition and spatial distribution of demand changed cumulatively and considerably after 1690, and by 1739 it was quite different from what it had been in the seventeenth century. Domestic demand probably remained predominant, though by a reduced margin, and this, coupled with the continued and perhaps increased availability of peat, must have been related to the ephemerality of exploitation at many sites, just as it had been in the seventeenth century. However, the growth of industrial demand and the decline in the proportional share of the domestic market accounted for by scattered rural households meant that the nature of coal demands were on the whole less conducive to discontinuous working. This, again, seems to have elicited the predictable response from the mining industry: colliery life-spans were certainly longer on average during this period.

The changing composition of demand was also clearly reflected in the emergence of strong distinctions between the mining industries of different areas of the coalfield. The new industries and the most rapid urban expansion occurred on and near the Mersey in the market areas of the collieries of the South and South West. The largest collieries of the coalfield, hugely different from any of the previous century, developed in these areas alone. Elsewhere, coal markets and the coal industry gradually evolved from those of the seventeenth century. By 1739 the coal industry of the Central area had fallen well behind that of the South West in terms of total output and the average size of collieries, because its domestic and small-scale industrial markets expanded less rapidly than the new industries on the banks of the Mersey.

These contrasting correlations between variations in demand and differences in the mining industry from area to area on the coalfield were also expressed in price trends. Generally, prices were highest throughout the period in the South West, where they rose from 1½d and 2d to 2d and 2½d per cwt during the course of these fifty years, although they seem to have remained steady until at least 1710 (figure 24). The detailed geography of coal prices in the southern limbs of the coalfield was quite complex (see figure 25): the highest prices were charged at collieries nearest to Liverpool

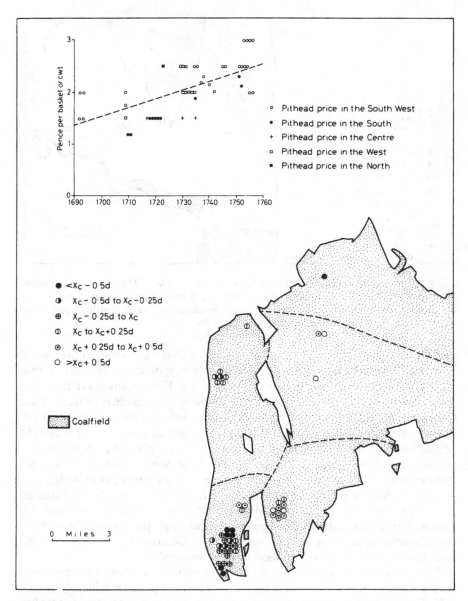

24. Pithead coal prices, 1690-1739. Prices from the South and South West are included on the graph until 1757, until when these areas were unaffected by water transport, to facilitate the construction of figure 25. Mapped symbols represent residuals from the regression line $X_c$ plotted on the graph.

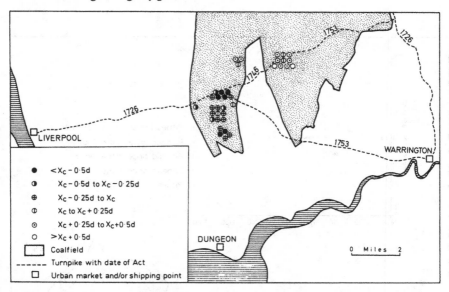

25. Coal prices in the South and South Western areas, 1690-1756. The map includes all prices recorded in these areas from 1690 until road haulage ceased to be the dominant means of getting coal to Liverpool. Symbols represent residuals from the regression line $X_c$ plotted on the graph of figure 24.

and Dungeon, with a saddle in between and a sharp dip away to the north east, a pattern which indicates a very precise response to the pulls of the shipping and consumption points on the Mersey. The evidence available on prices elsewhere on the coalfield is too sparse to yield a clear pattern. The Centre seems still to have been an area of low prices with ordinary coal 1d per cwt cheaper than in the South West. The price of small cannel at Haigh (2) drifted up from 2d per cwt in 1687 to 2½d in c.1730, when best 'round' cannel sold at 3d per cwt. The collieries of the West, insulated from the competition of more intensively mined areas, sold expensive coal, whilst the one price which has survived for the North is the highest on the coalfield in the 1720s. No inference can be drawn from this snippet of evidence, of course, but it may not be coincidental that this high price was followed by considerable expansion in the number of collieries in an area where population seems to have increased quite considerably.

Thus, as markets became more varied in nature and rates of expansion from area to area, so did the mining industry and the prices at which coal was sold. Demand obviously exerted a considerable influence on the geography of the mining industry. But it was not an all-pervasive, completely determinative influence: supply conditions must have changed in unison to allow the industry to respond. In the seventeenth century capital shortages, drainage problems and, in some districts, the fragmentation of coal reserves had limited the size to which individual collieries could grow. These constraints must have been relaxed to allow correlated changes in markets and mining. Furthermore, certain small-scale features of the geography of mining are contradictory to a simply hypothesis of demand controlled location. Given that most collieries worked sites opened in the seventeenth century and that the locations of the major centres of demand were the same as then (even though their growth rates were very different), there was still a lack of close correspondence between the situations of markets and the collieries that served them. Wigan continued to be supplied from pits two or three miles distant, and a location on the southern or western edge of the coalmeasures outcrop or on the Croxteth Park outlier would have given closest proximity to the Mersey, whilst the collieries of the South and South West were all eccentric to these positions. On a smaller scale still, there was virtually no exploitation at Rainford, where the village and mugworks provided a relatively large and continuous demand.

The chronology of changes in demand and in mining seem not to have been closely parallel, either. In the West there was population growth and hardly any expansion in the industry, and, even more discordant, there was a contrast between the apparently smooth acceleration of demand and the sudden upsurge in the number of collieries in the North in the 1730s, the sudden slump in the South in the same decade, and the contemporaneous meteoric rise of the South West. Moreover, the hypothesis that industrial demands led to greater stability in the industry, which holds firm for the coalfield considered as a whole, disintegrates somewhat when pursued onto a smaller scale because collieries were not, on average, any more long-lived in the South and South West, which served the main industrial markets, than they were elsewhere on the coalfield (see figure 17).

To explain these discrepancies requires an examination of supply factors, but, in addition, changes in transport methods, and therefore in transfer costs, also occurred in the early eighteenth century and their effects will be considered first.

## Variable transport costs and methods

Roads almost certainly worsened during the early eighteenth century and horses and panniers were substituted for carts and wains, which were immobilised by ruts and mire.[35] The roads which linked Wigan and Liverpool to their coal supplies two and five miles away became execrable. At Whelley, between Haigh and Wigan, mourners had 'comen with dead corpse women hath torne their shous of ther feet some have lost them in the mire' in 1712,[36] and Nicholas Blundell's coach 'was laid fast in yᵉ Rode being so deep, so we left it in the Leane all night and we went on with our horses to Wigan' as he neared the town from Liverpool in 1724.[37] The road from Prescot to Liverpool carried all the port's coal and most of its exports. It was almost impassable to carts and could normally only be used conveniently in summer. Coal prices were more than doubled by the five mile journey even when the road was at its best; 'the Inhabitants . . . have suffer'd much for want of getting their coales' and after bad weather in 1725 a horseload, usually 7d, sold for 2/6d.[38]

These conditions must, of course, have had an enormous effect upon mining. Potential demand could be transmitted only weakly even over short distances. A location at the market was more than ever desirable for colliery proprietors; when this was impossible the poor roads must have drastically limited the quantities that could be hauled, made supplies irregular and unreliable and frequently pushed up prices to prohibitive levels. It was probably this which restricted the growth of output in the South West to a niggardly thirty per cent from c.6000 tons a year between 1690 and the middle 1720s, despite the rapid growth of population and industry in Liverpool. A quarter of a century of lagging coal supplies, inexorable narrowing of the transport bottleneck and economic growth must have built up enormous pressure in Liverpool by the middle 1720s. Improved transport could suddenly release it, bringing massive and rapid expansion in output in the area served by the new route. Each sudden change in output in particular areas of the coalfield seems to have been triggered in this way.

The growth of mining in the South from the last decade of the seventeenth century to the third of the eighteenth was associated with its supply of industry on the upper Mersey estuary, made navigable in 1701, and its advantageous location for the overland supply of the Cheshire salt industry. The rapid decline of the 1730s occurred when new transport routes gave the South West easier and cheaper access to these markets. The Act of 1702 which prohibited the construction of further rock salt refineries on the Mersey kept the bulk of the salt-refining market within reach of supplies from the Southern area. Although coal from Prescot and Whiston in the South West could be floated cheaply across the estuary from Dungeon to Frodsham, unloading and reloading were expensive and the stretch of water was not long enough to compensate for the shorter distance between the saltworks and the collieries in Sutton (59, 72 and 73), Garswood (29) and Haydock (28). The northward shift of the salt industry to Northwich after the discovery of rock salt put it beyond the reach of Staffordshire coal, too, and by 1710 the collieries of the South were sending 'a full 3d part' of their output to the saltfield. Thatto Heath colliery (59), the largest on the coalfield in the early eighteenth century, witnessed the strength of this connection, being developed in 1691 by William Mascall of Frodsham and run until 1727 by Richard Turner of Middlewich.[39]

Turner pulled out of the colliery as the mining industry of the South declined sharply. At the same time, output surged ahead in the South West. These complementary shifts were caused by the construction of two new lines of transport. The Weaver Navigation was largely completed between

1728 and 1732, giving a complete water route from Dungeon to Northwich and a decisive advantage over the South to the collieries around Prescot and Whiston in the supply of Cheshire, which amounted to 8500 tons a year by 1740. The turnpike from Prescot to Liverpool was built between 1726 and 1732 and the rates for carriage upon it greatly favoured coal, penalising manufactured goods, thus encouraging the development of the latter in Liverpool itself.[40] The rapid growth of output in the South West, which can be precisely dated in the 1730s, must have been the product of these two stimuli, one releasing the long pent up demands of Liverpool, the other usurping the Cheshire market which had been monopolised by the South for thirty years.

The Douglas Navigation, the first waterway on the coalfield, was not completed from Wigan to the Ribble before 1742. The appropriate Act was passed in 1720, but the scheme foundered in the general collapse caused by the South Sea bubble speculations and work was not begun in earnest until 1738.[41] Although the undertakers made an agreement with an Orrell colliery owner in the same year, nothing came of it and the only impact of the Navigation before 1740 was off the coalfield in Bispham, where a colliery (91) was opened into the thin seams of the Millstone Grit in 1739. It closed in 1740.[42]

It is possible that the noticeable quickening of mining in the North in the 1720s and 1730s owed something to transport improvements, too. It is difficult to suggest any other explanation and an Act was passed for the turnpiking of the road from Wigan to Preston in 1726. Coal was being carted from pits near Wigan in the Central area to Preston by the 1730s when the market was sufficiently important to the proprietor of Haigh colliery (2) for him to commission a survey of competitors there.[43] Given this, it does not seem unlikely that mining in the North, the nearest coalfield area to Preston, owed its sudden spurt of the 1720s and 1730s to greater penetration of that market after the turnpiking of the road between them.

The incorporation of transport as a variable greatly improves the simple demand orientation model by providing possible explanations for the erratic chronology of expansion in the South, South West and North. However, its inclusion does not help to solve any of the other discrepancies between this model and actuality. These must have been due to variables operating upon the supply of coal.

## Supply

### Coal seams

The fringe of coalmeasures nearest to Preston, and the Croxteth Park outlier, which was closest to Liverpool, were in the *Lenisulcata* series where seams were few and thin, though of high quality. Borings were sunk at

Croxteth, but no workable coal was found before 1740.[44] Until then the outlier was barren. Two collieries definitely worked *Lenisulcata* coals in the North (see figure 17). Harrock Hill (86) was a drift or adit mine driven into the edge of the 500 foot erosion surface. The thinness of the seam caused rapid subterranean progress and the cost of getting coal to the surface soon became prohibitive.[45] The thin seam of Charnock Richard colliery (17), which was the nearest supply point to Preston, gave individual pits short lives and the geological distance between seams in the lower coalmeasures meant that workings must be confined to a single vein. In consequence, workings moved swiftly down-dip and had reached seventy-two yards, much deeper than anywhere else on the coalfield, by 1742.[46] Clearly, working and drainage must have been expensive in the lower coalmeasures of the North and South West and this was probably the reason why coal was hauled across them to Liverpool and Preston from more distant collieries in the main productive series.

In the West, too, difficulties in finding workable seams probably limited mining in districts with relatively favourable locations. The main workings in this area were on valley sides where superficial deposits only thinly camouflaged the underlying coal. At Newburgh (85) the outcrop was of lower coalmeasures, exuding thick ochre water which precipitated and blocked the drainage soughs, causing the closure of the colliery in 1739, just as the Douglas Navigation promised to provide bigger markets to the hitherto remote workings.[47] The more abundant and less troublesome coals of the main productive series were tapped in Lathom (56, 71 and 87), the longest and most extensively exploited district in the West before 1740. Elsewhere in the mossy plain of this area drift was thick and must have precluded the chance discovery of coal, whilst markets were insufficiently large to stimulate systematic searches. The apparent absence of seams certainly contributed to the lack of mining in the extensive Western area of the coalfield.

Numerous seams contorted by folding and faulting provided an abundance of outcrops and shallow reserves on the main productive coalmeasures. Even so, little new coal was opened in the early eighteenth century and most collieries continued to exploit sites first worked in the seventeenth century, despite increased outputs. This inertia was only possible where large reserves were available in plethora of shallow seams.

Greater depths were penetrated than had been customary in the seventeenth century because of this long continuance of mining at most collieries, although even some of the largest were kept shallow by skipping from outcrop to outcrop over a wide area rather than pushing down through the sequence of seams. Only in the North was it impossible to keep workings at less than forty yards; elsewhere, many collieries still operated at less than the seventeenth-century maximum depth of thirty yards, although their number became fewer as time passed.[48] Although the increase in depth was apparently small, the down-dip progression of pits left many waterlogged 'hollows' above current workings and it brought problems. Water caused difficulties almost everywhere before 1740. Wrightington colliery (55) in the North was closed by flooding in c.1710 and Shevington (49) in the Centre was drowned two years later.[49] Adlington colliery (22) seems to have been closed because of the expense of draining new coal in the middle 1720s, when the pits at Blaguegate in the West had exhausted crop coal and were 'overloaded' with water.[50] Even the Great Sough proved insufficient at Haigh (2), where a supplementary engine was built before 1728.[51] Pumps and a sough were required at Orrell (74) and at Edge Green (65) in the South in 1711, where the sough proved inadequate to drain workings that descended from twelve to forty yards in the first twenty years of the period.[52] The full spectrum of drainage experience was recorded at Haydock colliery (28) in the same area: in 1708 a jubilant auditor reported that 'the Spring being so very easy will be drawn by one horse with a Small Engine', but a steam engine was required before 1750.[53]

Fire-damp was an insidious new problem brought by increased depth. Because 'cannel is by many degrees ye more combustible' than ordinary coal,[54] the pits of Aspull (12) and Haigh (2) were most sorely troubled by it, particularly the latter, which were deeper and larger. Bellows were installed at one of the pits there in 1726, but in March 1737 the colliery caught fire. Pits were blocked and the sough was stopped to smother and drown the blaze and by June colliers were sent down to disperse any damp that remained in the charred workings. In October hewers were hired again, but later that month the owner was informed that 'ye Fire is still burning in yr. Cannel Mine'. It was extinguished by February of 1738, but flared up again in the following May, when the pits and sough were again stopped up. Local expertise was obviously inadequate and men were hired from Coalbrookdale and Newcastle. A fire pit was dug and the outbreak was at last, after nine months, contained, but it cost three quarters of the normal output in 1737 and 1738. There were 'Sev.[le] Instances in this neighbourhood of Wigan, of y[e] Difficultys y[t] have been found to Extinguish y[e] Fire in a Delf of Coal' by the time of the Haigh blaze.

Haigh (2) was the largest, oldest and deepest colliery in the Central area of the coalfield, but the rate of extraction was greatest and the collieries were largest in the South West and it was there that the consequences of deeper working were most prevalent. Tarbock colliery (60), on the southern tip of the coalfield, was the best situated to serve Liverpool and Dungeon Point. Water caused problems as soon as it was opened in 1692 and by 1695, when it was closed, the expense of unwatering the workings was consuming half the weekly running costs. The superb location of the colliery attracted adventurers from Staffordshire, who probably brought in colliers from Shropshire in 1716 and at least ten pits were subsequently sunk. But the seams were 'much troubled with water' and after the exhaustion of winnable coal in 1733, when borings stopped at fifty-one yards, nine yards above another seam, the colliery was again closed.[55]

To the north east, on the other side of the south western promontory of coalmeasures, the full range of seams in the main productive series outcropped *en échelon* between Prescot and the southern boundary of Whiston. This superabundance of crop and shallow coal was also partially drift-free, so that the huge increase in the rate of extraction in the district which supplied Liverpool and Cheshire after 1732 did not at first cause any problems: indeed, these geological conditions were no doubt an important factor in the rapid and sustained expansion. The reserves of Prescot Hall (5) were first worked in the sixteenth century and almost continuously exploited thenceforth. The colliery had reached a capacity of 15,000 tons a year by 1735, but still there were several crop pits and a number of others that were only twelve and twenty yards deep.[56] But even in this vicinity there were harbingers of future difficulties if seams were pursued very far along their steep dips. Huyton colliery (58) was hit by explosions of damp as early as the 1690s; a large engine was necessary to drain the deeper coals at Whiston (34) or Huyton colliery (75) in 1719, and windmills and gins were built to serve the same purpose at Prescot Hall (5) in the 1730s.[57] This inexorable trend culminated in the construction of the first steam engine on the coalfield at Whiston (34) in c.1729,[58] but drainage and damp problems were not generally too great in the South Western area as a whole before 1740 because of the wealth of shallow reserves.

The colliery distribution pattern was skewed away from demand orientation by the depth or absence of seams only in parts of the North and on the Croxteth Park outlier, and by the camouflage of drift in the West. Elsewhere, geological processes had ensured huge reserves of very shallow coal, greater than those of any other British coalfield. But the seams were wet; even relatively small increases in depth could cause expensive difficulties and had done so in the most heavily exploited areas of the coalfield by 1740.

*Colliery sites*

The little obsequent valleys which tumbled into the wider vales of the main streams had provided sufficient fall and power to drain the workings of the seventeenth century and exerted an important influence upon the siting of collieries. Most eighteenth-century collieries still occupied them (figure 26) and new coal was opened up in significant amounts only at Thatto Heath (59, 72 and 73), Tarbock (60), Newburgh (85) and Harrock Hill (86) in the early eighteenth century. But this does not mean that the geomorphological attractions which had originally drawn mining to the majority of sites continued to operate. Indeed, it was when collieries grew too deep and

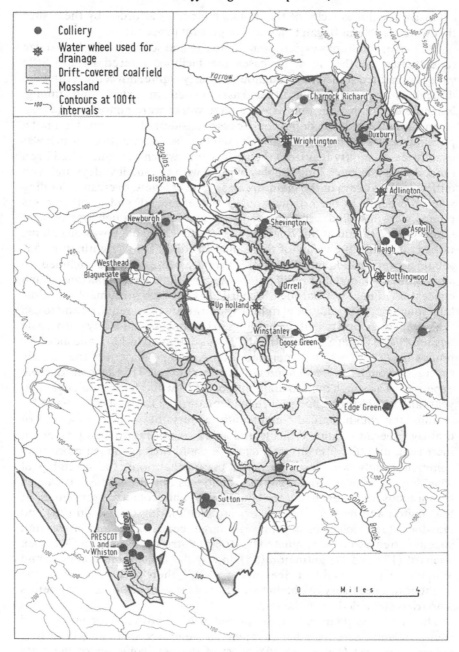

26. Relief and the distribution of collieries, 1690-1739. Only those collieries which can be located reasonably accurately in relation to relief features and drift are plotted.

extensive for the capacity of the drainage facilities afforded by these sites that water problems began to assume important proportions.

Of course, conditions varied from site to site and so did the depth and size of collieries, so that in some cases the facilities offered by the little valleys remained sufficient throughout the early eighteenth century, and the difficulties experienced in other cases varied in extent and timing. Nonetheless, difficulties and new drainage works were almost ubiquitous. Even a small increase in depth could cause considerable problems if it put a pit sump below the level of the sough outlet in the valley bottom, which was usually less than fifty feet from the surface into which the valleys had been eroded. In some cases long soughs, parallel with the small valleys and with outlets in the valleys of the main streams at lower levels, overcame flooding when the depth of the original soughs was exceeded, and in other cases pumps were installed as supplements. A new deeper sough was built at Garswood (29), and at Shevington (47, 49 and 77) a sough was dug along, rather than into, the valley of the Calico Brook.[59] The Wrightington (55) sough proved inadequate before 1710 and a lease of 1728 included an agreement to extend it or build pumps, and the same lessee amassed a hugh concession at nearby Charnock Richard (17) in 1712, presumably to allow long soughing, with additional rights to build an engine if required and to use the water courses on two of the holdings in his concession for any purpose he desired.[60] The Great Sough at Haigh (2) was extended in 1716 and aided by pumps by 1728,[61] whilst at neighbouring Aspull colliery (12) the already extensive soughs were lengthened further in c.1713 at considerable expense.[62] The sough built to drain the coal at Newburgh (85) took over two years and £100 to finish and proved inadequate after thirteen years of use.[63]

Thus, the sough, until the eighteenth century the preferred method of drainage, became more difficult and more expensive to build as depth increased, and the difficulty and expense could increase much more than commensurately with increase in depth. Furthermore, the assembly of soughing rights over large areas of ground was not always easy. It caused acrimonious disputes between landowners and colliery proprietors at Shevington and between the owners of neighbouring Garswood (29) and Haydock (28) collieries.[64] Although large concessions and associated soughing rights were accumulated by Robert Patrick Lancaster in Charnock Richard (17) and Wrightington (55) and by the Dicconson, Halliwell and Whitehead partnership in Shevington (49),[65] there is no record of the negotiations necessary to obtain them, nor of those of any other proprietors who tried and failed to do the same.

The small obsequent valleys also provided water power to run pumps, and such machines were installed to replace or reinforce small soughs at four collieries at least (see figure 26). Most of the new collieries of the early eighteenth century had drainage machinery, powered by horses if the fall of water was insufficient. Even the large Thatto Heath colliery (59), which raised 8000 tons of coal a year, used horse powered engines and so too, probably, did the colliery at Edge Green (65) which produced c.5000 tons a year.[66] At the largest collieries in the South West these expedients were inadequate; windmills powered the gins at Prescot Hall (5) in the 1730s and at Whiston (34) a steam engine provided the only solution.[67]

Thus, although there was still a reasonably strong correlation between colliery location and small, deep stream valleys, skewing the distribution pattern away from demand orientation, the causal link between geomorphology and the pattern had been largely snapped by the middle of the eighteenth century. Water power might have kept some pits there, but the reason for the siting of most of the workings in them seems to have been inertia – partly, no doubt, a rational response to fixed capital installation and accumulated knowledge but partly also, one suspects, simply because that was where the proprietor concerned had always mined his coal.

## Labour supplies

As production grew or shrank so the collier population must expand or contract. Labour mobility must have been an important concomitant of the changes in output between 1690 and 1739. The small numbers of baptisms of collier children on figures 19 and 20 represent quite substantial mining populations. In the North, thirty-six colliers were named in the registers of Eccleston church and Douglas chapel in the 1720s and 1730s; in the South, there were thirty-seven in the register of St Helens chapel in the same decade and thirty-one in that of Up Holland between 1700 and 1740.[68] It is rarely possible to trace as many as fifty per cent of the names in colliery accounts in the parish registers, so that the actual populations of colliers must have been at least double these totals.

The indexes in the published transcriptions of the registers of Up Holland and St Helens provide a means of discovering where colliers came from and went to as their numbers rose or fell in unison with output. Each person who was described as a miner on one or more occasions between 1690 and 1739 in these registers was traced forward and back through time, ideally to the terminal events of his baptism and burial, and changes in his occupation and residence were noted. The resulting data are by no means perfect representations of occupational and residential mobility. The register record is woefully incomplete (table 7),[69] and the family recruitment of labour, plus the custom of naming children after grandparents and uncles, sometimes make it impossible to decide how many people are represented by a particular name. In consequence, men could enter and leave the labour force without being noticed, their movement masked amongst the entries of others with the same name. Nonetheless, one can derive some reflection of the behaviour of miners from the registers. It contains clear signs of considerable mobility. This is summarised on tables 8 and 9, which exclude

TABLE 7. *Registration of vital events of sometime colliers in Up Holland and St Helens, 1690-1739*

|  | No. of men | Baptisms | Burials | Marriages | Other entries |
|---|---|---|---|---|---|
| Up Holland | 32 | 10 | 10 | 0 | 96 |
| St Helens | 37 | 0 | 14 | 1 | 192 |

TABLE 8. *Occupation mobility of sometime colliers registered in Up Holland and St Helens, 1690-1739*

|  | No. of men* | No. of entries | Occupations entered | | | % collier† |
|---|---|---|---|---|---|---|
|  |  |  | None | Collier | Other |  |
| Up Holland | 24 | 101 | 8 | 60 | 33 | 64 |
| St Helens | 31 | 196 | 71 | 114 | 11 | 91 |

* With more than one entry of occupation.
† % of occupations entered, not of entries.

TABLE 9. *Residential mobility of sometime colliers registered in Up Holland and St Helens, 1690-1739*

|  | No. of men* | Movers | | Moves | | |
|---|---|---|---|---|---|---|
|  |  | No. | % | No. | Per mover | Per collier |
| Up Holland | 28 | 8 | 29 | 8 | 1.0 | 0.3 |
| St Helens | 32 | 14 | 44 | 25 | 1.8 | 0.8 |

* With more than one place of residence entry.

those men who were registered only once. Men who were at some time described as miners almost equally frequently had other jobs in the vicinity of Up Holland. The most usual alternative was labourer, entered twenty-six times; weaver, husbandman and nailor were entered four times, twice and once, respectively. The St Helens figures present a stark comparison for there, it seems, there was little occupational mobility and only eleven out of 125 occupations entered against the names of sometime colliers represented jobs outside mining. However, the St Helens colliers were much more residentially mobile than those of Up Holland (table 9); a much higher proportion moved from one township to another in the area and the average number of moves made by particular individuals was much greater. Even in Up Holland, however, there was little occupational mobility after about 1730 and few men changed their occupations after they had once entered the pits. After the labour force of the new Holme colliery (74) in Orrell had been recruited the mineworkers were as tenacious of their occupation as those of St Helens, and the other jobs recorded were those from which miners came into the pits. In St Helens, where the industry was declining in the 1730s, colliers seem to have moved around locally to find jobs rather than change their occupation. Judging by the fall-off in the registration of colliers' children's baptisms and colliers' burials, many moved out of the area altogether and some from Sutton, Eccleston and Windle in the western part of the chapelry went to work in the expanding industry of Prescot and Whiston.[70]

The commonness of short distance movement to work is further testified by attempts to link names in colliery accounts with those in parish registers.[71] Four out of twelve Haigh (2) workers can be traced in this way, one to Haigh itself, two to adjacent Aspull and one to the neighbouring Scholes area of Wigan. Ten of the fifteen colliers named at Holme's colliery (74) in Orrell between 1744 and 1748 lived, in roughly equal numbers, in the contiguous townships of Orrell, Up Holland and Winstanley, four of them moving from one of these townships to another in the 1740s and 1750s.

Thus, it was quite common for miners working in one township to be registered in another and for registered places of residence to change. Some of this spatial mobility must have been effected through daily or weekly journeyings from home to work, some by lodging and some by changes of family residence, which was probably not uncommon. Although the borough regulations of Wigan made entry to the town difficult,[72] the Poor Law documents of Parr bear ample witness to family moves.[73] The majority were switches between adjacent townships of the kind revealed by parish registers (figure 27). Indeed, the operation of the Poor Laws encouraged residential mobility; they militated against a change of occupation in a township where a settlement had not been secured after 1740 and may well have had the same effect earlier. Employment at a colliery in a particular township for a full year, or for many years, did not give the right to a settlement in that township, at least in the absence of a signed bond. When a miner died in a township where he had no settlement his widow and children were immediately sent back to one where he had: if he became unemployed, he and his family were returned to his place of settlement,[74] whence he would presumably sally forth again to a colliery elsewhere if he could find no local work. The moves and subsequent losses of work would be punctuated by spells on the relief provided by his township of settlement. How the colliers and their families were housed in the places between which they shuttled is unknown, but the finding of work and housing must have been facilitated by kin networks such as that of the Baxindens, which extended through Coppull, Standish, Wrightington, Parbold and Lathom in the 1730s and 1740s.

Figure 27 reveals that the vast majority of moves to and from Parr fell well within the range of such contacts, but some passed beyond the two or three miles which must usually have circumscribed it. Almost invariably, these were contacts with other mining townships.[75] Parish registers are not generally useful for tracing long-distance movements because a migrant

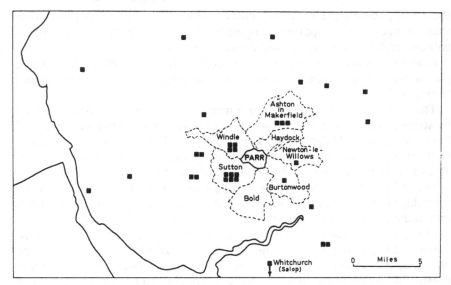

27. Pauper migrants to and from Parr, 1694-1739. Includes all those bringing settlement certificates and being removed to and from Parr. One such contact with a place is represented by a square symbol.

would already be resident somewhere in a parish before using the register and therefore be described as 'of' that local place. Nonetheless, there are six ascriptions of relatively remote places of residence in the registers of coalfield townships. A collier 'late of Blackrod' had a son buried in St Helens, where colliers of Blackrod and Winstanley were also interred. A Skelmersdale collier moved to Wrightington, and one from Pemberton was buried in Hindley. On a yet larger scale, colliers moved between Cumberland and Shevington and Yorkshire and Up Holland.[76] Whether these more widely separated points were the ultimate termini of long aimless wanderings, places of kin contact or, like the infusions of Shropshire men to Haigh (2) and Tarbock (60) and of Newcastle men to Haigh (2),[77] bridged by the contacts and cash of coalmasters, is not, of course, discoverable.

Migration and short distance mobility supplemented a labour force which seems to have been latent everywhere amongst the local labourers, husbandmen, weavers and so on, and amongst the sons and brothers of colliers who would have entered these occupations in the absence of jobs in the pits. Although the grip of the Lowes was loosened at Haigh (2),[78] and although no single surname or small group of them completely dominates the entries in any register, family recruitment obviously remained important everywhere. There were six or seven men named Welsh at work in the Wrightington collieries in the 1720s and 1730s and more than one Charnock, Turner, Baxinden and Howarth as well; in St Helens, there were four Adamsons, three Greenoughs and two Leylands, Hills and Tickles, and in Up Holland there were three Unsworths, three Andertons and two Birchalls.

The readily available labour force, the tendency to continue as a collier once men had entered the pits and their willingness to move after work suggest that mining was a relatively attractive occupation, despite the ever present threat and common incidence of injury and death.[79] The reason must have been relatively high wages. As in the seventeenth century, it is impossible to calculate how much an individual miner earned because of the multiplicity of piece rates for different jobs between which he might be shifted, compensatory payments for slack or difficult conditions, earnest and 'odd' payments and unknown proportions of idle time. The later accounts of Prescot Hall (5), Orrell (74) and Standish (41) show much more standardisation in payments than the earlier ones of Winstanley (6) and Tarbock (60), but payments for 'getting' coal were not accounted, even in the best of them, presumably because they were paid at standard rates by the auditor out of his weekly takings, and information on these standard rates is very sparse.[80] Nonetheless, some degree of comparison is possible and the rates paid at various collieries have been recalculated and presented in comparable statistics in table 10. Although these figures must be treated with some caution because of uncertainty about the exact contents of the measures in which piece rates were expressed, they should be reasonably accurate.

The day wage rates of 9d and 10d were, as one would expect, slightly lower than the daily amounts which could be earned in getting: assuming an annual get of 500 tons a miner and fifty weeks of six days each a year to cut it,[81] a rate of ½d per basket would yield about 16d a day, out of which the drawer must also be paid. It seems, then, that Lancashire colliers must have earned about 5/- for a six day week, perhaps ranging up to 6/8d, or between about £12 and about £17 for a fifty week year, excluding odd payments for ale, tobacco and celebrations, earnest and concessionary coal.[82] There were quite considerable variations around these sums besides those due to higher payments for time spent getting, which themselves varied with the depth of the work. At Standish (41), day rates varied systematically according to which collier received them: day wage earnings were calculated for each collier each week by the auditor and it is clear that some colliers were always paid 6d per day, others 10d and one 14d falling to 12d. The wages of auditors were also expressed in different ways at different collieries and it is impossible to convert them to comparable forms. The least equivocal statement was made at Adlington (68), where the auditor was simply paid £20 a year – considerably more than an ordinary collier.[83] At Haigh (2) the auditors got a standing payment of £10 per year in 1698 with, presumably, additional payments in proportion to the amount of coal mined, as at contemporary Winstanley (6), where the standing payment was only £4.[84]

TABLE 10. *Wage payments, 1676-1748*

| Colliery | Day wage | Getting per basket[a] | Winding or banking per basket | Tunnelling per yard | Other payments | Selling price per basket | Getting as % of price |
|---|---|---|---|---|---|---|---|
| Tarbock (60) 1692-3 | 10d | 0.58d[b] | 0.20d | 2/6-3/4d | earnest, odd cash | 2.0d[c] | 25 |
| Winstanley (6) 1676-96 | 9d | 0.31d[d] | 0.13d | 2/2d | earnest, odd cash, ale, tobacco, candles | 1.5d | 21 |
| Prescot (5) 1735 | | | | | | | |
|   deep coal | | 0.73d[e] | 0.20d | | | 2.5d | 29 |
|   bassett coal | | 0.53d[f] | 0.13d | | | 2.5d | 21 |
|   glass-coal | | 0.53d *or* 0.67d[g] | ? | | | 2.5d | 26 |
| Orrell (74) 1746-8 | | 0.38d[h] | | | 1/- earnest and 2/6d | 1.5d | 25 |
| Standish (41) 1746-7 | 4d-1/3d[i] | 0.50d[j] *or* 0.60d | | | | 1.5d rising to 1.75d | 33 / 29 |

a. Wherever drawing costs were given separately, they have been included with the getting cost because it seems to have been usual for the getter to pay the drawer out of his own wages.
b. 2/6d per work (of sixty baskets) or 10d per score (of twenty baskets) = 0.5d per basket for getting plus 3d per work for drawing.
c. A small quantity was sold at 1.5d per basket, probably due to degeneration from exposure on the pit bank.
d. The usual rate was 3.5d for getting and winding a load (of eight baskets) but occasionally 4d a load was paid, or 4d or 5d for getting and 1d for winding a load in low coal, soughs or endways.
e. 3/8d per work (of sixty baskets) 'to y$^e$ men underground' and 1/- for banking.
f. 2/8d per work 'to ye men underground' and 8d for banking.
g. 'The charges of getting, p$^r$ work . . . 3. 4', with no banking cost itemised. Perhaps glass-coals were only mined to order, or the cost of banking was included in the getting charge or inadvertently omitted.
h. 'Nine pence a score as usual'. The sale score contained twenty-four baskets at this colliery in 1769 but at an adjacent colliery at the same time there were twenty-six baskets per getting and selling score and each getting basket weighed 150 lb rather than the common 120 lb which was the weight of the sale baskets there. Coal got in driving the water lane was paid for at '12 a score' or 0.5d per basket in 1745.
i. The most usual rates were 6d, 10d and 12d.
j. Most coal was got at 1/- per score but each week very small quantities were also got at 1/2d, 1/3d or 1/6d per score. The hewed score of twenty baskets contained twenty-four sale baskets costing 1.5d each. Thus, the rate per getting basket was 0.6d and per sale basket 0.5d.

It is clear from table 10 that wages of all kinds, whether for hewing, day work, tunnelling or winding, were highest at the collieries of the South West and lowest, early and late in the period, at those of the Centre. As output and prices rose with demand, so too did the wages of miners, which did not vary much as a proportion of the selling price of coal. It seems, then, that there was some kind of market mechanism behind the fixing of wages. Its imperfections prevented their equalisation: the mobility which occurred in response to differences was not sufficient to eradicate them.

Thus, it is probable that the standard of living of miners varied over the coalfield. Industrial growth undoubtedly brought them higher wages and although the deep coal colliers of Prescot (5) in 1735 may not have earned almost twice as much as the miners of Orrell (74) in 1744, as the tabulated figures suggest, because of odd payments, earnest, differences in the extent to which the owners supplied candles and tools, concessionary coal and so on, they must have earned very much more: all the possible compensations cannot nearly have made up the difference. But it may have been considerably blurred by the greater opportunity for earning outside mining in the less intensively developed parts of the coalfield. Certainly, there were still 'farmers and miners' in Wrightington in the 1720s and 1740s and at Lathom in the 1690s. A number of miners obviously had a smallholding as well as their jobs in the pits and at least one Ashton collier had fustian looms. Styles and standards of living would obviously vary considerably, therefore, according to supplementary occupations as well as mining wages. Some colliers managed to accumulate small fortunes; Edward Greenough of Parr left £64 in goods and cash at his death in 1693, Thomas Brighouse of Lathom left £24 and William Brown of Prescot, who described himself as a collier but may have held part leases of the collieries in which he worked, left £217 in owed money alone in 1727.[85] But all except a very few colliers must have been amongst 'the lower people' with whom they were categorised in Prescot.[86]

The families of colliers cannot be successfully reconstituted because of the incomplete registration of their vital events. The information summarised in table 11 is all that can be abstracted. There is no single instance of information on both birth and marriage for any of the hundred or so men who were at some time colliers in the registers of St Helens, Up Holland, Eccleston and Douglas Chapel. On the very meagre evidence of age at the first child's baptism and the interval between marriage and its birth, marriage seems to have been relatively late even for colliers, who could, one presumes, reach their maximum earning power in their late teens or early twenties. Children followed in rapid succession, about two-thirds surviving in Up Holland, where the registration of burials seems least incomplete. The average number of children baptised by each collier in a parish must, because of mobility, under-represent the number he had; there must have been many instances of men who moved out of a parish before their family

TABLE 11. *Some demographic characteristics of collier families registered in Up Holland, St Helens, Eccleston and Douglas Chapel in the early eighteenth century*

| | Interval between marriage and first baptism | | Age at first child's baptism | | Interval between baptisms | | Baptisms per collier | | Interval between last baptism and burial | |
|---|---|---|---|---|---|---|---|---|---|---|
| | N | av. years | N | av. years | N | av. years | N | av. | N | av. years |
| Up Holland | | | 10 | 27.7 | 81 | 2.33 | 27 | 3.16 | 7 | 10.4 |
| St Helens | 1 | 1.0 | | | 142 | 2.53 | 36 | 4.37 | 5 | 10.3 |
| Eccleston and Douglas Chapel | 4 | 0.5 | | | 43 | 2.43 | 25 | 2.08 | 13 | 12.0 |

N = no. of colliers.

was complete, or into it after it was begun. This makes the St Helens figure of 4.37 baptisms per collier quite impressive. This average was exceeded by thirteen men and four had more than seven children baptised. In Up Holland and Wrightington, four and three men, respectively, registered five or more children. Fathers usually died before their last child was into its teens. The average interval between the baptism of the last child and the father's burial was barely ten years and even this is skewed upwards by a few men of far greater than normal longevity; ten men died within six years of their last child's baptism and six within two. In figures such as these, and in the constant threat of maiming, we can perhaps glimpse one of the reasons for the strength of kinship bonds within the mining labour force.

*Capital and entrepreneurship*

Profit data are available for more collieries in this period than in the seventeenth century (see table 12).[87] The figures for the smaller collieries are similar to earlier ones: about £100 was made on an output of about 1500 tons a year at a rate of about fifty per cent of receipts. However, there are some interesting novel features. Losses were recorded at Whiston (89) in 1737 and a rate of less than ten per cent at Tarbock (60) in the 1690s. At nearby Prescot Hall (5) high rates of profit on large outputs yielded an enormous annual income in the 1730s.[88] In the South West, at least, development brought more varied profits and, therefore, greater risk.

Although profitability did not change much over most of the coalfield, mining at greater depths increased the costs which must be incurred before income began to accrue and, therefore, the losses which would be incurred if a colliery failed or ran for a shorter than customary period.[89] At the larger collieries of the South and South West the costs and risks must have risen commensurately with output. The lengthening of soughs and their supplementation with engines was an almost universal call on extra funds. A

TABLE 12. *Colliery profits, 1690-1739*

| Colliery | Area | Year | Annual profit £.s.d | Annual output tons | Profit as a percentage of receipts | |
|---|---|---|---|---|---|---|
| | | | | | Before rent | After rent |
| Winstanley (6) | Centre | 1690 | 100. 9. 7½ | 1,485 | 54 | |
| | | 1691 | 100. 1. 8 | 1,239 | 65 | |
| | | 1692 | 114.12. 4 | 1,613 | 57 | |
| | | 1693[a] | 60. 0. 1 | 1,439 | 33 | |
| | | 1694 | 109.18. 3½ | 1,487 | 59 | |
| | | 1695 | 94.13. 0 | 1,269 | 60 | |
| Tarbock (60) | South West | 1693 | 118.16. 4 | 2,481 | 33 | |
| | | 1694 | 33. 2. 1 | 2,415 | 8 | |
| Winstanley (6) | Centre | 1716 | 100. 0. 0 | | | |
| Standish (41) | Centre | 1719 | 39.15. 0 | | | |
| Westhead (87) | West | 1731 | 108. 1. 2 | 1,279 | 41 | |
| Prescot Hall (5)[b] | South West | 1753 | (i) 350. 0. 0 | 6,000 | 44 | 28 |
| | | | (ii) 680. 0. 0 | 6,000 | 75 | 55 |
| | | | (iii) 180. 0. 0 | 1,500 | 71 | 58 |
| | | | (iv) 180. 0. 0 | 1,500 | 71 | 58 |
| | | | (v) 1187. 0. 0 | 15,000 | 54 | 38 |
| Whiston (89)[c] | South West | 1737 | *157.11. 9* | | 70 | |
| | | 1738 | 86.15. 1½ | 1,031 | 43 | |
| | | 1739 | 183.14. 7 | 1,703 | 57 | |

a. New pit sunk.
b. (i) Deep house-coal (ii) shallow house-coal (iii) deep glass-coal (iv) shallow glass-coal (v) all coal.
c. Losses in italics.

simple horse whimsey or windlass cannot have been expensive, but it usually required an extra pit so that the total cost of the cheapest form of drainage installation must have exceeded £100.[90] Moreover, their running costs could be high. A few shillings were paid each normal week for winding water at Standish (41) and Tarbock (60), but running costs increased alarmingly if water proved unusually troublesome: they were absorbing half the total running costs at Tarbock (60) just before it closed, hence the desultory profits of 1694.[91] At Blaguegate (71), too, where a whimsey was used, drainage caused 'great Expenses' through the necessity to continuously wind up water.[92]

More complex drainage equipment must, of course, have been more expensive to install. The Whiston (34) steam engine cost £1500 and although the cost of the windmills of Prescot Hall (5) and the water wheels of other collieries may have fallen far short of this drastic expenditure, they must have cost far more than a simple horse whimsey or the short soughs of the seventeenth century. Indeed, the cost of a water wheel proved prohibitive at the average sized Adlington colliery (22) in 1726: 'As for a new Water Wheel and Pit I take the charges to be so gr! that we should let 'em lye for those yt come after.'[93] Even when the most complex machinery was installed, drainage still consumed about ten per cent of weekly running costs.[94]

The only records of soughing charges which have survived relate to works in newly opened coal at Park Lane (40) and Newburgh (85).[95] These collieries were also small, averaging less than 2000 tons a year, so that the expenditures there of about £100 were probably pale reflections of the sums necessary to open long soughs at larger collieries in coal which had been worked for many years – in other words, at the typical colliery where sough drainage was continued in the eighteenth century.

The cost of pits also increased with depth and output. The method of working was still to dig many small shafts, each of which was worked for about three years, in rows or 'ranks' which gradually moved down-dip.[96] Although the use of wooden props reduced the amount of coal that must be left in pillars to support the workings in at least one colliery,[97] the capacity of a single pit seems generally to have remained at between 3000 and 5000 tons, or about 1000 to 1500 tons a year, so that most collieries in the early eighteenth century required more than one working shaft in addition to drainage pits, and must open new ones every two or three years. The largest collieries had many more pits; Thatto Heath (59) had seven coal pits and an engine pit in 1718 and Prescot Hall (5) had ten coal pits in 1735.[98] The cost of sinking must, therefore, have become appreciable and it must, too, have been an ever increasing burden as collieries moved down-dip at rates proportional to their size. Shallow pits cost between £5 and £10 at Prescot Hall (5) in the 1730s. This was about the same as the pits of the seventeenth century, but the deepest pits at this colliery, which was not among the deepest on the coalfield, cost £60.[99]

It is quite clear that mining became much more expensive even though the rate of profit was generally maintained at its high seventeenth-century level once collieries got under way. The Earl of Derby's collieries at Newburgh (85), Westhead (87) and Whiston (89) were slightly smaller than average with outputs of about 1500 tons a year, but they cost £274. 11. 11½, £161. 2. 6 and £171. 0. 9 to open up.[100] These are much larger amounts than had been required in the seventeenth century, but even so they were far exceeded at larger collieries or at collieries which tapped reserves that had already been extensively worked. The two pits of Aspull colliery (12), reopened in the early eighteenth century to exploit the Cannel seam and needing the renovation of an old long sough, rendered Thomas Gerard 'above £700 out of purse in getting them on foot'.[101] The large collieries of the South and South West must have cost about the same amount to bring to full capacity even without a steam engine, which would more than treble the expense.

The need for large financial rescources was not ended once a colliery was under way. Worsening drainage problems made decisions between closure and expensive new machines or soughs necessary, and it does not seem to have been customary to accumulate a proportion of profits to allow for this almost inevitable eventuality. Fire-damp explosions could cause expensive damage and put outgoings far in excess of income over long periods.[102] Moreover, the common practice of selling 'on trust' locked up much of the accounted profits anyway. At Tarbock (60) in 1693 about one-third of the coal was still owed for when the annual account was closed in 1693, and a 'trust book' was kept by the auditor of two Sutton collieries (72 and 73). Its compilation must have kept him busy because he was owed for 3000 tons in 1711, about half the collieries' output in the previous two years.[103] Industrial consumers were very slow payers in the 1750s and there is no reason to doubt that they had already developed this trait in the previous half century.[104] Cash reserves greater than the surpluses accumulated in annual accounts were thus vital for the long-term success of a colliery; the large differences which normally existed between annual running costs and annual receipts could not continually be syphoned off without risk of later financial embarrassment, and the first few years of any colliery's life were more of a drain than a gain to the finances of its owner.

Where then did the greater reserves of cash as well as the greater technical and financial expertise necessary to run and manage deeper and larger collieries come from? It is perhaps surprising, given their difficulties in the seventeenth century, that the gentry still maintained their numerical dominance, although the proportion of collieries for which they were responsible slipped below one-half (see figure 28). In the South West the gentry share rose considerably as Viscount Molyneux and the Earl of Derby opened a number of collieries, but everywhere, even in the South West, the collieries of the gentry were small and only that of the Bradshaighs at Haigh (2) produced over 3000 tons a year.

The generally Papist gentry were in just as bad financial straits as they had been in the seventeenth century, however, and many found their colliery operations a strain. The seventeenth-century difficulties of the Gerards of Ince at Aspull (12) and the Langtons at Hindley (13) continued into the eighteenth;[105] the Harringtons of Huyton (58) obtained an Act of Parliament to allow them to sell their estates to pay off debts in 1713; the Byroms of Parr, who worked Parr Hall colliery (43) between 1704 and 1707, were sued for bankruptcy on a mortgage taken out two generations earlier in c.1705 and sold their estates to the Claytons of Liverpool in 1711, and the Ogles of Whiston (34) sold out to the Cases of Liverpool in 1701.[106] Even those Papist families who held onto their estates had problems. The collieries at Standish (41), Shevington (47, 49 and 77), Wrightington (55), Charnock Richard (17) and, if it was working then, Eccleston (62) were all forfeited with the estates of their owners for Jacobite sympathies and complicity in the Standish Plot in c.1717. If the experiences at Standish (41) and Shevington (77) were typical, sequestration was very damaging: profits

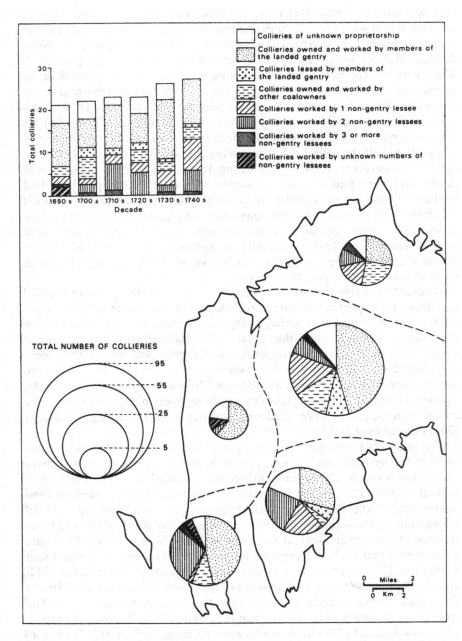

28. Sources of capital and entrepreneurship, 1690–1739. See figure 13 for explanation of volumetric symbols.

plummeted to about £40, lessees were difficult to find and a wall of uncommunicative resistance from colliery workers prevented anything beyond barely adequate administration. The more important Papist estates were milked to sustain the remnants of Jacobite power abroad. The Dicconsons, who had collieries at Wrightington (55), Shevington (47) and Wigan (79), sent £2000 a year to the heir who lived at the Pretender's court at St Germain, and Thomas Eccleston, who had become a Jesuit priest, 'is incapable of holding land except for the benefit of the Jesuit Order, which is a Popish or superstitious use'.[107]

Inevitably, the gentry found it difficult to capitalise mining properly. The Gerards of Garswood, also Papist Jacobites, leased rather than purchased coal reserves at Edge Green (65) and took into partnership Thomas Potter, who had the rights to adjacent coal under his own freehold, with expenses and profits shared equally.[108] The Dicconsons did the same at a Shevington leasehold colliery (49), where their yeoman partner claimed that Roger Dicconson, who administered the estate for his elder brother, 'being a man in mean circumstances has not the wherewithall either to pay the lord's rent nor any part of the workmen's wages'.[109] The Holmes at Orrell (74), who were not Papist, had similar trouble. They opened their mines after netting a dowry of £1500 with a Manchester grocer's daughter, but it took them two years to open two pits, a sough and an engine pit. During the course of the work they first leased out the unfinished colliery, then made an agreement that two local farmers would be paid in coal rather than cash to finish the sough into the engine pit.[110]

Quite obviously, such pervasive and chronic capital shortage must have seriously affected the efficiency with which gentry collieries were run and limited the size to which they could grow. Those who were not troubled in this way, such as the Earl of Derby (63, 71, 85, 87, 88, 89 and 91), the Bankeses of Winstanley (6 and 122) and the Bradshaighs of Haigh (2), whose cannel colliery provided £651 out of total estate receipts of £965 in 1728,[111] seem to have been a small minority. It may or may not be significant that these three were the main Protestant gentry families of the region.

The royalties asked by coalowners fell quite markedly as their own difficulties in exploiting their reserves mounted (see figure 29 and *cf.* figure 14). In the Centre and North, where growth was least and collieries grew little beyond their seventeenth-century sizes, rents fell to levels current elsewhere. During the seventeenth century no rent of less than twenty per cent of prices had been recorded, whereas in the early eighteenth century only three rents exceeded this figure, and they were demonstrably exceptional.[112] As a result, the working profits of lessee coalmasters were between forty and fifty per cent. The incentive to rent collieries was therefore, in the eighteenth century, strong over the whole coalfield.

But there had been little with which to respond to any incentive outside the gentry ranks of the region in the seventeenth century, when most of the

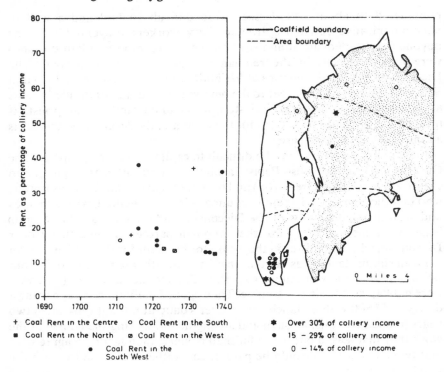

29. Coal royalties, 1690–1739.

lessees were themselves from the gentry. Although this tradition was continued in the Centre and the South, non-landowners came to dominate the leaseholding fraternity in all areas of the coalfield. These local lessees were 'gentlemen', yeomen and husbandmen except for one blacksmith, and frequently, as in the previous century, they banded together in partnerships in order, one supposes, to scrape together the capital required. An interesting element emerged amongst these leaseholders in the areas of slow development: specialist coalmasters took leases, almost invariably with partners, in widely scattered works. The Halliwells of Wrightington were partners in collieries in Wrightington (55), Aspull (12) and Shevington (77); Robert Patrick Lancaster of Coppull Moor had collieries in Charnock Richard (17), Wrightington (55) and Shevington (77); John Chadwick, sometime steward of the Dicconsons of Wrightington and ironmaster at Birkacre, partnered Halliwell at Aspull (12), ran the Dicconson's Bottling Wood colliery in Wigan (79) and worked coal on his own account at Birkacre (96) and Charnock (17).[113] 'Mr Halliwell' was often called in to give advice to other coalmasters, even those of Haigh (2) and Winstanley (6),[114] and it may be significant that all three men came from the North, first working coal in the deep collieries there, and that the collieries they leased in the North and Centre were all old works where more than usual expertise would be required to make profits.

For the first time, therefore, capital and skill were becoming mobile on the coalfield. But still, little was flowing into the Centre, North and West from outside. The only examples were an Ormskirk partner in Orrell colliery (74) and a Preston man who partnered William Halliwell and others in Shevington (77).[115]

In the more rapidly developing South and South West small local flows also supplemented landed finance, but the pools tapped by these channels were not deep enough to fund the largest collieries there. Thatto Heath (59) was opened up with £300 borrowed from Richard Turner of Middlewich, who eventually took over the colliery and ran it until it was acquired by a local glassmaker when the South lost its advantage in supplying the Cheshire saltfield.[116] The massive upsurge in the South West was in large measure based upon injections of capital from outside the coalfield. Knowsley colliery (63) was leased by Josiah Poole, a Liverpool alderman with interests in glass and pottery, until his bankruptcy in 1729.[117] The sodden mines of Tarbock (60) were leased by George Salt and Stanier Parrott, a Staffordshire man who mined at Griff in Warwickshire, where he introduced the first colliery steam engine, and who was responsible for many of the early steam engines in Durham, Northumberland and Cumberland.[118] The large Whiston colliery (34), where the first steam engine was built in Lancashire, was financed by the Case family who purchased Whiston manor with the proceeds of slave trading through Liverpool. They kept this interest after acquiring Whiston and it presumably at least helped to fund their colliery and a saltworks at Dungeon.[119] The other very large colliery of the South West was that on the Prescot Hall estate of King's College, Cambridge. This was an attractive concession. Not only were the seams numerous, thick and well situated for Liverpool, but the owners and the lessee from whom the colliery was sublet were distant absentees who found it difficult to look after their interests.[120] The origins of Thomas Cobham and Thomas Makin, who leased the colliery, cannot be traced. Cobham also owned a glassworks in Prescot and Makin was related by marriage to the Leafes, who had glassworks in Sutton and Warrington. Their colliery at Prescot was clearly the most important property of both Cobham and Makin and of their heirs.[121]

Contrasts in sources of capitalisation thus mirrored contrasts in the scale of development. In the North, Centre, West and, after c.1730, the South, the industry was not much different from that of the seventeenth century. Even so, because of greater depth mining was becoming a more expensive and risky adventure, beyond the pockets or propensity for speculation of most of the gentry. The men who took leases from them cannot have had much firmer financial bases, and the small size and discontinuous operation of the collieries of these areas was at least in part due to their flimsy capital backing.

In the South before c.1730 and in the South West, rapid growth was associated with an influx of capital from outside the industry and from outside the region, and larger markets allowed the operation of the classical formula of growth, ploughed-back profits and further growth once large collieries had been established.

### Competition between collieries

Competition affected the location pattern of collieries in two ways in the seventeenth century. Works producing peculiarly attractive fuel served much wider markets than others and this, plus the higher prices which their coals commanded, ensured longer continuance and larger outputs than normal. In addition to this competition through product differentiation, there was rivalry between collieries in equidistant markets, the mechanism whereby the colliery distribution pattern and coal prices were exercised towards a reflection of demand and supply conditions. Both kinds of competition became more influential agencies in the early seventeenth century.

Although the 'charring' of coal, presumably to make it more suitable for metallurgical use, seems to have been quite widespread, the naturally superior coal mined in the Smith seam at Winstanley (6) and Orrell (74) was still carted over considerable distances.[122] Despite its superiority over ordinary bituminous coal it was, like all the coals of the Central area, cheap compared with that of the South, South West and North (see figure 24) because of competition in the Central market of Wigan with other high grade coals and the distant or scattered nature of alternative markets. Cannel was superior even to smith's coal, selling at 2½d per basket of small and 3d per basket of round in c.1730, compared with 1½d per basket of ordinary local coal at the same date and smith's coal prices of 1½d per basket in the 1690s and 2d in 1745 (see figure 24). Despite its high price, cannel served very widespread markets through Standish to the west as far as Heskin, eight miles away, and to the north as far as Preston, twelve miles away, and beyond to the Fylde. About thirteen per cent of Haigh's (2) total output was sold in or through Preston in c.1730.[123] Clearly, the high grade of the produce of Haigh (2), Aspull (12), Winstanley (6) and Orrell (74) in the Central area explains why these collieries were bigger, longer lived and more assiduously capitalised than the ordinary coal-producing collieries of that area. However, the fact that only Haigh (2) grew much beyond its seventeenth-century size, and even then to only one-third of the output of the largest collieries in the South and South West, suggests that this aspect of competition did not increase greatly in importance in the eighteenth century.

Moreover, it was not axiomatic that collieries producing especially attractive fuel served wider market areas and sold at higher than normal local prices. The glass-coal of Thatto Heath (59) and Prescot Hall (5) was no more expensive than locally produced ordinary coal. Like the non-ferrous metalworkers of seventeenth-century Wigan, the industrial consumers were attracted into the normal market areas of the collieries which produced the coals for which they had a predilection, which fostered the opening of more collieries, local price cutting and cheap supplies.[124] Special coals only ensured high outputs and prices if their market was of necessity relatively scattered and the seam was not easily accessible.[125] The growth of specialist industrial demand during this period did not, therefore, necessarily mean that differentiation between the coal of different collieries became a more important influence on the sizes to which they grew or the prices at which they sold.

The large-scale impact of transport improvements has already been described, but it remains to examine their impact upon the competitive strategies of individual collieries. As coal haulage became easier and cheaper along certain narrow channels, transport costs became proportionately smaller as a component of the market price of coal from collieries located on them. This allowed pithead prices to be lowered to compensate for greater, but now relatively inexpensive, distance. This price response could only occur, of course, to the extent allowed by mining costs. However, the more distant a mining site was from a major market the less intensively it had usually been worked, so that production costs would generally fall with distance along a newly turnpiked road. Price manipulation of this kind seems to have been important in the South West and adjacent parts of the South, where single routes carried most of the coal to Dungeon Point and Liverpool. Figure 25 demonstrates how pithead prices fell regularly away from Prescot and Tarbock, with lower prices at Whiston, even lower ones at Eccleston, and the lowest prices of all at Thatto Heath and Gillars Green. The competitive situation in this part of the coalfield was even more finely tuned than figure 25 suggests, because the relatively low price at Huyton, nearer to Liverpool than Prescot, is largely illusory. It is a result of a high regression estimator, which occurs because of a crop of high Prescot and Whiston prices and an absence of low ones from elsewhere for the time at which it was recorded. The competitive situation was also more complex than figure 25 suggests. In 1735, 'There is a Colliery some distance from the Town (about 2 or 3 miles) where y$^e$ Coals are sold at 9s 6d p$^r$ Work, and from thence many of y$^e$ inhabitants of Prescot supply themselves.'[126] It thus seems that Prescot coal went to Liverpool and Dungeon and Thatto Heath coal to Prescot in a stepwise progression. Why Thatto Heath coal could not also compete in Liverpool can only be surmised: perhaps Prescot people did not cost the carting, done in their spare time, as a commercial haulier would have done, or perhaps the commercial suppliers behaved according to the principle of least effort, going only as far as the nearest reliable supplies. Whatever the rationale, the situation is a clear testimony to the way in which pithead prices were fixed by coalmasters in manoeuvres for competitive advantage.

Transport costs were still too high to allow competition through price reduction to be effected on a larger scale. The low prices of the Central area were not low enough to allow its coal to compete in the markets of the South and South West, except in the limited and exceptional cases of cannel and smith's coal. However, coal from Shevington (47 or 49) and Standish (41) did manage to compete effectively in Preston with supplies from Harrock Hill (86), Adlington (22) and probably Wrightington (55) and Charnock (17) in the North.[127] At these latter collieries, plagued with depth, water or thin seams, prices were probably necessarily high and the quantities which could be raised from the sparse coals of the *Lenisulcata* measures cannot have been great. Thus, coal from the Centre was probably necessary to satisfy Preston's needs irrespective of the prices charged at the pitheads of the North, but even so the geography of coal prices over the North and Centre, in so far as it can be inferred from the scanty evidence, again corresponds to one which would have been produced by conscious price manipulation by coalmasters at works in abundant and easy coal at the furthest distances from markets.

Finally, the relationship between pithead prices (and therefore, other things being equal, output) and competition was also evident in the isolated West. Coal from the South infiltrated the southern part of this area at Rainford, but further north in Lathom (71 and 87) the mosses and the scarp of the Billinge–Ashurst ridge were adequate protection from the collieries of the South and Centre. The price of 2½d per basket charged there in 1730 was as high as prices in the South West. When a competing colliery was opened on an adjacent holding in 1731 the local price fell to 2d and nearby Newburgh colliery, opened two years later, charged the same.[128]

The geography of coal prices demonstrates that much greater competition occurred over longer distances in this period than in the seventeenth century. The mining industry was becoming economically integrated on a much larger scale. The price structures of the South West and South, the Centre and North were coalesced by their supply of common external markets. Two distinct 'price regions' still existed, focused on the South West and the Centre, but they had expanded to incorporate adjoining parts of other areas and all parts of the coalfield except the isolated West swung in one of these two orbits by the end of the 1730s.

### The geography of the South West Lancashire mining industry, 1690-1739

In the seventeenth century there had been two mining regions on the coalfield, each integrated around points at which the industry was linked into wider national or international economies. In the first half of the eighteenth century these two regions became larger and more sharply distinguished as the factors which governed the development and character of mining became more spatially varied. Markets grew on the Mersey and Weaver, transport routes were improved between these markets and the South West, where shallow coal was abundant and where, like everywhere else, labour supplies also seem to have been abundant; capital flowed in from outside the coalfield and numerous collieries developed, some very large, arranged around the turnpike and Dungeon road into a system that was integrated through price competition and encompassed the whole of the South West and part of the South by the 1730s. Elsewhere development was much slower. The Centre, North and West were too far inland to feel the quickening which pulsed from the Mersey. The economy of Wigan and the Central area evolved slowly along the course set in the seventeenth century: the collieries remained small and undercapitalised, stuck by accumulated investment in relatively remote valley locations whose advantages disappeared as depth increased. Prices remained low, aggregate profits and wages were much less than in the South West, and technical changes such as steam power and the use of pit props seem not to have occurred there. Even there, however, the Preston to Wigan turnpike seems to have generated a wider economic integration between the industry in the Centre and North and collieries producing especially desirable coals, notably cannel at Haigh (2), were larger, more technically advanced and more heavily capitalised than they had been in the seventeenth century.

Clearly, demand was the prime mover. Barriers to growth such as poor transport and capital shortage were only removed where markets grew quickly and massively. Of course, there was considerable reciprocation between the expansion of industry, trade and population in the South West and the availability of easily increased, reliable and inexpensive coal supplies. To an extent, coal attracted and fostered the markets of the South West. But cheap and abundant coal alone was not enough for industrial development, as the slow growth of the Centre, with the cheapest and best supplies, bears witness. The navigable water of the Mersey and the Weaver and the huge maritime markets available through Liverpool were the catalysts of industrial development, and coal supplies were a subsidiary and mainly dependent variable.

The integration which came with development was not simply a linkage between points of demand and supply through flows of coal, nor between particular collieries and particular kinds of industrial consumption through ownership. Flows of capital and colliers, of coalmasters and skills, were necessary to lubricate the process of geographically varied rates of expansion. Numerous small movements of these kinds, as well as flows of coal along price gradients, bound collieries together into industrial systems within which such reciprocations were both causes and effects of development. There were still two quite distinct regional systems of this kind: that of the Centre and North and that of the South West and South, and there were still very few flows of any kind between them, although the links within them, even the former, were considerably increased and strengthened. Capital, colliers, coalmasters and coal flowed within but not between them; there are more records of movements from other parts of England and other English coalfields to the Prescot and Wigan areas than there are of contacts between the mining industries of those two areas themselves. Their methods of coal measurement were still sufficiently different to make it difficult to compare prices, wages and outputs.

The regional structure of the seventeenth century was thus made considerably more distinct by the growth processes of the early eighteenth century. One region was completely transformed, the other slowly evolved. Although Prescot and Wigan were only twelve miles apart in linear distance they were separated by an era in terms of economic development by the end of the 1730s. But a change in these relativities was presaged as the Douglas Navigation, the first coalfield waterway, was in the process of being driven from the mouth of the Ribble to Wigan.

# PART THREE:
# THE EARLY WATERWAY ERA

CHAPTER 6

# THE COLLIERY LOCATION PATTERN, 1740-99

Three waterways and a number of new turnpikes and extensions were built in the second half of the eighteenth century. To facilitate description of the consequent changes in the mining industry, I have split the period into three with breaks at the opening dates of the Sankey Navigation in 1757 and the Leeds and Liverpool Canal in 1774.

This procedure prevents the use of decades as units of time, an arbitrary methodological convention which is in any case of limited value at a time of numerous small-scale unsynchronised changes in different parts of the coalfield. Graphs of the numbers of collieries working the various areas of the coalfield in each year have been produced to summarise changes. Of course, they must to some extent under-represent the actual colliery population and the exercise is potentially hazardous because of the nature of the evidence on which it is based. Whether collieries ran the full length of leases, or whether the cessation of an account signifies closure, a change of account book or more lax administration, is not often known for certain and the isolated mentions which provide much of the evidence of mineworking are obviously difficult to convert into a continuous aggregate series.

Many arbitrary assumptions are necessary to produce the graphs of colliery numbers. It may be considered that the evidence has been stretched beyond its capacity in the process of making them. However, it is necessary to produce some kind of summary to condense the voluminous surviving evidence into something beyond a collection of disjointed anecdotes, and all the surviving information, however fragmented, should be included within it.[1] The independent testimonies of parish registers and canal accounts generally verify the chronology of change shown by the graphs, and switches in trend occur when they would be expected in response to new transport routes. The graphs can, therefore, be treated with some confidence as descriptions of the main shifts in the mining industry of the region in the second half of the eighteenth century, if not of the exact numbers of collieries which operated within it at any particular time.

135

**1740-56**

*The numbers and distribution of collieries*

The numerical expansion of the early eighteenth century accelerated after 1740 (see figure 30). The bulk of the increment occurred in the Centre after the construction of the Douglas Navigation from Wigan to the Ribble in 1742. The colliery population more than doubled there from six to fourteen in the next four years. Thenceforth, it remained steady until 1753, when it fell to eleven. Most of the new collieries of the Centre worked sites which had already been exploited and abandoned. Five of the nineteen which operated produced cannel (2, 12, 101, 102 and 104) and five others worked in Winstanley (6), Orrell (1, 74 and 93) and Pemberton (30) in places where only the high quality Arley and Smith seams outcropped. Thus, over half of collieries developed in the Centre definitely produced fuel of exceptionally high grade and many of the others might have done so from the more numerous seams which underlay their sites.

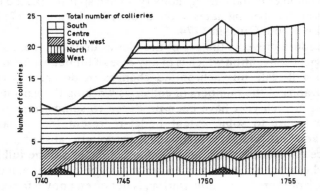

30. Numbers of collieries, 1740-56.

The apparently extinguished industry of the South slowly revived in the 1740s and surged to a total of seven in the 1750s. As collieries increased in number, they spread eastwards across the coalmeasures promontory from St Helens (100) to Haydock (28). In the North, too, there was apparent quiescence in the early 1740s and resurgence in the 1750s, although the trend was less marked than in the South. In the South West there was almost absolute stability at three or four collieries, whilst the West remained almost completely unexploited.

*The duration of collieries*

No more than twenty-five of the forty-one collieries which worked in this period were open in any one year and only five seem to have worked continuously from its beginning to end (figure 31). A further seventeen operated without interruption after they had opened during the course of the period, but nearly half of the collieries recorded between 1740 and 1756 closed down before 1756, a quarter of them after having opened since 1740. The lives of most of the collieries in the Centre seem to have begun and ended within the span of seventeen years. Of course, much of this apparent ephemerality must be due to the nature of the information from which it derives, but this is not completely the case. Royalties were received from four Haigh collieries leased out by the Bradshaighs (90, 97, 98 and 103) for only eight years on average,[2] and even the collieries at Prescot Hall (5) and Whiston (34), which have been reckoned as continuous, only lasted because large new reserves were opened up nearby: the old collieries there ceased.[3] Generally, it seems that mining was less fitful in the South and South West than in the Centre.

*The size of collieries*

Figure 31 demonstrates that many of the numerous collieries of the Centre were small as well as short lived. Three of the four leasehold collieries at Haigh (97, 98 and 103) raised only a few hundred tons a year and were small even by the standards of the late sixteenth century.[4] The fourth Haigh leasehold colliery (90) was bigger, with a capacity of 1000 or so tons a year, and accounted outputs at Park Lane (40), Kirkless (104) and Standish (41) were between 2000 and 3000 tons a year,[5] only slightly above the average for this area in the preceding half century. The two pits and seven or eight hewers of Holme's Orrell colliery (74) put it into the same category.[6] Throughout this period in the Centre, only Haigh (2) raised more than 4000 tons a year. Output there seems to have risen to about 5000 tons a year by the middle 1750s (see Appendix 4).

The collieries at St Helens (100) in the South and Gillars Green (99) on the edge of the South West were of about the same size as the typical works of the Centre.[7] The Earl of Derby's colliery at Whiston (89) was slightly larger with an average output of over 3000 tons a year, whilst the well situated Tarbock (60) workings, with between eight and twelve colliers, was probably about half as big again.[8] Thus, the 'normal' colliery of the South West seems to have been larger than the 'normal' colliery of the Centre. The largest ones there were far bigger than elsewhere. Until 1744 the 15,000 ton a year Prescot Hall (5) 'went on slowly', but 'from that time the Works were Carried on much faster than before'. Representatives sent up by a distant and suspicious King's College in 1750 found it 'far greater than the College ever aprehended' and variously estimated its output at between 18,000 and 24,000 tons a year and 140 works a week (c.21,000 tons a year) 'besides what are sold to the Glass Houses at Liverpool'.[9] In 1753 the colliery

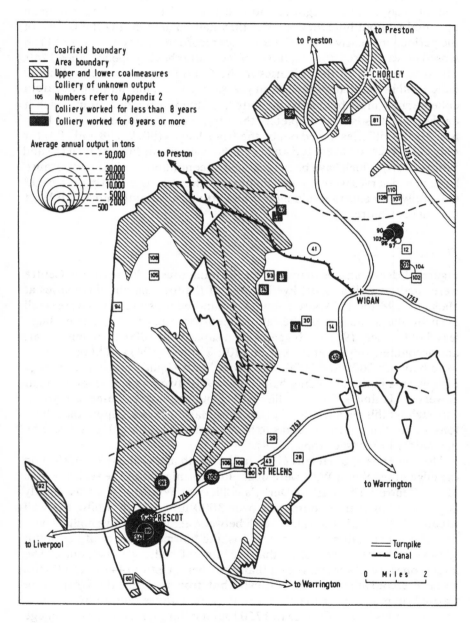

31.  Distribution, duration and size of collieries, 1740-56.

produced about 24,000 tons.[10] The Whiston colliery (34) of Jonathan Case, the bitter rival of Cobham and Makin at Prescot Hall (5), had a second steam engine by 1747 to lay dry reserves extracted through his New Paradise workings.[11] The annual output of the Case works was never recorded, but its profits were great enough in 1744 to provide £80 a year to Jonathan Case II, £50 a year or £11 out of every £80 profit, whichever was greater, to Thomas Case and contributions to a fund of £400 to cover unforeseen colliery expenses, with any residium to be spent at the discretion of Jonathan and Thomas.[12]

The colliery location pattern changed in a number of reasonably distinct ways between 1740 and 1756. There was rapid numerical increase in the Centre, but the collieries were generally small and short lived. In the South and North, collieries increase in number in the 1750s; although only one output figure has survived for these two areas together, there are no indications of any kind that the collieries there were any larger than those of the Centre. In the South, numbers remained steady and small, but two of the collieries there were probably eight times larger than most others and the rest were nearly as big as Haigh (2), the clear champion of the remainder of the coalfield.

## 1757-73

### The numbers and distribution of collieries

The colliery population increased from twenty-four to twenty-six over these seventeen years, reaching a peak of thirty in 1769 (figure 32). There was

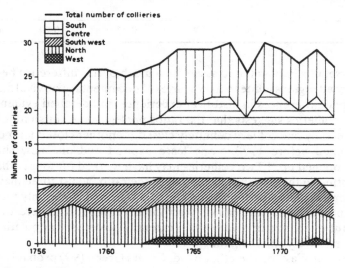

32. Numbers of collieries, 1756-73.

numerical stability in all areas, ranging from four to six a year in the North, three to five in the South West and nine to thirteen in the Centre. Even in the South, newly penetrated by the Sankey Navigation, the colliery population hardly increased: the rate of numerical increase after 1757 was slower than it had been earlier in the decade and only carried the total from five to eight in the canal's first two years, with no further change before 1774.

## The duration of collieries

The stable totals did not subsume an unchanging population of collieries: openings and closures were common throughout the period, especially in the South. Table 13 gives an indication of the degree of flux in the various coalfield areas. A ratio of unity would indicate that the same collieries worked in every year of the period and the lower the ratio the greater the fluctuations in the colliery population. If the low ratio of the West is ignored because of the very small numbers involved, a contrast is apparent between the Centre and North, with high ratios and therefore relatively constant populations, and the South West and South, with low ratios and unstable populations.

TABLE 13. *Ratios of total recorded collieries and average number working in each year, 1757-73*

|  | West | North | South West | Centre | South |
|---|---|---|---|---|---|
| Number recorded | 2 | 7 | 9 | 16 | 18 |
| Average per year | 0.35 | 4.88 | 4.23 | 11.23 | 7.12 |
| Ratio | 0.18 | 0.70 | 0.47 | 0.70 | 0.40 |

Of course, it is impossible to assess how much of this difference between the two broad regions is due to the vagaries of documentary survival and how much to real differences in historical experience, but detailed evidence for particular collieries suggests that the ratios possess some verisimilitude. In the Centre, only one colliery definitely closed,[13] another closed temporarily and reopened before 1773,[14] one lease expired and one new site was definitely opened up.[15] Two collieries may have worked for longer than the spans allotted to them in the calculation of the ratio, whilst three may have worked for less. Thus, openings and closures were not commonly recorded in the Centre and the ratio can only err slightly, probably underestimating the amount of change to a very small extent. The colliery population was definitely relatively stable there.

In the South West, on the other hand, Gillars Green (99) probably closed in 1757 and Windle Ashes (116) was definitely freshly opened in 1758.[16]

Thenceforth, the same four collieries (5, 54, 60 and 116) worked continuously until about 1770, when there was wholesale change. Prescot Hall (5) closed after almost two centuries of uninterrupted production,[17] and so did Tarbock (60) and Windle Ashes (116).[18] Jonathan Case opened Carr colliery (127) in 1769 and closed either it or his other older Whiston workings (34) in 1772.[19] Two other collieries certainly opened at about the same time in Whiston (133) and Prescot (136).[20] The low ratio of the South West thus reflects a sudden complete change in the composition of the colliery population in about 1770.

The lowest ratio of all was that of the South along the new waterway, and it is probable that the pattern on the previous new route into the Centre was repeated on the Sankey in this period in a more extreme form. True, five collieries on the Sankey may have worked for longer than the one or two years in which they were recorded, and which they were allocated in the calculation of the ratio. However, there is considerable evidence that many collieries there did have short lives. Some (126, 129 and 134) did so because they were opened towards the end of the period on the branching extensions of the canal beyond St Helens to the Thatto Heath area of Sutton, built in 1771. Even one of these (126) was closed by the end of 1773,[21] however, following in the wake of earlier collieries to the east. Half of the collieries which opened on the Sankey before 1769 had closed by that date; only two collieries at work in 1757 lasted until 1773 (28 and 43) – indeed, these were the only two at work as late as 1769 which lasted to the end of the period. Of the four collieries opened to the west of Parr and Garswood before the canal was extended, one worked for fourteen, one for eight and one for four years, and the other was also probably ephemeral.[22] Four collieries at the eastern end of the coalfield stretch of the canal also closed within the period, one (28) after a very long, if much interrupted, existence, one after eight years and two after what were probably very short lives.[23] It seems reasonably clear, therefore, that the successful collieries on the Sankey were all on the eastern coalfield edge before the 1770s and that closures were common even there, with only two collieries lasting through the full span of the period. Collieries opened further west before 1770 all closed before 1773, as did one of the three sunk on the canal extensions after 1770. The stable colliery total of the South thus masked an amount of churning change unprecedented within an area's mining industry.

## The size of collieries

The South and South West may have contained a large majority of the short-lived collieries, but they also contained by far the largest. Many of the outputs plotted on figure 33 are based on estimates rather than accounts and are therefore not completely reliable. This is, unfortunately, true of all those along the Sankey Navigation and their derivation will be described in some

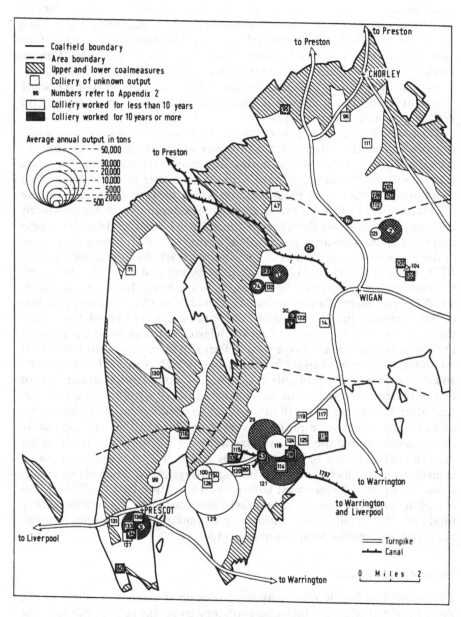

33.  Distribution, duration and size of collieries, 1757-73.

detail in order to justify the stark impression of their relative enormity conveyed by the map.

Receipt totals have survived for the Gerards' colliery at Garswood (29) between 1766 and 1777, but there is no indication of whether they represent takings at the pithead or at the colliery's Liverpool office,[24] to which coal was carried in the Gerards' own flatts or boats. The receipts have been divided by averages of local and Liverpool prices to produce annual output estimates, which fall continuously from about 18,000 tons in 1767 and 1768 to about 9500 tons in 1773 and 6000 tons in 1777. Output may have been even higher before the account began because 'in the year 1765, and 1766, the great supply of coals from Garswood reduced the price and caused an export of more than 8000 chaldrons from Liverpool'.[25] The nearby Parr colliery (121) of John Mackay, opened by 1763, was probably even larger than Garswood in the middle 1760s. If the twenty longwall hewers for whom he advertised were recruited they would have produced about 25,000 tons a year.[26] The colliery at Laffack (118), which could raise '40 Works a Week or more if Occation' in 1760[27] (about 6000 or 7000 tons a year), was small compared with the later Garswood (29) and Parr (121) works but bigger than anything else on the coalfield outside the South West and South.

Even less firm evidence lends support to the impression given by these estimates. Clayton's Parr colliery (43) worked two seams through pits linked by wagonway to the canal bank as soon as the waterway reached it in 1757.[28] By 1759 its coals, like those of Garswood, were taken in the colliery's own flatts to a coal yard in Liverpool and when it was sold up in 1778 it included 'horses, waggons, and other carriages, railed roads, windlasses, materials and utensils thereunto belonging, particularly a remarkable good FIRE ENGINE, lately erected and in thorough repair' and 90,000 tons of winnable coal.[29] The Legh works at Haydock Wood (28) was the nearest to Liverpool on the Sankey and tapped three seams drained by a fire engine by 1758.[30] In 1760, coal was being sold so rapidly that labour shortage was the only possible threat to continued expansion.[31] The colliery was closed, exhausted, in 1767 and Legh replaced it with the nearby Florida colliery (125) a year later. Within a few months the successor was producing '165 tons' through four pits (about 8000 tons a year), gradually coming into full production with a number of new pits 'Air'd and in fine order to raise a large quantity, whatever quantity we get there is sure to go', by August of the following year.[32] The Clayton and Legh collieries in Parr (43) and Haydock (28 and 125) were linked with those of Gerard and Mackay in a number of projected cartels, which suggests that these four were of about the same capacity and between them dominated the Sankey coal trade. They were the only ones to work throughout the period in the South and the only ones still open at the eastern end of the Sankey in 1773.

John Mackay opened a colliery at the western end of the canal which undoubtedly rivalled these four leviathans of the east. He acquired coal rights in Sutton and Eccleston in the middle 1760s and in February 1770 advertised for 'any number of good COLLIERS or SINKERS' at Thatto Heath (59). Coal was for sale on the banks of 'Hatto Heath and Ravenhead' in May 1771, and two years later 'THIRTY COLLIERS With able DRAWERS' were required at Ravenhead (129).[33] If only this number was employed at the colliery, they would have produced about 40,000 tons a year working longwall as at Mackay's Parr colliery (121).[34]

It is impossible to say whether any other collieries on the Sankey approached these five in size, but it is probable that none did. The exclusion of other coalmasters from price rigging and cartel negotiations suggests that no one else commanded a similar proportion of total output. All other collieries for which records have survived experienced greater or lesser degrees of trouble before prematurely closing and this, too, indicates their inferiority. The St Helens colliery of Leigh and Mackay (120), opened in 1762 and steam drained by 1765, was subject to a minimum royalty equivalent to 6000 tons a year, but it was not so successful as Mackay's own works judging by his partner's lament in 1765 that 'I heartilly wish ways & means may be found out to make the Concerns in Collearys More Agreeable & the Sooner the better'.[35] Peter Berry's engine drained works in St Helens (80) were advertised for sale in 1759 and 1761 and never again recorded,[36] and Jonathan Case's Sutton colliery (126), though subject to a minimum royalty which was equivalent to 6000 tons a year, never raised more than 2000 tons and lasted only five years.[37] Most of the other collieries apart from those discussed were situated at some distance from the canal and probably relied to a great extent on land sale (see figure 33), which could not have sustained outputs as large as those on the canal bank.

In the South West, the output of Prescot Hall (5) fell substantially to an average of 16,500 tons a year between 1760 and 1768 and the colliery petered out in 1770.[38] Its proprietors replaced it with Hillock Street colliery (136), but this was still drained by horse gins and had not come fully into production by 1773.[39] Windle Ashes (116) had a steam engine but cannot have been very large because it relied on land sale from a disadvantageous location (see figure 33),[40] as did nearby Gillars Green (99), which never produced more than about 4000 tons per annum.[41] Tarbock colliery (60) also seems to have been quite small.[42] Two other collieries of the South West might have matched the size of the largest in the South. If all eleven pits at Gildart's steam drained Whiston (133) colliery had worked simultaneously and produced the same as similarly equipped Orrell works, then about 30,000 tons a year would have been raised.[43] Jonathan Case's Whiston colliery (34) may have been even larger; after 1769 his own 'immense' reserves were supplemented by those previously tapped through collieries at Prescot Hall (5) and on the estates of the Earl of Derby (88 and 89).[44]

The collieries of the Centre and North appear to have been much smaller. Haigh Park (123), Pemberton (30) and Arley (112) definitely raised less than 2500 tons a year on average and Winstanley (6) and Orrell (74) were less than twice as big (see figure 33).[45] However, Haigh (2) and Orrell Hall (1) were, although much smaller than collieries in the South and South West, bigger than anything that had previously operated in the Centre. They produced 8000 and 5000 tons on average, the latter with the aid of a steam engine.[46]

Although evidence is sadly insufficient, it is possible to discern that the beguiling constancy of the colliery populations of the coalfield areas between 1757 and 1773 masked considerable change and instability. Even in the Centre, where change was least, certain collieries grew appreciably and the steam engine, symbol of modernity, was introduced. Despite the demise of Prescot Hall (5), the South West almost certainly remained the scene of large-scale workings, some newly opened, and the South experienced a continual turnover of collieries around the stable monoliths of Parr (43 and 121), Garswood (29), Haydock (28 and 125) and, late in the period, Ravenhead (129). It is almost certain that these six were as large as any collieries that had previously worked on the coalfield and that some of them far surpassed their largest predecessors. Their more numerous ephemeral neighbours were almost certainly much smaller, though still, perhaps, larger than collieries anywhere outside the South and South West.

## 1774-99

*The numbers and distribution of collieries*

The stability of the colliery population apparently continued through the 1770s (figure 34). The small flurry which immediately followed the construction of the Leeds and Liverpool Canal quickly subsided and significant and sustained increase only began in the early 1780s. The number

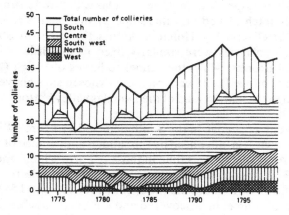

34. Numbers of collieries, 1774-99.

of collieries on the coalfield increased from twenty-six in 1780 to twenty-seven in 1783, then surged upward to forty-three in 1793 before falling slightly in the last few years of the century.

It is possible that this apparently clear trend is due to characteristics of the source materials rather than the progress of the industry. The increase of the early 1780s corresponds almost exactly to the number of collieries listed in the Land Tax Assessments of Orrell, Up Holland, Shevington and Parbold, which begin then (see Appendix 4), but were not previously recorded elsewhere. One of them (144) was referred to as 'late Berry's' in 1782 and must, therefore, have been open before then. The Hesketh mines in Shevington (49) might, too, have been exploited between the expiry of a lease in 1752 and their next record in the Land Tax Assessment of 1782. On the other hand, Orrell Post colliery (142) was definitely not worked for long before 1782 because its reserves were advertised for sale in virgin condition in 1777, and Dean colliery in Orrell (137) was opened in 1775, which perhaps suggests that the colliery through which the same proprietor worked adjacent reserves in Up Holland (143) did not antedate its first mention in 1782 by much either.[47]

Thus, the upsurge of mining on the Leeds and Liverpool Canal in the Centre may have been earlier and more gradual than figure 34 indicates, but it cannot have occurred immediately after the canal was opened. The gradual decline of the colliery population of the Centre in the 1790s is testified to by the relatively reliable evidence of the Land Tax Assessments, so it can be concluded that the number of collieries in this area fell back to its pre-canal size by the end of the century, after climbing to a plateau thirty per cent higher between 1783 and 1793.

Within the area, mining was heavily concentrated in the adjacent townships of Up Holland, Orrell and Pemberton which lined the southern bank of the canal (see figure 35). Collieries seem to have been situated on the top of the steep slope down into the Douglas valley and strung like beads down the deep notch carved into this slope by the Dean Brook along the boundary of Orrell and Up Holland. Across the canal in Standish and Shevington, the collieries were similarly fixed in the little valleys of the Calico and Mill Brooks, whilst the cannel collieries of Haigh (2), Kirkless (104 and 135) and Ince (172) were on more exposed sites above and below the slope down to the marshy land of the Douglas valley bottom to the east of Wigan.

The chronology of change in the South depicted by figure 34 is also open to question because it is influenced by the sudden appearance of collieries in the Land Tax Assessments, this time in those of Parr, Windle and Golborne in 1778 and 1779. It is possible that five of the thirteen works recorded there were in operation before that time (150, 151, 152, 106 and 154). How long before cannot, of course, be discovered and it is even possible that some of them worked before 1774, although this is unlikely in the cases of (150) and

(154).[48] Thus, it is again only possible to conclude that there was some numerical expansion between 1774 and 1792, most of it probably after the middle 1780s, and a subsequent small decline which began, like that of the Centre, in 1793. In the South, the collieries were stretched out along the whole coalfield length of the canal, as they had been in the first few years of its existence but not in the intervening period, when the two ends had been completely dominant.

The Whiston Land Tax Assessments include collieries from 1783. All that are recorded there are also known from other sources and the number of collieries in the South West definitely remained quite steady throughout the period. The increases registered in the North and West after 1790, on the other hand, might well have been less late and sharp because they correspond with the beginnings of colliery registration in the Assessments of Blackrod, Lathom and Skelmersdale. The numbers in these areas were so small that possible inaccuracies there cannot seriously affect the trend for the coalfield as a whole.

*The duration of collieries*

There were undoubtedly some colliery closures during the new canal's first few years, whether or not numbers actually declined. Collieries at Kirkless (102) and Haigh Park (123) were shut at about the same time that the canal was finished and Orrell Hall (1), Standish Highfield (171) and one of the Orrell works (132) closed within the next few years.[49] Nonetheless, there was much greater stability on the Leeds and Liverpool Canal than there had been on the newly opened Sankey Navigation. Only four collieries definitely closed in the Centre during the rest of the century. If the workings in Pemberton are ignored, fourteen out of twenty-seven enterprises worked continuously after they were opened along the Leeds and Liverpool and only five begun after 1780 closed before 1799.[50] It is true that a continuous record exists in the Land Tax Assessments, whilst the sources used to reconstruct the earlier pattern on the Sankey were extremely fragmented, but even so there is more evidence of actual colliery *closures* on the Sankey between 1757 and 1773 than on the Leeds and Liverpool during the next twenty-five years.

On and near the older canal, too, there seems to have been greater stability after 1774. Four collieries definitely closed and another stopped work temporarily, but eleven out of eighteen worked through from 1774 or their opening until 1799. Twenty collieries were recorded on the rest of the coalfield. Twelve of them definitely worked continuously from their beginning until 1799; no colliery seems to have closed in the South West after 1774, and only one of the four opened in the West after 1778 did so.[51] In the North, two collieries out of the six that worked during this period closed before it expired, another had an interrupted life and the remainder were continuous.[52]

By concentrating upon evidence for the closure of collieries, the effect of the continuity of the Land Tax Assessment Returns on conclusions about the duration of workings, compared with conclusions based upon the fragmented evidence of earlier times, should have been nullified. Collieries were definitely longer lived and the turnover within the colliery population was much smaller after 1774 than previously. This difference was highlighted by the situation on the Leeds and Liverpool Canal compared with that on the Sankey Navigation during the earlier period.

*The size of collieries*

The picture displayed by figure 35 is in stark and clear contrast to that of the preceding period. However, most of the outputs mapped onto it are derived from estimates. Accounts have survived for only nine collieries over these twenty-five years: a mere two stretch for more than a few years, only three extend into the 1790s, and nearly all are for collieries in the Centre.

The accounts lend their feeble support to the hypothesis that growth was at first slow on the Leeds and Liverpool Canal. Orrell (74) and Orrell Hall (1) raise on average no more than 5000 tons a year or so in the 1770s.[53] They were thus little larger than the collieries that worked before the canal was opened in the Centre. Dean colliery in Orrell (137) was probably even smaller in 1775, and the two collieries in Standish (41 and 171) definitely were.[54] Later accounts are fewer, but they record considerably larger outputs. The Holme's colliery in Orrell (74) had a capacity of 8000 tons by 1780 and the cannel colliery at Kirkless (135) increased in size continuously from 5000 tons in 1791 to 11,000 tons in 1799.[55] Blundell's colliery in Orrell (93) dwarfed them both, growing from 18,250 tons in 1788 to 42,480 tons in 1799.[56] Thus, outputs at individual collieries, as well as the total number of collieries, seem to have climbed strongly through the 1780s and continued to do so as numbers declined in the 1790s. The manpower figures recorded for Orrell collieries in the Land Tax Assessments display the same trend (see figure 36) and so, somewhat surprisingly, do the accounts of Skelmersdale colliery (105) in the West, where output increased from 1201 tons in 1793 to over 8000 tons in both 1798 and 1799.[57]

Unlike accounts, indexes of output are relatively abundant for the last twenty years of the eighteenth century, and the volumetric symbols on figure 35 are largely derived from them. Most of the indexes are pit and hewer numbers from the Land Tax Assessments, requiring productivity data for conversion to estimates of output. That which has survived from 1740-99 is summarised on tables 14 and 15. The average outputs of cannel pits at Kirkless (102 and 135) were reasonably similar and span those for Haigh (2) earlier in the century.[58] The Kirkless (135) figure can therefore be used with reasonable confidence as a multiplier for the pit numbers recorded at Haigh (2) and Kirkless (104), which were, like it, drained by steam engines. All

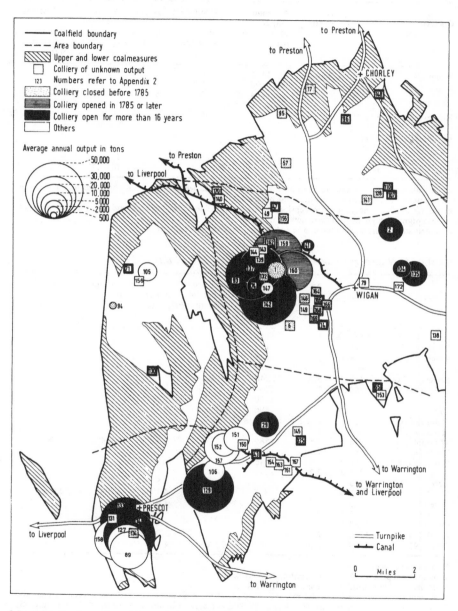

35. Distribution, duration and size of collieries, 1774-99.

TABLE 14. *Numbers of hewers and productivity at cannel pits, 1725-1801*

| Colliery | Years | Av. no. of pits p.a. | Av. annual output per pit in tons (o) | Coefficient of annual variation of (o) | Av. no. of hewers per pit | Av. output per hewer in tons (g) | Coefficient of annual variation of (g) |
|---|---|---|---|---|---|---|---|
| Haigh (2) | 1725-55 | 3.46 | (i) 1441 | 32%[a] | | | |
| | | | (ii) 1490 | 14%[b] | | | |
| | 1748/9-1754/5 | 4.14 | 1174 | 33% | 5.1 | 227 | 21% |
| Kirkless (104) | 1750/1-1754/5 | 2.2 | 1214 | 45% | | | |
| Kirkless (135) | 1792-3, 1795-6 and 1799-1801 | | 1801 | 21% | 8.3 | (i) 217 | 44% |
| | | | | | | (ii) 273[c] | 15% |

a. Calculated from five year moving means of receipts data which have been corrected for prior wage payments, and divided by current selling prices.
b. As (i) but ignoring years of exceptionally low output due to an underground fire and two years of exceptionally high output due to new pit openings.
c. Calculated on the basis of the number of days actually worked by hewers.

TABLE 15. *Numbers of hewers and productivity at coal pits 1737-99*

| Colliery | Years | Av. no. of pits p.a. | Av. annual output per pit in tons (o) | Coefficient of annual variation of (o) | Av. no. of hewers per pit. | Av. output per hewer in tons (g) | Coefficient of annual variation of (g) |
|---|---|---|---|---|---|---|---|
| Haigh (90) | 1737-56 | 1 | 1132 | | | | |
| Haigh (97) | 1744-7 | 1 | 480 | | | | |
| Haigh (98) | 1745-51 and 1757 | 1 | 144 | | | | |
| Eccleston (100) | 1747-9 | 5.3 | 540 | 67% | | | |
| Prescot (5)* | 1750 (i) | 2 | 2800[a] | | | | |
| | (ii) | 7 | 3120[b] | | 4.0 | 780 | |
| Orrell (74)* | 1763-76 | | 2696 | 64% | | | |
| Standish (41) | 1764-84 | 1 or 2 | c.1000 | | | | |
| Standish (171) | 1764-76 | 1 | 902 | 60% | | | |
| Winstanley (6) | 1766 | 2.5 | 1589 | 35% | 3.95 | 384[c] | 21% |
| Haigh (123) | 1768 | 1 | 1192 | | | | |
| Haydock (28)* | 1769 | 4 | 2145[d] | | | | |
| Orrell (1)* | 1771-2 | 2.7 | 2629 | 14% | | | |
| Orrell (137) | 1775 | | | | 3.5 | 531[e] | |
| Orrell (74)* | 1782-90 | 1.5 | 2887 | 22% | 4.1[f] | 704 | 23% |
| Orrell (142)* | 1783-5 | 4.7 | | | 4.9 | | |
| Orrell (137)* | 1783-5 | 2 | | | 4.17 | | |
| Orrell (93)* | 1783-5 | 4 | | | 4.17 | | |
| Orrell (147) | 1785 | 2 | | | 3.0 | | |
| Orrell (93)* | 1788-99 | 6.6 | 5568 | 23% | 4.1[f] | 1357 | 23% |

* Steam engine drained.
a. Glass-coal pits, which were not continuously worked.
b. House-coal pits, which were continuously worked.
c. Composed of two large pits which raised 2125 and 1920 tons p.a., had 4.0 and 4.29 hewers on average, and where the hewer outputs were 433.74 and 447.63 tons p.a. respectively; and one small pit, intermittently worked by 2.65 hewers at an output of 271.29 tons per hewer. Calculated from a forty-six week weekly account × 52/46.
d. Calculated from the statement that four pits raised 165 tons (per week?).
e. Calculated from a thirteen week account.
f. Estimate comprising the average hewer complement of pits at six Orrell collieries, 1783-5.

three cannel collieries were, according to these estimates, of about the same size and together raised about 30,000 tons of this choicest of fuels in the late 1790s. Coal was also raised at the main colliery in Haigh (2), through three pits in 1786 and one or two in the late 1790s. However, productivity seems to have been much more variable at pits producing ordinary coal than at those producing cannel and the data merit more discussion before being used as multipliers.

Table 15 makes this variability clear. Even at collieries with more than one pit, where each would presumably be worked to capacity, annual outputs ranged from 540 to 5568 tons and production per man year from 384 to 1357 tons. The differences were not patternless, however. Small collieries working through one or two pits had uniformly low figures similar to those of earlier times, whilst large collieries with steam engines had much higher ones. The reasons for this impressive doubling and trebling of productivity per pit and per man can only be surmised, but there are certain aspects of large-scale steam drained mining which would obviously be conducive to it.

Continuous interruption by flooding could severely reduce the output of individual pits, as it did in the 1740s at Eccleston colliery (100), where five or six pits were used intermittently as the level of water within them allowed and none managed to produce much more than their average of 540 tons a year.[59] Steam drainage removed the continual interruptions and therefore increased pit outputs. Moreover, the number of hewers employed in the pits of the larger collieries also increased from the two or three that were normal before 1740 to between four and five. In Orrell in the 1780s, the largest collieries usually employed more than four men per pit, the smaller ones three or four.[60]

Only five pieces of evidence about hewer productivity have survived, so it is difficult to make even tentative generalisations. The figures do, however, fall into a reasonably distinct sequence, ranging from the small sough drained Winstanley workings (6) and the first few months of Longbotham's Dean colliery in Orrell (137), through Holme's equally small though steam drained colliery (74) and the large, steam drained but early works at Prescot Hall (5), to the exceptionally high 1357 tons per man year at Blundell's Orrell colliery (93), one of the largest recorded before 1799 and working in the last decade of the century.[61] Why should steam drainage, size and lateness combine to have this dramatic effect? Steam drainage meant that pits were continuously active, and sheer size meant that hewers could spend all their working time cutting coal, with sinking, soughing, tunnel driving and other miscellaneous jobs being done by their own specialist labour force. On the evidence of figures for Ince cannel colliery (135) the elimination of such absenteeism from getting by hewers could contribute an increase of twenty per cent or so to their annual production.[62] But this is not enough to account for the difference between the Blundell figures and those for smaller and earlier collieries. It is possible that the remainder was due to the innovation of more efficient methods of working. Longwall mining was certainly introduced in the large collieries on the Sankey, and the 'Lancashire system', combining features of both pillar and stall and longwall, was developed in Orrell and used at Blundell's colliery (93) in the late eighteenth century.

It seems reasonable to suppose, therefore, that large steam drained works would produce as much per man day as the Blundell colliery (93), small ones as much as at Holme's colliery (74). This assumption was used in the conversion of manpower figures from the Land Tax Assessments of Orrell and Windle into the estimates of output plotted on figure 35. The two estimates for the South besides those for the Windle collieries were derived from receipts at Garswood (29)[63] and statements about industrial coal consumption at Ravenhead. Ravenhead colliery (129) had a monopoly of the supply of a glassworks, which alone would have sustained its output at about 20,000 tons a year in 1786, and a local copper smelter. In 1795 these and one or two smaller industries nearby consumed 700 tons of coal a week, giving the colliery an output of 35,000 tons, or even 50,000 if the tons used for reckoning supplies were all, like those of the copper works, of 30 cwt.[64] Ravenhead colliery (129) was thus comparable in size with the large works on the Leeds and Liverpool Canal in the 1780s and 1790s, although the Windle collieries were less than half as big.

The estimates of outputs at the collieries in the South West are based on shaky foundations, but these works were undoubtedly still large and some impression of their size is desirable. If the twin drainage engines of Case's colliery at Whiston (34) are taken as an indication that it was of a parity with Ravenhead (129),[65] and if the liabilities entered into the Land Tax reflected, when averaged, the relative sizes of the Whiston collieries, then their outputs were as given on figure 35, ranging from 20,000 to about 35,000 tons a year. These figures seem about right considering the earlier sizes of the collieries of this area and the fact that still in the 1780s they commanded a large share of the Liverpool market, where their competition was feared by the proprietors of the works along the Leeds and Liverpool Canal.[66]

Even if all the outputs entered upon it were unimpeachable, figure 35 would still provide a somewhat misleading impression of the geography of coal production in the last quarter of the century. The mapped data relate to randomly scattered dates through the period, so that the picture it gives mixes up some features of the 1770s pattern with others of the 1780s and 1790s. In the Centre at least, there was considerable directional change of which the map gives no inkling. The small outputs of Orrell collieries (1 and 137), of those at Standish (41 and 171) and Kirkless (104) all derive from the 1770s. All these collieries had closed by the 1790s except Standish (41), which was linked to the Leeds and Liverpool by an undergound canal, its huge reserves worked by a consortium which included the proprietors of some of the largest Orrell collieries.[67] Two of the Shevington collieries (47

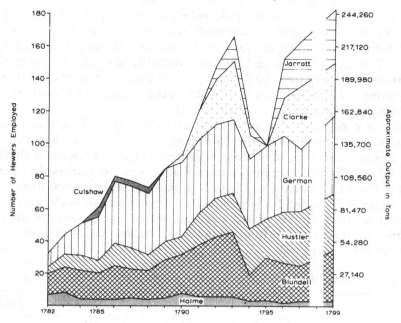

36. The Orrell collieries as recorded in LRO, Land Tax Assessment Returns, 1782-99. See Appendix 4 for conversion rates of hewer numbers to tonnages.

and 155) and many of those of Up Holland, Parbold, Pemberton and Winstanley, too, were in the hands of the same entrepreneurs by the last few years of the eighteenth century. All of them could have been as large as those in Orrell by the end of the period, but they could only have become so just before its close. Moreover, the collieries of Orrell itself had relatively small outputs in the early 1780s, expanding gradually but continuously to their maximum dimensions, achieved in the 1790s (figure 36). Thus, the size of collieries quite definitely increased greatly after the early 1780s, and after it abated in Orrell the expansion spread to other townships before the close of the period, if entrepreneurial changes indicated changes in the scale of exploitation.

In Windle on the Sankey, on the other hand, the number of men employed in each colliery, as well as the number of collieries, fell after the early 1790s.[68] Although this is not enough to allow a conclusion that output declined along the whole of the Sankey Navigation, it does indicate that the continuously cumulating expansion along the Leeds and Liverpool was not matched there.

## Summary of main trends, 1740-99

There were eleven collieries in operation in 1740 and thirty-eight in 1799, with a peak of forty-three in 1793, when twenty-five per cent of all the collieries that existed in the seventeenth and eighteenth centuries were at work. Information on size and duration points the same way. Except in the South West, collieries before 1740 had invariably been small, surpassing 5000 tons a year rarely and 8000 tons never. By the end of the century there were groups of long-lived collieries in the Centre, South and South West with outputs of over 30,000 tons within each.

Table 16 contains estimates of annual production in the three main areas at various times. They are obviously somewhat rough and ready, based upon estimates of output at collieries for which no quantitative evidence at all has survived as well as the sometimes questionable estimates discussed earlier, but they should fall within a margin of error that is acceptable for statistics of this vintage.[69] The ninefold increase of production between 1740 and 1799 far exceeded the growth in colliery numbers, so that expansion in the size of collieries contributed most to it. Growth was not smooth, nor was its rate constant from area to area.

TABLE 16. *Output in the three main areas of the coalfield, 1740-99 (in tons)*

| Area | 1740 | 1760 | 1773 | 1799 |
|------|------|------|------|------|
| South West | 50,000 | 55,000 | 60,000 | 100,000 |
| South | 8,000 | 115,000 | 125,000 | 200,000 |
| Centre | 20,000 | 29,000 | 36,000 | 380,000 |
| TOTAL | 78,000 | 199,000 | 221,000 | 680,000 |

In the already important South West, output probably doubled with the main increase coming in the last quarter of the period, although a categoric statement is impossible because of the shortage of information about the size of collieries there after 1750 or so. In the South, the chronology of growth was erratic. Production was undoubtedly small before 1750, and then, before the canal was built, the number of collieries began to increase. This did not continue far after the Sankey was opened in 1757, but collieries became much larger. Many of the works were ephemeral before the middle of the 1770s: all those opened away from the very eastern end of its coalfield course closed, except the works in Ravenhead and Thatto Heath (59) begun in and after 1770. There were casualties even in the east and amongst the late flurry in the extreme west, and it is noticeable that the few long-lived collieries were also quite definitely the largest before the mid-1770s. Through these traumas, the progress of total output seems to have been quite slow. Colliery output figures have only survived from the middle 1760s, but by then the collieries of the Sankey seem to have been as large as they ever became before the early 1780s.

The Douglas Navigation had a much smaller effect seventeen years before. Collieries increased rapidly in number, but not in size, and the output of the Centre probably did not even double as a result of its opening. Neither did the Leeds and Liverpool Canal, built into the same area in 1774, have an appreciable immediate effect. All the indications of colliery numbers, durations and outputs are that relatively little expansion had occurred by the early 1780s, when the largest collieries produced, on average, less than 10,000 tons a year each. But expansion was thenceforth cumulative except for a sharp fall in 1794 and 1795. By 1799 there were six collieries in Orrell producing nearly a quarter of a million tons between them, one alone raising 60,000. There may well have been others of this size in Up Holland, and in the very last years of the century collieries in Standish, Shevington and Pemberton might have been developing towards a similar size. On this evidence, 380,000 tons seems to be a conservative estimate of the output of the Centre in 1799.

After about 1780, growth seems to have occurred throughout the coalfield and there was considerably more continuity of operations, even in the South. A pattern of sustained expansion seems to have settled over the coalfield. The staggered and sometimes stunted bursts of growth in the main producing areas over the previous forty years had gelled into a process whose course was steady by the last twenty or so years of the century.

## Supplementary evidence

Unfortunately, only in the 1780s and 1790s do sufficient of the parish registers contain occupations to yield maps of the frequencies of baptisms of colliers' children. These generally corroborate the summary conclusions just outlined. A comparison with the corresponding map for the 1720s (figure 19) is salutary. Then, Prescot and Whiston in the South West had the highest frequencies of collier children's baptisms with between five and ten per annum, and only four other townships had two or more a year. In the 1780s the three main mining areas stood out starkly, and by the 1790s the Centre was clearly predominant amongst the three, with averages of five or more in at least six of its townships.

The pursuit of correlations onto a finer spatial scale is difficult because of an absence of occupations from some registers in each of the areas of the coalfield. But even bearing this in mind, it seems that the geography reconstructed from mining manuscripts is largely supported by the evidence from the registers. Colliers were concentrated in Prescot and Whiston in the South West and in Parr, Windle, Sutton and Haydock in the South, the townships where nearly all the collieries for which records have survived were also located. In the Centre, too, most colliers were registered in the townships for which records of mining have survived, both being particularly concentrated in Orrell. By the 1790s in the Centre, townships eastward along the canal, particularly Wigan, also had large numbers of collier children's baptisms. Whether this increase was commensurate with the late

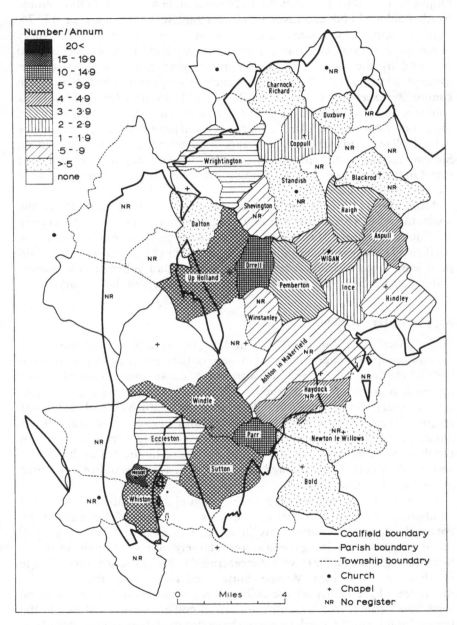

37. Distribution by township of fathers recorded as colliers in parish baptism registers of the 1780s. See figure 19.

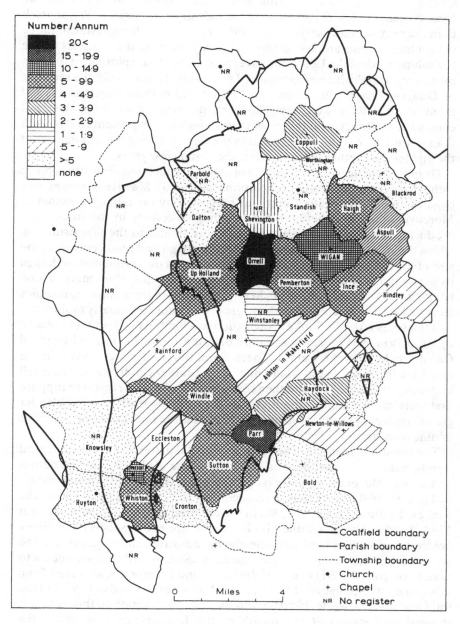

38. Distribution by township of fathers recorded as colliers in parish baptism registers of the 1790s. See figure 19.

spread of mining into Pemberton and its accelerated growth in Standish, Ince, Aspull and Haigh, or whether the industry was larger in Wigan itself than the very meagre surviving records for the borough suggest, cannot be determined. There are few records of occupations in the registers of the townships north of the Leeds and Liverpool Canal; those plotted on the map represent 'strays' who were registered elsewhere. In the 1780s the registers of Douglas Chapel, Eccleston and Coppull did contain occupations, and numbers of colliers were few. At least, the register evidence does not contradict the conclusion that mining was much less important to the north than it was to the south of the canal in the Centre during the last quarter of the eighteenth century until, perhaps, the closing few years.

Graphs of changes in the numbers of baptised collier children also confirm rather than contradict earlier conclusions (figure 39). Massive increases only began in the 1780s in the Centre, as suggested by the mining documents. Moreover, the ratios between the numbers of colliers in the three main producing areas in the 1790s are reasonably similar to the proportions in which their outputs were estimated. On this rough and ready evidence, the size of the industry in the Centre might have been overestimated (although its newer industry may well have had a younger and therefore more fecund labour force). Two features of these graphs differ from the chronologies surmised earlier. According to them, growth was not particularly rapid in the first ten years after the Sankey Navigation was built in the South, but picked up strongly in the 1760s, just as it did a decade after the Leeds and Liverpool Canal was built in the Centre. There were, too, quite large swings in the graph for the South West, where numbers of collier children's baptisms fell between 1782 and 1794, and then rose again. There is no reason to suppose that these more complex trends are further from the truth than the blander growth patterns proposed on the basis of partial and non–randomly selective mining records.

The register evidence can be related to spatial patterns and chronological trends, but not to actual quantities. There is little evidence of actual output on the coalfield, but what there is again supports the estimates made earlier. In 1771, 90,000 tons of coal were, according to Pennant, shipped down the Sankey. I estimated that 125,000 tons were produced in the South in about 1773,[70] and this is compatible with Pennant's figure if, as seems not unlikely before the growth of local coal-consuming industries, three-quarters of the area's output was taken out down the canal. Statistics of coal shipments to Liverpool from the Wigan end of the Leeds and Liverpool Canal exist from 1781 onwards (see figure 41). Only 31,401 tons were moved in 1781, but the total had quadrupled to 202,185 tons by 1799. In addition to this, coal was shipped to Tarleton at the mouth of the Douglas, by now under the ownership of the canal company. Accounts of tonnages sent to both destinations have survived for 1784, 1785 and 1788, when sixteen per cent of the total went to the northern destination. If the proportion remained the

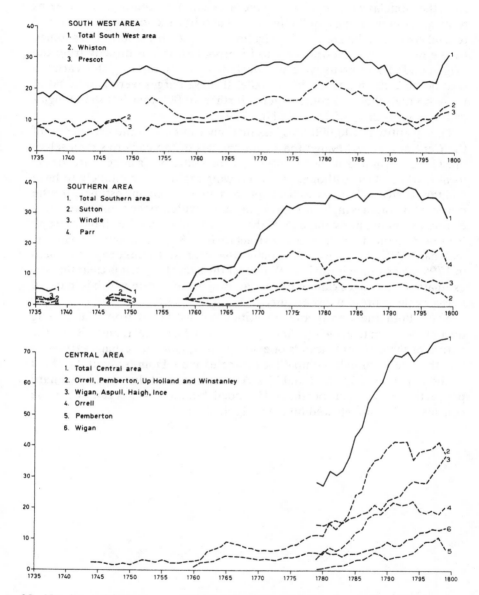

39. Numbers of fathers recorded as colliers in parish baptism registers of certain coalfield townships, 1740-99. See figure 20.

same in 1799, 243,000 tons would have been shipped along these two routes from the Douglas valley collieries in the last year of the century. Aikin seems to suggest (his meaning is not fully clear) that thirty-six boats were employed to haul coal from the Douglas Navigation – presumably where it was joined by the new canal at Newburgh – to Liverpool as well as directly from the Wigan end. These boats were obviously not simply carrying the Tarleton coal because the tonnages do not tally. If these barges were additional to those on the other two routes, then a further 40,000 tons left the Douglas valley, making 283,000 tons in all in 1799.

This quantity is 100,000 tons less than my estimate of the production of the Central area. Although the steam engines of the collieries themselves must have consumed at least 10,000 tons, this still leaves a huge share for the local market which, although it was growing rapidly, was unlikely to have had this capacity. Perhaps, as the baptism figures hinted, my estimate for the output of this area is high, though on this same evidence the Central area was clearly pre-eminent by the end of the century. On the other hand, Joseph Priestley's estimate of the rate of extraction in 1788 gives a tonnage which is less than half that hauled on the canal in that year, and if this proportion held in 1799 my estimate is too low. Whatever the exact figure, it is clear that my estimates of tonnages cannot be too far away from a truth which is in detail inaccessible to present day enquiry.

Thus, both the trends and quantities reconstructed from the mining documents seem to get within hailing distance of historical reality. Given the nature of the different strands of evidence it is, indeed, perhaps surprising that they yield mutually compatible estimates at all. Doubts about the effect of the appearance of the Land Tax Assessments and the sparsity of hard production data are, of course, well founded, but their assault cannot rob the conclusions of enough credibility to cripple them.

# CHAPTER 7

# FACTORS AFFECTING THE LOCATION PATTERN OF COLLIERIES, 1740-99

## Demand

The main changes in the geography of the industry in this period resulted from linking different parts of the coalfield with external markets through three waterways.[1] The effects of each on both the inter- and intra-area scales will be dealt with at the beginning of this chapter before the nature of changes in the pattern of consumption on the coalfield itself.

### The waterways

*The Douglas Navigation.* The Act to make the river Douglas navigable from Wigan to the Ribble estuary was passed in 1720 but the project languished until 1733 when it was revived by Alexander Leigh and Roger Holt of Wigan. The complete line of sixteen miles, falling seventy-two feet through eight locks, was finished in June 1742, although boats could reach the Millstone Grit seams on the fringe of the coalfield by 1740. Holt died before the work was completed and Leigh seems to have been mainly responsible for raising the initial capital of £7000 and the additional sums which were necessary continually to improve the channel. Modifications were still in progress in 1771 when the navigation was sold to the Leeds and Liverpool Canal Company.[2]

The Douglas was thus opened slowly, laboriously and imperfectly. But the proprietor's initial hopes were high. He anticipated that 'ye Coal Trade will be very Considerable, Not only because we Can Furnish all ye Coast with Cannel . . . but also For yt we hope to Establish a Trade with the Kendal Cartmel and Ulverston Side'.[3] Enquiries were made about the demand for cannel in London,[4] and the Irish Sea coasts, with their burgeoning cities of Dublin and Liverpool, lay open from the river's mouth. But a 'Disinterested Burgess of Wigan' wrote in 1753 that 'I always thought it a wild and impractical scheme . . . likely to end in nothing but the Ruin of the Projector. The Success answered my Expectation; and I am sorry to see the Misfortunes it has involved' Mr Leigh,[5] who himself admitted in 1768 that his navigation 'do's not admit of any Trade but at Spring Tides, nor of vessels more than 40 or 50 Tons burden'.[6]

The main reason why trade was sluggish on a river penetrating an area with abundant high quality coals was that it opened into an economically undeveloped estuary. The Douglas below Tarleton, the Ribble, the Fylde coast and the Wyre comprised the port of Poulton. Douglas coals could be taken anywhere within this flat and mossy hinterland without attracting the duty on sea-borne coal, but only Preston provided a market of any size there and even this could only be held from the competition of landsale collieries in the North and in the Blackburn area by keeping the canal dues down to 1/- per ton rather than the 2/6d allowed by the Act.[7] Without a reduction of the duty on water-borne coal the shippers on the Douglas could not compete in the Irish Sea market with the coastal collieries of Sir James Lowther at Whitehaven or in Liverpool with landsale coal from Prescot and, after 1757, with coal from the Sankey Navigation. If the duty was not reduced they 'must be shut up within our own river' and its unprepossessing estuarine and coastal extensions.[8] Leigh applied what pressure he could upon parliament through Sir Roger Bradshaigh of Haigh MP, for whom Leigh acted as attorney and estate superintendent during Bradshaigh's absences in London and whose own collieries would greatly benefit from an extension of the markets accessible through the Douglas. They tried to get the duty removed from Douglas coal bound for London directly or via Liverpool and reduced on coal moving between the Ribble and Mersey estuaries; they attempted to get hold of the farm of the duties of the Port of Lancaster. All their schemes were baulked by the more influential and wealthy Lowther, who seems to have succeeded in his 'design [which] is no less than to Monopolise the Coal Trade upon this coast by . . . prohibiting all others from Dealing in that Article'.[9]

The only coals which could penetrate markets outside the river were the high quality produce of the Arley, Smith and Cannel seams. Douglas coal 'bore 4 p Tun more price than the Whitehaven' in Dublin because of its superior quality[10] and competed successfully in Lancaster despite the port being 'over-stocked'.[11] The boats bound from Tarleton to Dublin shipped at least as much cannel as bituminous coal[12] and cannel must have continued to satisfy those wealthy enough to discriminate in Preston, Warrington and Liverpool after the Douglas was opened just as it had done before.[13] Winstanley (6) smith's coal certainly found its way to Liverpool and Warrington and managed even to penetrate to Ulverston, deep in Lowther's territory (figure 40).[14] A number of colliery proprietors had their own flatts on the river and others sold to independent haulage concerns.[15] Some of the latter acquired interests in collieries and Leigh himself had flatts, three coal yards and a colliery partnership before the 1760s.[16] It is perhaps significant that the colliery proprietors with flatts, like Leigh, all worked areas where the Arley, Smith or Cannel mines outcropped.

The Tarleton Trading Company was founded by James Winstanley and Thomas Hatton in 1752.[17] Within two years the partnership had expanded

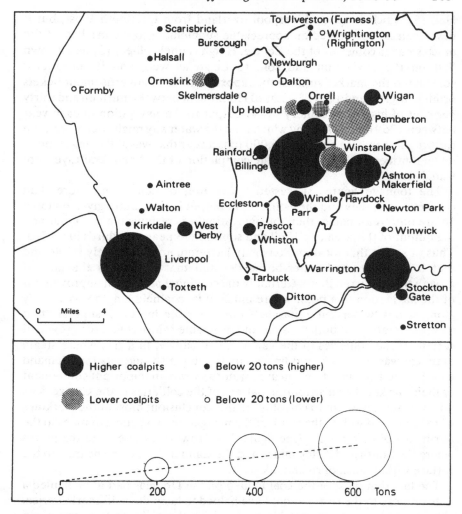

40. Coal sales from the higher and lower coalpits at Winstanley (6) in 1766. Based on LRO, DDBa/ Account of Sundry Trusts . . . 1766.

to six members and controlled two coasters and two river flatts which shipped between seventy and a hundred tons out of the Douglas valley each month: in other words, this firm alone could have sold the entire output of a small colliery. Like Leigh, the Tarleton traders had shares in collieries, Winstanley and Hatton in Kirkless cannel works (104) and James Bradshaw in Orrell (1).[18] At the Tarleton Company's shipping rates the forty flatts on the navigation in 1758 could have moved about 15,000 tons a year out of the Douglas valley.[19] This was, of course, less than half the output of Prescot

Hall (5) which supplied Liverpool overland from the South West, but it nevertheless represented an appreciable supplement to the markets of the much smaller collieries of the Centre. Haigh cannel colliery (2) had its own flatt on the navigation and sold to Leigh as well, and its proprietor considered the market sufficiently important to have advertisement tickets printed in 1760 and 1761.[20] Richard Holme sent between thirteen and thirty per cent of his Orrell colliery's (74) output to the navigation in each year between 1769 and 1773,[21] by which time the waterway might well have been carrying somewhere near the 26,000 tons a year that was still being shipped down it fifteen years or so after the completion of the Leeds and Liverpool Canal alongside.[22]

The tonnage figures suggested above are, of course, no more than conjectures. If they and the estimates of output in the Centre given on table 16 are both reasonable, then the navigation was responsible for about half the output of this area in c.1760 and somewhere near two-thirds in c.1773. Thus, it seems that local markets may have remained relatively stable and that it was increased water-borne demand that stimulated the modest growth that occurred there. Certainly the continual piecemeal improvement of the navigation to a point where much of its coalfield length was already along a deadwater cut separate from the river by 1773 suggests that increased traffic was both possible and profitable.[23] Moreover, the growth of certain of the collieries in the valley, albeit slowly, to a size where steam drainage was necessary and financially feasible also suggests that demand concentrated at a point on a nearby waterway was an important component of their market. The distribution pattern of the collieries in the Centre does not, at first sight, seem to corroborate this conclusion; most of the workings were strung out well to the south of the navigation or a couple of miles to the north east of its terminus (see figure 31). However, these were the places where the Smith and Arley mines and the Cannel seam came nearest to the surface within reach of river bank sales.

The fact that most of the coal loaded on the Douglas had to be hauled a mile or more to the river cannot have helped to keep down its market price. This was not an insurmountable problem for these choice fuels but it can only have reinforced the difficulties of finding easy markets for Douglas coals. It was these intertwined influences which kept down the size of the maximum to which collieries grew in the Centre before 1773 to less than a third of that of the large workings on the Prescot to Liverpool turnpike and the Sankey Navigation. The relatively small demand, certain only for high grade fuels accessible at relatively long distances from the river, must also have been conducive to the early plethora of tiny workings in the vicinity of the established larger works at Haigh and Orrell: speculation was possible, but its fruits could not be large and they were by no means certain.

It seems quite clear, then, that the nature of the market opened up by the Douglas Navigation was responsible for a number of the characteristics of the changing colliery location pattern in the Central area in the three middle decades of the eighteenth century. The gradual growth, the small collieries eccentrically distributed and the rapid turnover in the colliery population during the early years were all consistent with its influence. Because the river did not give the collieries of the Centre any advantage in markets previously served by the pits of other areas its effects were not felt beyond that area.

*The Sankey Navigation.* The Sankey Brook Navigation Act of 1755 authorised the improvement of natural river channels from the Mersey into the middle of the Southern area of the coalfield with termini in Ashton, Windle and Sutton. In fact, a deadwater cut was built under the authority of the Act and the Sankey was opened in 1757 as the first industrial canal in Britain. Eight locks took the waterway's main channel of eight miles to the point at the Old Double lock where it split into three branches, completed to their termini in 1759, 1762 and 1771.[24] This was the major coalfield waterway until the Leeds and Liverpool Canal was opened in 1774. Its impact until that date will be considered in this section of the chapter.

The Sankey was actually financed and built by Liverpool capital and skill, unlike the Douglas Navigation which was originally projected from Liverpool but constructed and run by Wigan proprietors. The Liverpool interest ensured the speedy completion of an access to the coalfield. It is easy to understand. The port was dependent on supplies from the Prescot to Liverpool turnpike, 'which of late Years are become scarce and dear and the measure greatly lessened, to the great imposition and Oppression of the Traders manufacturers and Inhabitants of this Corporation'.[25] Ordinary coal was 3d per cwt at Prescot and 5½d in the city compared with 1¾d for ordinary coal and 2d and 3d for best smith's coal and cannel in the Central area of the coalfield. This situation developed because the tremendous growth of Liverpool exerted pressure on the monopoly supply prices of the South West. Its population probably trebled to over 40,000 between 1720 and 1750 and by the third quarter of the century the port contained, on the evidence of craftsmen's wills, far more coal-consuming industry than the other towns of South West Lancashire: indeed, it probably accounted for as much as existed in the whole of the rest of the region (see figure 43). Liverpool itself was not the only point of massive coal consumption in which the projectors had an interest. The Cheshire saltfield was already drawing in supplies up the Weaver to supplement those from Staffordshire, but Lancashire coal from the South and the South West required expensive overland haulage, whether direct to the Weaver through Warrington or to Dungeon Point on the Mersey whence it was floated to Frodsham Bridges at the Weaver's mouth. In addition to these two fast growing markets of large potential, Warrington on the upper Mersey estuary presented a not inconsiderable demand which could be tapped by coal leaving the mouth of the Sankey a few miles downstream of the town on the Mersey.

Supplies from the Sankey allowed continued acceleration in the growth of these markets. After it was opened salt and sugar refining, copper smelting, brewing, distilling and potting burgeoned in large works in Liverpool which must, by 1770, have possessed one of the largest congeries of coal-consuming industry in Britain besides having a population, and therefore a domestic coal market, increasing by about a thousand persons a year between 1750 and 1770.[26] The export of white salt down the Weaver doubled to 40,000 tons between 1760 and 1780 and this must have required a roughly commensurate increase in the consumption of coal on the saltfield.[27] Full runs of statistics of coal shipments down the Sankey to these markets are not extant, but sufficient data exist to suggest their magnitude and rate of growth. The tonnage hauled up the Weaver from the Sankey increased from c.16,000 to c.37,000 tons between 1760 and 1773 (figure 41). In 1771 total coal shipments down the Sankey were 90,000 tons per annum, of which 45,568 tons went to Liverpool, c.27,000 tons to Cheshire and c.18,000 tons to Warrington and elsewhere.[28] Thus, if the output estimate for the South in the early 1770s is correct, seventy-five per cent of the area's production was exported down the Sankey Navigation.

Given the size of this market and its importance to them, one would expect the collieries of the South to be drawn to the canal bank. After the initial period of flux with numerous sporadic openings and closures which lasted until about 1757, the workings were strung out along the length of the canal before becoming markedly concentrated on the edge of the coalfield where the waterway distance to the markets was least (see above). However, this apparently rational location pattern changed comprehensively after 1767. Thenceforth until 1774, the terminal stretches of the canal became

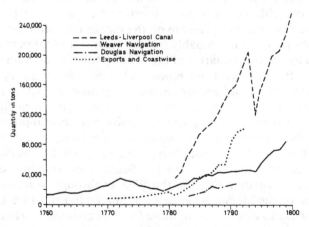

41. Coal tonnages hauled on the Weaver and Douglas Navigations, the Leeds and Liverpool Canal, and exported in the late eighteenth century. Based on sources and calculations given in Chapter 7, notes 49 and 51.

dominant in terms of coal output: only three collieries remained at the market end of the canal by 1773 with four, including the largest on the coalfield, near the termini (see above). The more recent completion of the end branches into virgin coal may have had something to do with this shift, but it cannot have been the sole reason because large reserves of high quality fuel remained in Haydock, Parr and Garswood nearer the market. Other variables than transport alone must, therefore, have been responsible for it.

Unlike the Douglas Navigation, the Sankey brought fresh and abundant reserves into communication with large markets which had previously been supplied from elsewhere on the coalfield. The shattering of the monopoly of the South Western collieries in Liverpool and Cheshire was anticipated with great foreboding at Prescot and Whiston. The mediate tenant of the Prescot Hall reserves anticipated that the South Western collieries 'must stop for want of a proper demand' and claimed soon after the canal was opened that 'the Success of the Liverpool Navigation . . . will lay a load on the other works, that they must sink under'.[29] One of the Prescot Hall (5) agents informed the lessors in March 1759 that 'The Navigation now affects us much . . . Our Sale . . . is very Slack, which if it does not mend in a little time . . . I would advise the parties to give [the lease of the colliery] up'.[30] In the following month there was 'a monstrous quantity' on the bank at Prescot Hall (5) due to 'want of sale'[31] and a correspondent to a Liverpool newspaper wrote in 1762 that 'The Country Carters, at the opening of the Navigation, found that there would not be the Demand for their coal that there formerly had been; therefore, as their Livelihoods chiefly depended on that of leading Coal, they were obliged to convert their Carts to other uses'.[32]

However, these lamentations of gloom were not disinterested observations of the state of affairs in the South West. They all had definite ulterior purposes. The newspaper correspondent was concerned, like many others who published in the Liverpool press, to demonstrate the 'evils' resulting from the construction of the canal. The Prescot Hall (5) correspondence had a more tangible objective: to obtain terms as favourable as possible in the new lease which was under negotiation because of the expiry of the existing one in 1763. A draft lease of 1761 reveals that 195,432 tons of coal would still remain ungot on the expiry of the earlier lease in 1763 plus other reserves 'supposed to be considerable', and a willingness to pay an entry fine of £2565 in lieu of six years rent.[33] This hardly suggests that the demise of the colliery was imminent. Indeed, Prescot Hall (5) and the other collieries of the area enjoyed long continuous subsequent careers at outputs which although no greater than before 1757 were certainly not much less. The apparent slight decline in the output of the area from between 60,000 and 65,000 tons in the early 1750s to about 55,000 tons in the latter part of the decade was probably due to competition from Sankey coal in Liverpool, but output had probably reached its earlier level again by 1773. Moreover, Prescot Hall (5) and Whiston (34) collieries continued to be amongst the largest and technologically most advanced on the coalfield.

One of the reasons for the survival of the industry of the South West almost undiminished was an immediate defensive cut in pithead prices from 3d to 2½d per basket for best coal and 2½d to 2d for inferior coal when the Sankey was opened. However, South Western coal must still have been more expensive in Liverpool than that from the Sankey (figure 42), where pithead prices were significantly lower than around Prescot (see figure 45). South Western coal was able to penetrate the Liverpool market at its higher pithead and customer prices because of conditions on the Sankey itself and the policies pursued by the coal producers and dealers there.

Although much better than the Douglas, the Sankey waterway was by no means perfect. The flatts must negotiate a lock every mile and a bank at Fiddler's Ferry where there was only five feet of water at highest spring tides.[34] Once in the Mersey they encountered 'the many dangers, delays and uncertainties of . . . conveyance thro' the tideway'.[35] 'The vessels sometimes make three trips, but more generally no more than two in a month' and 'the supply of coal is very uncertain'.[36] Thus, the vagaries of the canal link

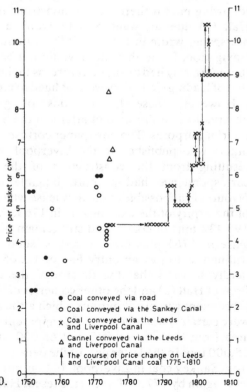

42. Liverpool coal prices, 1752–1810.

prevented a completely adequate meshing of the virtually unlimited demands of the Weaver basin and Liverpool and the virtually limitless supplies of the South. Even though Prescot coal was more expensive, supplies were more regular and could reach Liverpool within a day, so that the collieries there were more easily able to benefit from short-term fluctuations in demand and to guarantee the constant provision of consumers who required large regular deliveries.[37]

These inherent disabilities were exacerbated by the policies of the merchants who dealt in Sankey coal. 'The place of delivery is excessively dirty, there's no seeing the quality or measure before they are delivered, and no stocks ever kept on hand.'[38] The second of these strictures was keenly felt by the customers. Prices increased from 7/- a ton to 8/6d a ton between 1757 and 1762 and in 1772, 9/- was charged 'for what they call a ton, but their "ton" had been measured and weighs 15½–17cwt., according to the quality of the coals'.[39] This problem was chronic and the source of numerous complaints during the first fifteen years of the Sankey's existence.[40] The country carters who dealt in coal from the South West seem to have been paragons of virtue in comparison. Thomas Lyon, who carted coal to Liverpool from Carr colliery (127) in Whiston, claimed that persons had 'tried promiscuous loads which weighed 27 Hundred Weight, without slack, and measured 12 Tubs, which is One Third more than the Navigation measure: He will uphold every 20 Baskets One Ton and a Quarter'.[41] Whatever the actual price difference between equal quantities of coal from the two sources it is clear that it was by no means as much to the advantage of the Sankey as the ostensible prices imply. Moreover, the colliery proprietors there did not necessarily themselves benefit from the chicanery of the Liverpool dealers; they spent the first fifteen years in a vicious coexistence of bitter rivalry and price cutting which left the quantities supplied from the Sankey much below those which would have been sent down in more benign competitive conditions. Coal was still 'very often scarce' in Liverpool in 1771.[42]

The impact of the Sankey on the geography of mining was massive: the Southern area rose swiftly to prominence as it gained access to Liverpool and Cheshire. But by the early 1770s the influence of the canal on the location of collieries within the area had become muted and, owing to the policies of coal producers and coal dealers and the vagaries of the channel to Liverpool, the South did not completely oust the South West from competition in the Liverpool market.

*The Leeds and Liverpool Canal.* In October 1774 'the Grand Canal from Leeds to [Liverpool] was opened amidst the acclamations of thousands of spectators – it is now navigable to the great collieries at Wigan.' The first traffic into Liverpool was 'several boats loaded with coals &c' which were greeted by 'a band of music', 'a grand salute of 21 guns' and 'universal joy

... in the countenances in the crouds which were assembled.'[43] Rising Sankey prices (see figure 42) and chronic short supply fired this enthusiasm for another coalfield link to Liverpool. Yet the original projected line of the Leeds and Liverpool did not penetrate the coalfield and access to Wigan was only achieved by the acquisition of the Douglas Navigation, which was initially joined to the canal by locks at Newburgh. By 1774 the navigation had been converted into a deadwater cut from Newburgh to Gathurst in Orrell, the township containing the largest number of collieries. Until 1780 the old river navigation was still used from Gathurst to Wigan, but in that year the canal was finished to the town. It was thenceforth linked to Liverpool directly by canal, by river to Newburgh and then by canal, and to Tarleton by the river, which was locked into the canal at Newburgh.[44] Although in the same ownership, the two waterways perhaps had separate accounts: coal bound for Liverpool down the navigation, presumably through the locks at Newburgh, was listed separately from that carried on the canal in some statements of tonnage.[45]

Coal from the Douglas valley became the lifeblood of this important and profitable waterway, but the traffic did not at first grow quickly. There was continual friction between the canal proprietors with interests at the Liverpool end, who wanted to keep down dues on coal to increase traffic, and those with interests at the Leeds end, who wanted to raise dues on the western part of the canal in an attempt to maximise canal revenues.[46] A year after the canal was opened, the coal duty on the western end of the canal was reduced from 2/4d to 1/6d a ton.[47] Even with this low duty and the completion of the new waterway to Wigan in 1780, coal shipments had only reached 43,259 tons on the canal and about 12,500 tons on the river by 1782, not much more than half of the Sankey tonnage of a decade earlier and little more than was carried up the Weaver alone from the Sankey in the same year (see figure 41). Thenceforth, shipments increased dramatically; a determined attempt by the Yorkshire interest to increase dues back to their original level was eventually fought off in 1787[48] and the climb in coal shipments continued, with a sharp interruption in 1794, reaching over a quarter of a million tons by the end of the century.[49]

This chronology of canal shipments explains the pattern of output growth in the Centre: slow until the early 1780s then accelerating, with an interruption in the early 1790s, until the end of the century. The pre-eminence within the area of the Douglas valley collieries is also, of course, due to this canal-based demand. Unlike on the Sankey, however, the collieries of the Centre were never strung out along the bank of the canal itself but tended to stand away to the south of it and to the north east of the terminus. As on the Sankey, the pattern changed quite markedly as time progressed; in this case, the areas to the north of the canal in Standish and to the south in Pemberton, east of Orrell, grew newly and massively in the last few years of the century. This was not related to the completion of the canal eastwards to Wigan, which occurred fifteen years earlier. Clearly the detailed pattern of location around the canal was due to factors other than the nature of the waterway itself.

With the completion of the Leeds and Liverpool Canal there were three local sources of fuel for Liverpool as well as the ever present threat of supplies from the Bridgewater Canal if prices rose sufficiently high. The rapid rise of supplies from the Centre after the early 1780s suggests that the Leeds and Liverpool took much of the market from the Sankey and the Prescot turnpike. But output in these other two areas did not decline – indeed, production increased in both during the last twenty years of the century at the very time when supplies from the Leeds and Liverpool were growing most rapidly. This eloquently illustrates the speed of Liverpool's growth. Its population had reached 82,000 by 1801, when it was the third largest city in the kingdom after London and Dublin. Its processing industries grew commensurately and the coal consumption of the smelters and refineries of the port must have been prodigious: Rowe's copper smelter alone had thirty-five furnaces burning between 10,000 and 12,000 tons of coal a year in the early 1790s.[50] On top of this, exports grew considerably. Coastal shipments from Liverpool increased fifteenfold between 1770 and 1792 and exports abroad twelvefold (see figure 41). According to figures compiled by Gregson in 1792, the equivalent of more than half of the coal brought into Liverpool by the Leeds and Liverpool Canal was being shipped out again in that year.[51] The assertion made in 1787 that coal only left as ballast and that there was no specialist shipping engaged in the trade can hardly have been correct.[52] By the early 1790s there was concern that 'the export of this useful article from this port, which is now considerable, may prove in the end detrimental to the Town.'[53]

The cause of this concern was continued pressure on supplies in the city itself despite the massive injections from the Leeds and Liverpool Canal and the sustained growth of output in the South and South West. In 1787 'the *Liverpool* Market has not been fully supplied with Coals for some Years past' and in 1789, after massive expansion in coal tonnages on the Leeds and Liverpool, the port was still 'short supplied often no better than half supplied'.[54] In 1792, after further huge increase in canal tonnages, 'on account of the scarcity and dearness of coal the copper Works are under contemplation of being remov'd'.[55] In the same year, 'there was not sufficient Number of vessels to convey the Goods to and from Liverpool along the canals, which caused great inconveniences and frequent interruptions to trade'.[56] The rate of expansion of the Liverpool coal market was clearly beyond the capacity of the turnpike and two canals which linked it to the coalfield.

The strong pressure of demand upon supplies did not cause prices to rise. They were remarkably stable until the late 1790s (see figure 42). This was not simply because of competition between the three areas of the South

West Lancashire coalfield in Liverpool but also, and perhaps mainly, because of potential competition from the Duke of Bridgewater in Liverpool itself and from the collieries of Flintshire, Cheshire and Cumberland in the Irish Sea trade. The Duke 'offered a supply to the town' in 1776; in 1785 it was claimed that he 'can whenever he pleases deliver coals at Liverpool at a lower price than the Leeds and Liverpool Canal Proprietors can; but at present, having a better market for his coal, he does not interfere or bring any of his coals to Liverpool', and in 1789 the canal proprietors 'will not permit the colliers to raise the price of coals at Liverpool if they did it would permit the entry of the Duke of Bridgewater's coals'.[57] In the Irish market, coal from the Leeds and Liverpool was at a disadvantage of 3/- a ton, because of the cost of haulage to the Mersey, compared with the collieries at Workington and Whitehaven in Cumberland and Neston and Parkgate on the Dee in Cheshire.[58] The markets available in and through Liverpool might be enormous, but they could only be held from competitors from other coalfields in the last quarter of the eighteenth century by keeping prices down to a level beneath which they could not penetrate.

Competition between the Centre, South and South West had the same effect. Immediately the Leeds and Liverpool was opened in 1774 the main coalowners on the Sankey published in Liverpool a joint advertisement of prices and a declaration of their intention to deal fairly and mete full measures.[59] Prices of coal from different collieries varied slightly, but the top price of Sankey coal was to be 4½d per cwt of 120 lb. Leeds and Liverpool prices of the following year were higher: 4½d was the lowest price charged for their variety of house-coal, and cannel and smith's coal were 8½d and 6½d per cwt of 120 lb for all but the last, which was in cwts of 140 lb.[60] When the Leeds and Liverpool tonnage dues came down in 1776 to the level prevailing on the Sankey, only cannel fell in price.[61] The Douglas valley collieries held their Liverpool price stable for nearly twenty years, but even so it was said to be higher in the middle 1780s than that of the collieries supplying the port from around Prescot.[62] It was only really in the last two years of the century that Leeds and Liverpool prices began the giddy upward spiral which marked the early years of the nineteenth century (see figure 42).

The high quality of the coals brought from the Douglas valley along the Leeds and Liverpool Canal guaranteed them a sale despite their slight price disadvantage. The relative merit of the coal from Prescot at its slightly lower price was nicely balanced in the 1780s[63] and, though unremarked in the discussions of the time, coal from the Rushy Park and Florida seams mined in Ravenhead and Haydock on the Sankey was also of unusually high quality.[64] Nevertheless, Gregson concluded in 1792 that 'owing to a superior article, which the country affords through the borders which this canal passes, all competition has yielded in favour of the Leeds canal'.[65] But the victory was by no means a rout. In 1785 the collieries around Prescot

supply above one half of the inhabitants of the town . . . with coals, upon lower and better terms than the Leeds and Liverpool Canal dealers do . . . And as to the public breweries, of which there are a great number in Liverpool, none of them make use of the Leeds Canal coal; nor does any of the Sugar refiners (except one) . . . and as to the Salt works, they are wholly carried on with Sankey Canal coal; and the copper works, which consume a vast quantity of coal, are carried on and worked with Sankey coal only.[66]

The only industrial market served by the Douglas collieries seems to have been that of the ironworks and anchorsmithies which, no doubt, derived full benefit from the coal mined in the Smith seam in Orrell and Winstanley. The important export market was given advantageous terms by dealers in Sankey coal. They also had the advantage of communicating with the port through the Mersey itself, whilst it cost 8d a ton to get coal to ships from the inland terminus of the Leeds and Liverpool Canal.[67] None the less, it seems to have been the latter which took the lion's share of the considerable export market.[68]

Thus, the rapid and sustained expansion of output in the Centre was clearly related to the success of its collieries in capturing an increasing share of the growing Liverpool market during the last quarter of the eighteenth century. The collieries on the turnpike in the South West maintained a share of the house-coal market in Liverpool, hence the stable production of the area. The Sankey could get coal to Liverpool householders at under the Leeds and Liverpool price and its entry into Liverpool was advantageous for sale to shipping and to the dockside and riverside locations of the major industries of the port. Yet Sankey supplies held only the last of these consumers in the closing twenty years of the century. Of course, this was a growing market, but it is doubtful whether its expansion could have compensated the loss of the household and export markets. Moreover, just at the time when the Leeds and Liverpool Canal penetrated its main market, there was also a slump in demand for coal in the Cheshire saltboiling industry, the Sankey's other main consumer. After climbing from nearly 13,300 tons in 1759/60 to 35,286 tons in 1772/3, shipments up the Weaver began to slide, precipitated by the technical developments of Chryssel which about halved the amount of coal needed to boil a ton of salt.[69] A nadir was reached at just over 18,000 tons in 1779/80, but buoyant demand for salt, especially from the Liverpool soap industry, carried Weaver coal tonnages back to their pre-slump level by 1784/5 and on to twice as high before the end of the century (see figure 41). It is perhaps significant that it was only after this climb began that shipments to Liverpool on the Leeds and Liverpool Canal began to move strongly upwards; perhaps the Sankey suppliers fought hard in Liverpool as their Cheshire market dwindled and then opted for their safe monopoly on the Weaver when demand there picked up, leaving Liverpool to Prescot and the Douglas valley. Whatever the exact nature of the case, the collieries of the South must have suffered a decline in the demands pulsed along the Sankey Navigation from Cheshire and Liverpool in the 1770s and early 1780s.

The shifting colliery location pattern, such a marked feature of the early development of mining on the Sankey, was related to these fluctuations of demand from the canal. The openings at the western end in the early 1770s, as far as possible from the markets, occurred in conditions of buoyant rising demand; the subsequent closures and lack of growth until the 1780s corresponded with a slack canal market, and the later expansion through collieries strung like beads along the canal bank occurred after the resurgence of demand in Cheshire.

*Other developments in transportation*

The growth and changing geographical patterning of the mining industry in the second half of the eighteenth century was clearly and firmly linked to the opening of canals and changes of demand upon them. However, alone the canals could not have effected these rapid transformations: the haulage of large amounts of coal to the canals from the collieries presented problems, and these would be particularly acute along the Douglas Navigation and Leeds and Liverpool Canal where the collieries stood back at considerable distances from the waterway (see figures 31 and 33). Road surfaces suffered badly; indeed, the impact on nearby roads was one of the arguments used against the Douglas Navigation Bill, an impact which would be severe 'because most of the Coal and Cannel Mines lie at some Distance from the River, and those Commodities must be carry'd in Carts and heavy Carriages to the Banks of it'.[70] The demands made on the supply of draft animals were also insatiable at certain times of year. At Haydock (28) in 1756, just before the Sankey was opened, coal sales stopped when ploughing began and in 1769, when the canal was presumably the colliery's main market, sales dipped during haytime as agriculture took its prime claim on the available horses.[71]

Getting coal from the pithead to the cheap routeway to the market was quite a significant problem. At Prescot Hall (5), Whiston (24) and Kirkless (104) causeways were built between the pits and the turnpike and some kind of paving must have been laid at every colliery to prevent carts becoming completely bogged down.[72] But the massive traffic along single lanes generated by the large canal-serving works required a surface which was both more durable and more efficient in the use of traction power. Wagonways provided such a surface and diffused with canals as their essential feeder links. In the Douglas valley collieries were situated at considerable distances up slope from the waterway and canal but the topography to some extent nullified the effect of distance. Two collieries serving the Leeds and Liverpool Canal were linked to it by wagonway within two years of the establishment of the link to Liverpool;[73] when the deadwater cut was extended all the way to Wigan on the northern side of the river, the wagonways of the Orrell collieries on the south bank were extended across the river navigation to intersect with the new line. By the end of the eighteenth century, Orrell, Pemberton, Up Holland and Winstanley were linked to the canal by four arterial wagonways with numerous feeders and cross links.[74] No other district had anything comparable with the Orrell network. Ambitious projects to link Haigh (2) with the canal terminus two miles away and 300 feet below were never completed[75] and no wagonways were recorded on the northern slope of the Douglas valley, although the massive concession taken in Standish (41) in the 1790s was joined to the Leeds and Liverpool by an underground canal from the bottom of Taylor pit.[76] Collieries were generally sited much nearer to the Sankey than to the Leeds and Liverpool and the valley was not so deeply incised. Even so, some of the collieries there had wagonways, the longest being that which linked the Florida colliery in Haydock (125) to the canal.[77]

Without these feeders to allow easy and relatively efficient passage from pitheads to canals the impact of water transport would have been considerably muted. They were not the only transport development other than canals to affect mining: the turnpikes were continually widened, improved and extended during the second half of the eighteenth century to provide certain large collieries with significant parts of their markets. The turnpike from Liverpool to Prescot was extended to St Helens in the South after the permitting Act in 1746.[78] It must have been this extension which stimulated the efflorescence of mining in the South in the years immediately before the Sankey was opened. A further extension from St Helens to Ashton on the main north–south turnpike in 1753 was meant to bring Liverpool into contact with 'several Ranks of Cannel pits' in the vicinity of Wigan so that dealers in the port might be 'certainly and regularly supplied with that Fuel which is greatly wanted by them'.[79] The hope for the new road implies the failure of the Douglas Navigation fully to supply Liverpool, but in the event the turnpike served no better and its effect was probably limited to hauling supplies of smith's coal from Winstanley (6) to the Sankey Navigation after 1757 and to Warrington (see figure 40).[80] The effective range of coal haulage on this eastward turnpike from Liverpool seems to have been St Helens, but after the Sankey was opened the collieries to the west of its terminus along the road to Prescot seem to have been badly hit, stuck in a limbo between the limits of effective competition to the canal at St Helens and the turnpike at Prescot.[81]

As the earlier discussion of Liverpool's coal supply demonstrated, the collieries of Prescot itself were able to continue competing in the port after the Sankey was built. In 1760, 'colliers carts and horses . . . fill the road all the way to Liverpool'; lamps were installed at toll booths to facilitate collection from carts during the night, and still in the 1790s the coal carts were 'no small nuisance at particular times on that road' as they plied between Prescot, Whiston and Liverpool.[82] Continual improvement and widening were necessary to sustain this heavy traffic. Before 1759, the carriageway was four and a half yards; in 1759 it was widened to six yards, and in 1783 it was extended by another four feet.[83] Not all the coal which left Prescot and Whiston by turnpike was bound westward for Liverpool. After the building of the section between Prescot and Warrington after 1753, some went eastwards and down to Northwich overland.[84] Whether or not this traffic continued after the Sankey was opened in 1757 is unknown, but it is certain that some of the collieries on the canal itself also served markets via the turnpikes and other roads to the south and east. The proprietor of the Haydock collieries seems to have been as assiduous in pushing his interests in road haulage as he was towards his traffic on the canal.[85] Roads clearly provided markets to the collieries of Haydock and Parr as well as the Sankey Navigation: 'Warrington is well supplied with coals from the pits of Haydock and its neighbourhood [and only] partly by the Sankey canal'.[86]

For the collieries of the North, roads provided the only means of getting coal out to Preston and the thickly swelling industrial population to the south of it. The roads between Wigan and Preston and Manchester and Preston leading northwards from the coalmeasures were execrable before turnpiking in 1726 and 1753, 'scarcely allowing room for the passage of the gangs of pack horses carrying coals and lime [which] were frequently obliged to make way for each other by plunging into the side road . . . out of which they found it difficult to get back upon the causeway'.[87] The small expansion which occurred in the North must have been due to some extent to the improvement of these roads: certainly the Douglas Navigation found it as difficult to compete with landsale coals in Preston at the end of the century as in the 1740s.[88]

*Local industries*

It is clear that the massive expansion of output in some areas and the more modest growth of others was due in large measure to the construction of low cost high capacity transport links into the coalfield to join it with large external markets. In addition, these same transport links allowed other raw materials into the coalfield to be processed with local fuel, and allowed the easy dissemination of goods produced by coalfield manufacturers. The growth of a coal-based manufacturing economy was not restricted to areas off the coalmeasures.

None the less, there was a marked decline of the small-scale workshop type of industries on the coalfield as the economic axis of the region swung inexorably towards the Mersey: water transport, not local coal, seems to have been what fostered expansion. After 1740, the brass, pewter and iron finishing trades, which had provided a significant component of demand in the Centre, seem to have declined as the glass, pottery and brewing industries of Liverpool and the surrounding area grew. The decline of the

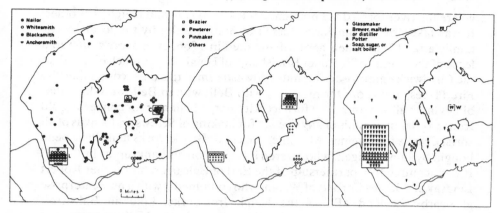

43. Distribution of coal-using tradesmen who left wills, 1741-80. Based on 'Indexes to wills'.

Wigan non-ferrous metal industries was quite spectacular (see figures 22 and 43), and although a small residuum remained, it was of no significance in comparison with the coal consumption of the large-scale industries which developed where coal and navigable water impinged upon each other.

During the second half of the eighteenth century the smelting of iron, glass and copper grew rapidly and massively on the coalfield lengths of the canals. Despite a shortage of local ore, iron smelting was introduced onto the coalfield in the second and third quarters of the eighteenth century, mainly to supply the nailors. Water power seems to have been the most significant early location factor and the early works were on the fast flowing reaches of the Yarrow south of Chorley, the Douglas north of Wigan and the tributaries of the Sankey Brook to the north east of St Helens. John Chadwick I built the first forge on the Yarrow at Birkacre in 1724 and extended his works after 1740. A slitting mill had been added by 1747 and by 1755 a second forge had been built by John's son, Thomas Chadwick I. By 1774, the plant comprised 'FORGES, containing two FYNERYS, two CHAFERYS, two LIFT and one TILT HAMMERS, one AIR FURNACE for Balling, and another for Casting Hammers, Anvils &c. &c. . . . . the higher pool, the Middle pool and the lower pool'.[89]

In 1755, the Chadwicks opened a slitting mill in the Scholes area of Wigan and a forge at nearby Bottlingwood, both on the Douglas. The forge, leased by a partnership of Furness ironmasters between 1757 and 1780, comprised 'two large water hammers, and three fire places, the bellows go by water, and there are several shops for manufacturing of pans, bols, bills, &c. &c. [and] very good warehouse room' in 1773.[90] These works were still in operation in 1780 but they were probably sold under the instruction of Thomas Chadwick's will soon after. The other works on the Douglas at Brock Mill, where the river formed the boundary between Wigan and Haigh, was older. It was taken over from John Russell of Staffordshire by the local Morris family after 1740. When her husband died in 1766, Ann Morris offered a lease of the works to Sir Roger Bradshaigh of Haigh. It was much bigger than the Chadwick enterprise a few miles downstream at this time, containing 'Six Fire Places, and room for more . . . with Bellows and Bellows Wheel that blows them all at a time . . . a furnace with Forge Hammer for converting old Iron into Bar or Moulds of any size . . . for making of Screws and Anvils of all sorts, Cranks for Engines, Machines &c.&c'.[91] Brock works must have been bought by James Wigan, who offered it for sale on his bankruptcy in 1775.[92] It was acquired by a partnership of the Earl of Balcarres, his brother Robert Lindsay and James Corbett of Wigan, who combined it with a new enterprise at nearby Leyland Mill as the Haigh Ironworks. Like all Balcarres' enterprises, this was a most ambitious concern. By the 1790s it comprised two furnaces, a steam engine, moulding works for shot and shell and shops for the construction of steam engines, sugar mills and other complex heavy machinery; the total cost of the ironworks not including a foundry in Wigan itself also run by the Earl was claimed to have been £40,000 and it employed four hundred people.[93]

The other ironworks on the coalfield were on the Sankey. A partnership including men from Coalbrookdale took over the old charcoal furnace at Carr Mill in 1759 and immediately built the first coke iron furnace in Lancashire.[94] The adjacent Carr forge and nearby Stanley slitting mill were run as an integrated concern by a partnership of Chowbent nailors, a local ironmaster and a Liverpool merchant after 1775, and in 1783 a new partnership of a local ironslitter and Liverpool merchants was formed and maintained its interest until the end of the century.[95]

The three ironworking complexes on the Yarrow, Douglas and Sankey were obviously oriented towards water power. Ore was not a significant factor: ironstone mined at Haigh was of very poor quality, much inferior to the lumps collected in the river for the preliminary experiments, and though iron ore leases were granted in Parr the ores of the Sankey district were neither abundant nor of suitable quality.[96] Fuel was also important. Coking coal only occurs in Lancashire in a few seams in the very lowest coal-bearing geological strata. These coals surfaced only in a few places, amongst them Burgh, where the Chadwicks had a colliery by their ironworks to mine a coal 'that cakes well' from the Arley or Bone seams which outcrop in the Yarrow valley.[97] At Haigh too, the lowest strata are thrown to the surface and there the King seam, considered by Balcarres to be superior even to the Arley, was opened up specially to serve the needs of the ironworks, although cannel seems also to have been used.[98] The coals of the Sankey valley were inferior to those of the Wigan area, and the Carr furnace had considerable trouble in getting supplies; Laffack colliery (118) was used, but the provision of suitable fuel required careful selection and enormous wastage of rejected coals which were difficult to sell elsewhere.[99]

What the Carr and Wigan ironworks possessed which the Yarrow works did not was access to water transport. The Chadwicks used the Douglas Navigation to bring ore from Furness and Sweden and to dispatch finished goods, and the sale advertisement of 1774 claimed that the works lay 'within six post miles of the Grand Canal from Leeds to Liverpool, on which all heavy goods may be conveyed on the most easy terms'.[100] This distance must have proved disadvantageous for a works requiring considerable heavy haulage and it was converted to a spinning mill in 1775 and sold to Richard Arkwright as his first Lancashire mill.[101] The Haigh ironworks and the other works on the Douglas in Wigan were only a mile from the terminus of the Douglas Navigation and later the Leeds and Liverpool Canal. Brock forge made 'Goods for Foreign and Home Consumption' in 1776 when the cost of carriage to Liverpool as well as the prices of coal and cannel at the forge were entered in the sale advertisement.[102] The nearby Chadwick works in Scholes supplied 'the West Indies and North American Trade [along] the Sankey and Leeds canals' in 1773[103] and the Haigh ironworks depended on the canal for ore and for the sale of its shot, shells, engines and milling machinery; 10,000 tons were carried upstream to the furnaces alone and goods bound eventually for destinations as far as Mauritius and the West Indies began their journey down the canal.[104] That the Carr ironworks on the Sankey were equally dependent upon water transport is testified by shipments of ore down the Weaver and up the Sankey and by the inclusion of Liverpool merchants in each of the partnerships formed after the completion of the canal in 1757.[105]

The ironworks symbolised the new industrial economy which grew after the completion of canals into the coalfield but they were not the only examples of it. The Sankey positively spawned large works after the early 1770s, and St Helens, 'its increase ... owing to the various works established in its neighbourhood', burgeoned amongst and around them as the first of the industrial revolution towns.[106] Copper and glass were its main manufactures, supplemented by brass, chemicals and pottery as well as ironmaking. Rapid growth began in the traditional local specialism of glassmaking on the completion of the British Plate Glass Manufactory in 1773.[107] It was a huge venture, second only to the glassworks at St Gobain in France. It cost £60,000, possessed the largest industrial building of the period in its casting hall and employed between three and four hundred full time workers by the middle of the 1790s. A smaller glassworks was opened in 1782 by George Mackay, West and Co., who introduced the crown glass process into St Helens. Like the ironworks, the glass manufactories initially had trouble with the relatively low quality of St Helens coal, but it was 'overcome by great industry and care in the choice and use of coals employed'.[108] The former stratagem must have involved settling on the coals of the Arley seam (locally known as the Rushy Park) which later became renowned as the best glass-smelting fuel, and the latter probably comprised the development of closed or 'coped' pot-smelting techniques, perfected at the works. Whatever the early problems with coal quality, its price was considered extremely attractive and larger steam engines than were absolutely necessary could be installed because 'coals are cheap at Ravenhead'.[109]

The copper works at St Helens, like the glassworks there and the Haigh ironworks, were of international importance. With the discovery and exploitation of the Anglesey copper reserves and the demand in Liverpool for copper sheathing and small copper currency bars and trinkets, the Mersey became an extremely attractive location. The Roe works were built in Liverpool itself in 1767 and another works existed at Warrington.[110] This market orientation was attractive because of the lack of duty on sea-borne copper ore and the availability of coal nearby. However, with the massive expansion of the industry and increased demands for fuel a location actually on the coalfield with canal access through the Mersey to Anglesey ore and foreign markets became desirable. Roe complained about the price of coal in Liverpool, whilst at the terminus of the Sankey 'coals are of little value'.[111] The Stanley works were built in Garswood in 1772 by the Warrington Company who expanded from their original location in search of cheaper coal.[112] The Greenbank works, which came to dominate completely the English copper industry, producing one-sixth of total national output, was built in 1780 in Ravenhead at the end of the Sankey Navigation adjacent to the plate glass works.[113] The proprietors owned the Parys Mountain ore deposits in Anglesey and by 1795 20,000 tons of Parys ores were being smelted annually at Greenbank and Stanley. The Greenbank works alone consumed 400 tons of coal a week in 1784 and 700 tons in 1795 – the equivalent of the output of one of the largest collieries on the coalfield and three times the consumption of the Roe works in Liverpool. The copper smelters also had trouble with the quality of their coal supplies. In 1787 they complained that coal from the St Sebastian seam contained sulphur and other impurities which 'retards the Process of Smelting the Copper and injures both the Metal and the Furnaces'. They demanded coal from the Ravenhead Main Delf, elsewhere known as the Wigan Four Foot, which was everywhere a good coal and at its best in the St Helens area.[114]

The four main towns of South West Lancashire thus became centres of smelting and heavy engineering in the late eighteenth century as large works grew at Liverpool and Warrington on the Mersey and at the termini of the Leeds and Liverpool Canal at Wigan and the Sankey Navigation at St Helens. The last was one of the foremost, if not the foremost, smelting centre of late-eighteenth-century England, its direct access to the Mersey and its cheap and, with care, adequate coal supplies making it a prime industrial location. The huge quantities of coal required at all these iron, glass and copper works made it imperative to guarantee continuous sufficient and suitable supplies. This was done by the vertical integration of coal and smelting in the iron and glass industries. The Chadwicks had their own pits into the Arley and Bone seams at Burgh and Birkacre (96) and at Bottlingwood (79); the Earl of Balcarres had his own colliery producing King, Arley and cannel coals at Haigh (2); the Ravenhead plate glass works was opened by John Mackay, who had collieries into the Rushy Park and Ravenhead Main seams in Parr (121) and Ravenhead (129), and Thomas West, partner in the crown glassworks, ran a colliery in Windle (157). The copper smelting industries of St Helens were not vertically integrated with mining. They were established by people from outside the coalfield and, perhaps in consequence, they bought in coal from separate concerns. However, each of these works had strong links with particular collieries which granted them special consideration in return for a monopoly of their coal supply.[115]

### Local households

Lancashire had one of the most rapidly increasing populations of any English county in the eighteenth century. By 1799 it was the most densely populated of any coalfield county. Most of the coalfield had a density of over 200 persons per square mile (figure 44) and it contained, in St Helens and Wigan, two of the fastest growing non-coastal British towns. By 1801, there were urban populations of over 117,000 on the coalfield and within the supply area of its collieries. Apart from the coagulation of population into urban masses, the most noticeable difference between the population density maps of 1667 and 1801 (cf. figures 8 and 44) is the emergence of the coalfield as the largest area of dense population in the region, with particularly marked thickenings in the South and South West, around Wigan and in a belt running northwards from the south east of Wigan through Chorley to Preston. All but the last of these areas owed most of their population increases to the expansion of mining and the industries associated with it. The northward-running belt of high density did not: it was really a part of the region dominated by cotton textiles which overlapped at Wigan and Chorley with the coal and metals economy of South West Lancashire proper.

This large increase in population must have entailed at least a commensurate growth in the domestic coal market. Being heavily urbanised, domestic consumers would now be generally less able than earlier to substitute turf for coal, though turf was still heavily exploited in the West, in Bickerstaffe as well as Rainford.[116] The same fastidiousness which characterised the industrial buyers and the household buyers of Liverpool and Preston must also have been apparent in local domestic markets. The people who came to the knowledge of Pococke in the Chorley area certainly showed it: 'they have coal in many places . . . but for their rooms they use the

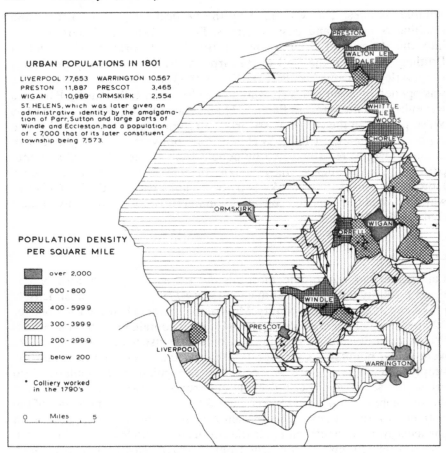

URBAN POPULATIONS IN 1801

LIVERPOOL 77,653   WARRINGTON 10,567
PRESTON   11,887   PRESCOT    3,465
WIGAN     10,989   ORMSKIRK   2,554

ST. HELENS, which was later given an
administrative identity by the amalgama-
tion of Parr, Sutton and large parts of
Windle and Eccleston, had a population
of c 7,000 that of its later constituent
township being 7,573

POPULATION DENSITY
PER SQUARE MILE

over 2,000

600 - 800

400 - 599.9

300 - 399.9

200 - 299.9

below 200

* Colliery worked
  in the 1790's

0    Miles    5

44. Distribution of population in 1801 and collieries of the 1790s. Based on *Census of Population, 1801: Parish Registers Abstracts*, British Parliamentary Papers, 1801.

Kennel or Candle coal'.[117] The Earl of Derby bought coal from various collieries in the South West and South in the 1740s, 1750s and 1760s as well as using the turf dug on his own manors; coal from Winstanley (6) was hauled over a considerable distance for the smithy, coke and coal (presumably for coking) was bought from Charles Dagnall, who mined the Rushy Park (Arley) seam in Eccleston (100), specifically for drying malt, a purpose served by the coal of Windle colliery (109) after Dagnall's bankruptcy.[118] Of course, not everyone had the wherewithal or the varied requirements of the Earl of Derby or Pococke's acquaintances, but presumably all were able to recognise that some particular coal would best satisfy their own requirements. Certainly, it was generally recognised by the end of the eighteenth century that 'the kinds of coal are as various, as the quantity is abundant'.[119]

## Coal markets and the colliery location pattern

During the second half of the eighteenth century coal and metal economies developed in two areas at the termini of canals to Liverpool. By 1800 too, all areas of the coalfield were engaged in competition in common markets, the North and Centre in Preston, the Centre, South, South West, and even the West in Liverpool. The only monopoly export market left was the salt industry on the Weaver served by the collieries on the Sankey in the South, but even here there may well have been competition from the South West down the Warrington turnpike or from Dungeon Point and the Willis' coal staithe on the Mersey. The markets in Liverpool and Cheshire were prodigal consumers of coal in smelting, refining, boiling, metalworking and domestic furnaces, ovens and hearths. The almost tenfold expansion of output on the coalfield between 1740 and 1799 was obviously due to the development of these markets and lines of cheap transport to them. Different parts of the coalfield expanded as they were joined to these external markets at different times or had their own local industrial markets engendered by the construction of cheap high volume transport routes.

The North and West were least firmly linked with the nexus of new economic forces, hence their relatively slow growth. There was no water transport in the North until the southern end of the Lancaster Canal was opened in 1799, linking Haigh in the South with Preston through Chorley.[120] In consequence, the Northern collieries had neither easy access to large external markets nor, after the demise of the Chadwick ironworks, large local coal-consuming industries of the type that canals brought elsewhere. There was considerable economic development and associated population growth in the area but it was based upon the textiles industry. The modest growth of local domestic markets and the two turnpikes to Preston, where there was keen competition with coal from the Centre brought down the Douglas, meant that the expansion of coal output in this area was really only a quickening evolution of what had gone before. The unspectacular increase in colliery numbers and the apparent lack of large outputs reflected the progress of demand in the North.

The West was also largely out of the swing of external markets or heavy coal-consuming industries, being without water transport. In addition, there were large reserves of peat in the area, busily exploited in Rainford and Bickerstaffe and probably other townships. The Leeds and Liverpool Canal skirted the northern fringe of the area and it must have been demand from this waterway, distant though it was from the pits, which stimulated the rapid expansion of output at Skelmersdale (105).

By the 1780s, the South, South West and Centre were in full competition in Liverpool. The South West had only a turnpike link, but it managed to keep a healthy share of the port's market and, perhaps, to take a part of the Cheshire market from the collieries on the Sankey. Although the expansion of the industry in this area was not so rapid and did not reach the same total tonnages as in the South and Centre, it did grow and the area continued to contain some of the largest collieries on the coalfield, a state of affairs commensurate with the large concentrated demand that it maintained. Certainly, the turnpike did not attract heavy industry in the same way that canals did: Prescot glassworks closed during this period and although the town was 'built over coal pits',[121] its other industries were mainly precision engineering related to watchmaking and some potting, although Liverpool was by far the largest centre of growth for the latter industry (see figure 43).

The scale and relatively complex chronology of growth in the mining industry of the South can be clearly related to the development of the markets it served via the turnpikes, the Sankey Navigation and locally. The early expansion in the late 1740s and early 1750s occurred as the Prescot turnpike was extended to St Helens. The development of large collieries and the large entry and exit ratio of the late 1750s and 1760s were responses to the completion of the Sankey and the jockeying of coalmasters for shares of the large markets it opened up. The turbulent flux of the 1770s occurred when demand in Cheshire fell off and coal from the Central area gained access to Liverpool, and the subsequent recovery and smooth expansion came with resurgent Cheshire markets and the massive growth of local coal-consuming industry. There was a similar correspondence between the chronology of change in mining and markets in the Centre. The Douglas Navigation gave access to relatively small markets, mainly for high quality coals, hence the smallness of the collieries and their skewed distribution within its catchment. The Leeds and Liverpool Canal injected a stimulus of much greater magnitude, but one which again was only fully active upon high grade supplies. At first gradual, then accelerating and eventually massive expansion of output in the Centre followed the pattern of coal tonnages sent down to Liverpool, even to the severe hiccough in the early 1790s when there was a slump in the port.

Thus, there were large collieries and rapid growth where large market potential was communicated through a cheap transport route, whether it came directly to the collieries or indirectly through canalside coal-consuming industries. The relationship is evident, too, on the intra-area scale, with most collieries clinging quite closely to the transport routes along which their produce flowed (see figures 31, 33 and 35).

The geography of pithead prices also changed in unison with the changing pattern of demand, but not always in consistent ways. Before 1757 the pithead prices charged by the collieries in the North selling overland to Preston were higher than those at the pits in the Douglas valley, selling to the same market along the navigation (figure 45). This pattern repeated itself

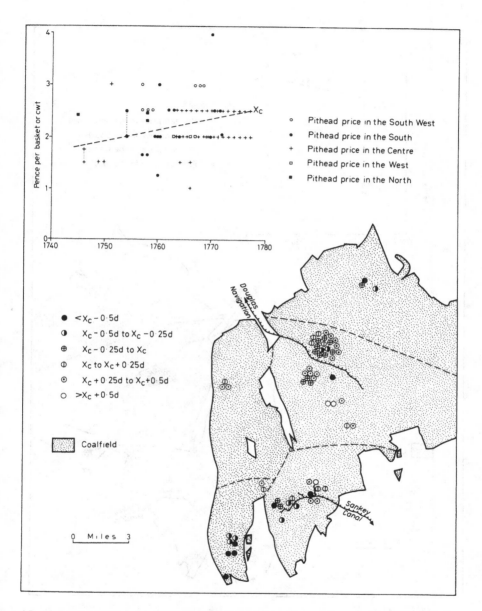

45. Pithead coal prices, 1740-73. Landsale coal prices in South and South West before 1757 are omitted (see figure 25). Symbols on map represent residuals from the regression line $X_c$ plotted on the graph.

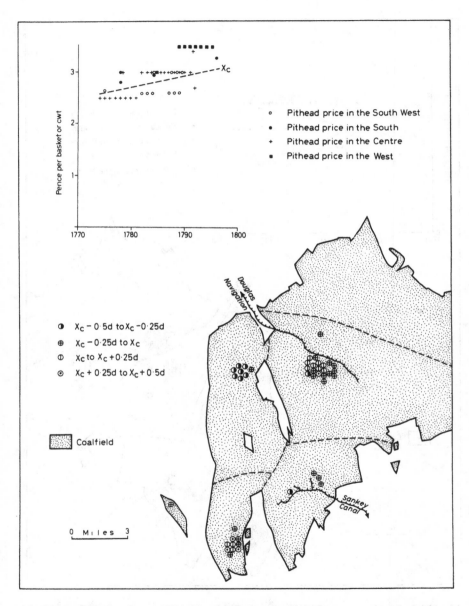

46. Pithead coal prices, 1774-99. Symbols on map represents residuals from regression line $X_c$ plotted on the graph. The high prices for the Western area were added after the estimation of $X_c$.

between 1757 and 1773 when the prices charged at the landsale collieries of the South West were higher than those at the pits on the Sankey which competed with them in Liverpool. Even within the South, prices were higher at Dagnall's colliery which sold via the turnpike than on the canal,[122] where there was a bewildering variety from colliery to colliery and through time.[123] Generally, within this wide range, prices were lower on the Sankey than anywhere except the Centre: the areas with water transport and the best coal thus sold at the lowest pithead prices. In the decade before 1774 there was a range of prices from 1½d to 4d per cwt at the pithead with considerable variation within both the Centre and the South. This variety seems to have been sharply curtailed on the opening of the Leeds and Liverpool Canal and between 1774 and 1792 all recorded pithead prices were at or between 2½d and 3d per cwt (figure 46). Low prices disappeared from the Centre and the South; uniform prices came to bear in particular areas and changes were made in unison after long periods of stability. Again, however, in the latter part of the century, it seems to have been the relatively remote collieries at Croxteth (92) and Skelmersdale (105) which charged the highest prices at the pithead.[124]

Two general and consistent aspects of these kaleidoscopic price patterns are striking. First, the collieries serving an external market through land sale, or with a long haul to a canal in the case of Skelmersdale (105), seem invariably to have charged higher prices at the pithead, although the prices of the South West were caught by those at the canalside works by the 1780s. The higher pithead prices of the South West were, too, translated into higher prices in the Liverpool market (see above). If this higher market price of the landsale works was general it must have been one of the reasons why they lost a share of those markets to water-borne coal. It might have reflected the lower costs possible at large steam drained works using new hewing methods in virgin coal compared with smaller collieries or the works in Prescot and Whiston, where heavy extraction had been long continued. But if canalside works could undersell landsale collieries and survive profitably at their lower prices, why did they not either monopolise the markets or force prices down in the landsale areas? If higher landsale prices were not always reflected at the market, then why did the landsale collieries not expand at the expense of those on the canals: why was investment so heavy in areas where receipts and, assuming equal costs, profits per unit of output were relatively low?

If the answer to these questions lay in an inability to raise larger quantities at a constant low price on the canals, plus lower unit costs on the canals at the quantities the collieries there did produce, then this must, obviously, have been due to conditions of supply in the various areas of the coalfield. Furthermore, when water transport came its effects on pithead prices varied. Prices started low on the Douglas Navigation, even though its coals were of the highest quality, with a wide range around a rising trend which carried them up to the coalfield average by 1773. On the Sankey, prices began very low and fluctuated wildly until the early 1770s; on the Leeds and Liverpool they started off at the level which had prevailed before the canal was built and then remained stable, changing only once and in unison before 1790. Again, an explanation of these differing experiences can only be sought in variations in supply conditions on the various waterways.

Thus, dominant and elemental though it was, the relationship between demand and the development of the mining industry was not all-pervasive. There were also discrepancies in both locational and chronological patterns. The locational discordances are apparent on figures 31, 33, and 35. In neither the South West nor the North were the collieries located as near as possible to the coalfield edge which gave them closest proximity to the markets they served, nor were they as near to the turnpikes as they could have been in either of these areas. Many of the collieries of the West, Centre and South serving the Leeds and Liverpool Canal and the Sankey Navigation were also considerable distances from the waterways. The reason for some of the discordance in the South was undoubtedly an orientation towards the turnpikes, and in the Centre it was to some extent the result of a pursuit along wagonways of the high quality coal that market penetration through the Douglas Navigation and Leeds and Liverpool Canal required. But these are by no means complete explanations of the lack of a total locational correspondence between demand and mining. Some of the chronological discordances are equally, if not more, jarring. The locational shift to the West and the rash of closures and ephemeral workings along the Sankey may have been exacerbated by a dip in Cheshire's requirements and a falling share of the Liverpool market in the face of competition from the pits of the Douglas valley, but it began in the late 1760s, before either of these causes began to operate. On the Leeds and Liverpool growth was slow before the early 1780s. A partial cause may have been continued desperate competition in Liverpool from the Sankey collieries before their local industrial markets came fully into production and before the resurgence of demand in Cheshire. However, if it was the quality of their coals that made the Douglas valley collieries eventually successful, this was an advantage they had always possessed over the collieries of the Sankey and their initially slow growth cannot have been due to a failure to drive Sankey coals out of Liverpool. Clearly, the situation was not a simple switch of markets by the Sankey collieries allowing Douglas coal into Liverpool. Similarly, the long constriction of mining along the Leeds and Liverpool Canal to the townships of Orrell and Up Holland, and its sudden efflorescence into Pemberton and Standish in the late 1790s, must be explained in terms of local supply conditions and can have had little to do with the nature of the demand transmitted from Liverpool, which apart from continuing to increase changed little. Supply factors must be considered to account for these discrepancies and for the patterns of prices. But their treatment is necessary for a more fundamental reason: they were the mechanisms which linked the stimulus of increased demand with the response of increased output and which allowed the appearance of spatial correspondences between markets and mining. Only an examination of these mechanisms can provide an account of the *process* of development.

## Supply

### Coal seams

The barrenness of the lower coalmeasures still precluded mining on the nearest parts of the coalfield to Preston and Liverpool and the nearest coalfield stretch of the Leeds and Liverpool Canal to Liverpool had the same disability. The island of coalmeasures at Croxteth was again prospected and perhaps the Iron Delf was fitfully mined. John Mackay, the dominant coalmaster on the Sankey, was keenly interested in taking out a lease on these nearest coals to Liverpool and royalties and leasehold covenants were carefully formulated by their owner, Lord Molyneux.[125] But the thinness of the two seams in the lower coalmeasures there must have prevented any but the most fitful and tentative exploitation: there is no record of commercial mining at Croxteth in the late eighteenth century nor of leases actually being taken out. In Dalton and Parbold, the nearest coalmeasures townships to Liverpool on the Leeds and Liverpool Canal and to Preston on the Douglas, the seams were again thin and few in the *Lenisulcata* geological stratum. There was intermittent mining in the Parbold area (91, 140 and 170), but in negotiations over a lease in about 1794 it was recognised that there were only two seams, one of which was claimed to be two feet and the other four feet and one inch thick.[126] In 1815 Parbold colliery worked seams of only sixteen and twenty inches, the main seam varying between fifteen and sixteen inches where it was actually being mined at the time.[127] In Wrightington in the North, also on lower coalmeasures, 'there was such a stream of Water as . . . those little Mines would not pay to draw out' in 1802.[128]

The contrasting superabundance of known seams on the main productive coalmeasures was stark. Nowhere there was a shortage of reserves yet a problem. At Ravenhead (129) seven seams, ranging from four to seven feet in thickness with a total workable thickness of twenty-nine feet, were exploited in 1783[129] and it was common for leases in that area to grant rights to coal from five seams in the same concession.[130] Makin's colliery (136) in Prescot had two seams, one of four and the other of eighteen feet in 1772, when the neighbouring colliery of Jonathan Case (34) worked two seams of three and nine feet and that of James Gildart (133) seven seams, including the Smith, with a total thickness of about thirty feet.[131] The long continued extraction of large quantities of coal for Liverpool from the seams of the South West obviously presented no problem of potential exhaustion with such numerous thick seams accessible at even the relatively shallow depths worked then.

In the Centre the outcrops of very high grade seams in the lower bands of the main productive series attracted exploitation to sites which were not optimum in terms of location relative to waterways, especially in Haigh, Aspull and Ince where the Cannel seam outcropped, and in Up Holland, Orrell and Winstanley where the Smith and Arley seams came to the surface. This latter group of townships were the nearest to Liverpool on the canal; their high grade seams must have made them much more attractive than the townships on the hardly exploited northern bank of the waterway. Even in the townships where these low seams of the main productive series were exploited, and where the measures were thus heavily truncated in comparison with Prescot and Ravenhead, there seems to have been little danger of complete exhaustion in the eighteenth century. In Orrell, Up Holland and Winstanley only the two bottom seams in the series existed and extraction was at the rate of twenty-five acres a year by the late 1780s.[132] Perhaps the spread of mining into Pemberton and Standish in the late 1790s may have owed something to the limited capacity for expansion in Orrell, where reserves must have been carefully husbanded and could not support a long continued massive expansion of output. At Haigh, Aspull and Ince all eight seams between the Cannel and Arley were present. At Haigh colliery (2) the abundance was such that supreme fastidiousness was possible: only the Cannel, Arley and superb King and Yard coals were ever touched and other seams of up to nine feet in thickness were left alone.[133]

It is quite clear that reserves were sufficient on the main productive measures for the heaviest exploitation except, perhaps, at the end of the period in Orrell and Up Holland where only the Smith and Arley seams existed. Nevertheless, because of the rapidly quickening extraction, the shattered and tilted nature of the strata and the pursuit of only the best coals, the depth of mining increased considerably everywhere. Pits of thirty and forty yards were common before 1740 and the deepest were those of seventy-two yards required to tap the *Lenisulcata* coals at Charnock colliery (17) in the North. By 1780 collieries were commonly twice as deep as the pre-1740 norm; ninety yard shafts were needed even in the plentiful and thick seams of the South West at Whiston (34) and Prescot (5), and at Haydock (125) and Garswood (145) on the Sankey 120 yards was reached by the early 1780s.[134]

The gas and water problems already evident before 1740 increased more than commensurately with the depth of pits. The Cannel seam continued to be the most sorely troubled with gas although there was no repetition of the catastrophic conflagration at Haigh (2) in the 1730s. Fire-damp outbreaks were recorded at Haigh (2) in the 1740s and 1770s,[135] but the pits at nearby Kirkless (104) seem to have been the most severely menaced. A bad outbreak occurred in the 'Close Hot Sultry Weather' of May 1751 when

'Damp [was] strong in ye pitts, fir'd & burn'd two Lads & Prevented the men working.'[136] Pococke described the problems and attempted remedies endured at Kirkless (104), 'much troubled by what they call fiery air', in the early 1750s.[137] Fire-damp again stopped the colliery in 1774 and 1775; its much interrupted life, the wide fluctuations in weekly output, and its complete closure in 1776 may well have been due to its susceptibility to fire-damp and flooding. The old deep workings of the South West were also hit periodically by damp; both the fire and choking kinds killed colliers in Prescot and Whiston.[138] The same problem occurred in the South on the Sankey. A clause stipulating nullity of contract on the appearance of fire-damp was inserted in at least one colliery lease there[139] and a severe outbreak occured in Haydock (28) in 1767, when smothering damp invaded Killbuck Lane pit and in Davies pit 'the damp & fire together is so strong that no man living can bear in it . . . the coal bleeds so very fast in every vein and likely lobe'.[140] Two men were burned before the pit was shut. At Haydock (28) fresh pits were sunk at some distance from the gassy ones to 'air' the workings and the use of bellows to create underground draughts between 'downcast' and 'upcast' pits may have become common after 1740.[141] Perhaps, then, the technology was available to handle fire-damp after its first eruption at most places and probably only Kirkless (104) was forced to close by it. None the less, it was still a dangerous and horrible menace and its eradication must have increased the cost of mining, particularly in the notoriously gassy Cannel seam.

The water problem was less spectacular but much more common. Flooding was experienced or threatened at every colliery for which relevant information exists. The low productivity per pit at Dagnall's Eccleston colliery (100) and at Orrell Hall (1) in the 1760s was almost certainly caused by the necessity to keep open numerous pits which could be worked intermittently as the water levels in them allowed (see table 15). As the size and depth of collieries increased, so did the hazard of flooding; but so too did the financial turnover which brought a remedy within reach. The steam engine was the only means of unwatering the deeper and more extensive collieries of the second half of the eighteenth century and it diffused rapidly over the coalfield in the wake of the waterways and turnpikes (figure 47).

Before the installation of steam engines a number of large collieries were overwhelmed by water. The Haydock (28) pits were closed by flooding just before the damp attacks in 1767 and in the following year the rueful Haydock steward reported with some relish that 'they have a fair chance of being fill'd with water' at neighbouring Parr colliery (121).[142] Flooding cost hundreds of pounds' worth of damage at Dagnall's Eccleston colliery (100) in 1757[143] and expansion was severely curtailed at Prescot (5) and Whiston (34) before the installation of steam engines: two were needed, running simultaneously, to keep the last free from water in the 1790s.[144] At Hawarden's colliery (30) in Pemberton it was necessary in the late 1750s to

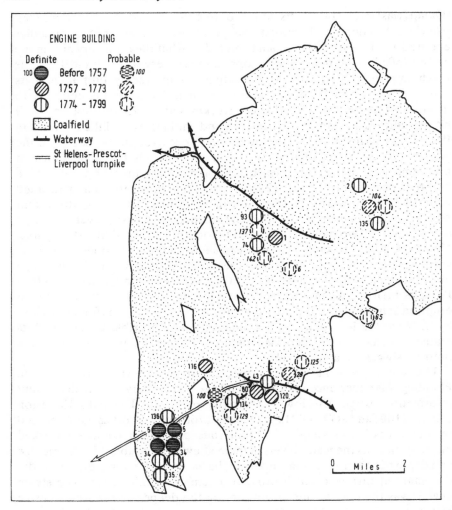

47. The spread of colliery steam engines.

'build a great dam from the top to the bottom of the Works [six feet high] and 48 yds. long caulked with Okam', yet still 'a constant stream issues forth' and four 'horse engines' and a 'water engine' must be worked constantly.[145] The nearby Jackson colliery in Orrell (93) was temporarily flooded out in 1769,[146] as was Kirkless Hall (104) in 1755 and 1756.[147] Steam engines were eventually installed at all these collieries except Hawarden's, which closed, and they were a standard item of equipment, 'the only Doctor on the deep', at the large canalside works in the South and Centre from the very beginnings of their exploitation.

Physical inaccessibility was not a severe constraint upon the distribution of mining and operated only in parts of the South West, North and Centre where the fringe of coalfield nearest the market comprised lower coal-measures and was therefore almost barren. However, the increased gas and water problems caused by deeper and larger-scale working caused low productivity, reduced output and intermittent working if unremedied. Solutions were available, hence the longer life-spans and higher productivity of the large collieries of the late eighteenth century. But solutions to both the damp and water problems involved the installation of heavy fixed capital equipment. As collieries grew and deepened their cost increased dramatically.

*Colliery sites*

With the diffusion of the steam engine, influence on the location of collieries by topographical configurations which were conducive to drainage declined. In the North, where development was small and the steam engine absent, the influence of geomorphology was still apparent in the late eighteenth century. The collieries on the turnpikes were situated where the roads crossed the deeply notched valleys of the Yarrow (17, 96 and 146) and Syd Brook (95). Those which were not on the turnpikes along which their coal must have been hauled seem to have been pulled away by similarly attractive sites on the steep valley sides of the Douglas (107, 110, 128 and 141) and Yarrow (81 and 96). Thus, all the Northern collieries seem to have been located according to considerations of drainage and strongly influenced by geomorphology, the pull of steep deep valleys often drawing them away from locations on the turnpikes themselves. Skelmersdale colliery (105) in the West may have been similarly influenced, situated as it was in the Tawd valley.

The clustering of collieries on and near to the boundary of Up Holland and Orrell to the south of the Douglas Navigation was also to some extent due to the value of the deep notch of the Dean Brook as an outlet for soughs. The collieries remained there after the Leeds and Liverpool Canal was built and after the widespread use of the steam engine at these sites: the Dean Brook valley and its fringes was by far the most productive area of the coalfield in the last quarter of the eighteenth century (see figure 35). However, by then its attractions in terms of drainage were irrelevant and the collieries remained and flourished there because the existence of smith's coal was known and considerable fixed capital had already been accumulated in pits, tunnels, working areas and wagonways. Geomorph-ology had little direct influence on the location of the collieries in the areas where output was growing rapidly in the second half of the eighteenth century.

*Labour supplies*

Increased productivity resulting from the more regular, continuous and specialised working associated with larger-scale mining and steam drainage, and from the innovation of the longwall and Lancashire systems of getting, meant that the labour force did not expand at the same rate as output. Neither was there any requirement for sudden large influxes of labour after the opening of canals because the early rates of growth on both waterways were relatively shallow. At first sight, it seems possible that an inability to recruit miners in sufficient numbers was to some extent responsible for this pattern of output expansion on the canals.

This may have been the case on the Sankey, which penetrated a thinly populated area without a pre-existing town. The steward at Haydock reported in 1760 that 'We pick up all [the men] We Can Meet With' at Stocks (113) and Laffack (118) collieries, but even so labour shortage threatened to close the latter. Later problems with faults and water at Haydock were in part blamed on the inexperience of the men, who 'was all strang in that buseyness when they Undertook it', although none the less 'bigeted in their one Opinins' for that.[148] 'Colliers are their only want' at Ravenhead (59) in 1771 and advertisements for miners there continued to appear in Liverpool newspapers until 1773.[149] An agreement 'about ye workm' (presumably not to employ each others') was made between the proprietors of three neighbouring cannel collieries in the Centre in 1750[150] and the bond system of hiring was used at Winstanley (6) and Orrell (74) in the 1750s as well as at Tarbock (60) in the South West.[151] These equivocal symptoms are the only evidence of labour shortage in the central area: there is none at all for the period after the opening of the Leeds and Liverpool Canal. Perhaps the existence of an already large wage earning population in Wigan and the handloom weaving villages around it mitigated any difficulties of worker recruitment.

It seems, from the very scarcity of the evidence, that labour shortage cannot have been a very significant cause of the slow initial growth of output on the canals. Furthermore, the numbers of colliers' children baptised in the townships bordering both canals surged ahead a decade or so after each was opened (see figure 39). There is no reason to suppose that this rate of recruitment had been impossible earlier.

Once canals became fully effective, the rate of recruitment along them was quite unprecedented. The five year moving mean of collier children's baptisms in the South around the Sankey increased from 12.8 in 1766 to 32.2 in 1774; in the Centre around the Leeds and Liverpool the sharpest increase was from 28.6 in 1779 to 69.4 in 1791. In Orrell in the Centre the number rose from 3.8 in 1770 to 6.6 in 1782 to 22.2 in 1791 and the number of hewers recorded in the Land Tax Assessments of that township increased from 36 in 1782 to 165 in 1794 and 180 in 1799.[152] The complete collier

TABLE 17. *The registration of vital events of sometime colliers in Up Holland and St Helens 1740-99*

| | No. of men* | Baptisms | Burials | Marriages | Other entries |
|---|---|---|---|---|---|
| Up Holland | 41 | 22 | 4 | 1 | 140 |
| St Helens | 123 | 49 | 26 | 20 | 604 |

* Includes five in Up Holland and eight in St Helens who began mining before 1740 but were also subsequently entered in the registers.

TABLE 18. *Residential mobility of sometime colliers registered in Up Holland and St Helens, 1740-99*

| | From birth to first adult entry | | | During adulthood | | | | | |
| | | | | Movers | | | Moves | | |
| | No. | Movers | % | No.* | Movers | % | No. | per mover | per collier |
|---|---|---|---|---|---|---|---|---|---|
| Up Holland | 18 | 8 | 44.4 | 31 | 10 | 32 | 16 | 1.6 | 0.6 |
| St Helens | 46 | 28 | 60.1 | 105 | 48 | 46 | 70 | 1.5 | 0.7 |

* Number with more than one entry of place of residence

population of Orrell must have been between double and treble these figures: a local coalowner claimed that 500 miners attended a meeting held there in 1792 excluding coal and cannel colliers from the pits to the east of Wigan.[153] Even with increased productivity, the doubling of output between 1780 and 1799 must have required about 350 new hewers and about 1000 new colliers in all on the coalfield as a whole.

Perhaps surprisingly, a large proportion of these new recruits were of narrowly local origin. About one half of the Up Holland colliers whose families have been reconstituted and about forty per cent of those of St Helens were baptised within the chapelry where they later worked (table 18).[154] Furthermore, many of these men were the sons of miners. In St Helens thirty-four of the forty-two colliers whose fathers' occupations are known were the sons of miners, and in Up Holland six out of twenty were, whilst the labourers and husbandmen might also, in fact, have worked in the pits[155] (table 19). The trees of three St Helens mining families have been drawn up from the reconstitution forms to illustrate this process of family labour recruitment (figure 48). Of course, accurate reconstitution is made difficult by the very commonness of the process and by the custom of naming children after close relatives: the pecked lines of the diagram joining children to more than one possible father illustrate the amount of guesswork required. None the less, it is apparent from the family trees that the often numerous sons of colliers who survived to child-bearing age were almost invariably miners themselves by that time. No son from these three groups of families had children baptised in St Helens whilst following an occupation different from that of his father.

This kind of recruitment pattern was almost inevitable given the way that work was organised in the pits. Children went to work with their fathers from

196

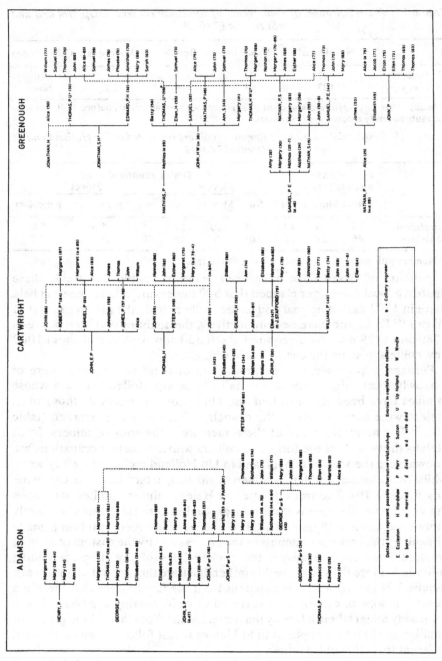

48. Three eighteenth-century collier family trees. Bracketed dates after a single
name are dates of birth unless otherwise indicated, and dates of marriage if inserted
after the surname of a marriage partner. These charts were constructed from forms
used to reconstitute collier families from parish registers.

TABLE 19. *Occupations of colliers' fathers and other occupations followed by sometime colliers in the eighteenth century**

| | Fathers | | | Colliers | | |
| | Up Holland | | St Helens | Up Holland | | St Helens |
| | pre 1740 | post 1740 | post 1740 | pre 1740 | post 1740 | post 1740 |
|---|---|---|---|---|---|---|
| Collier | 2 | 6 | 34 | | | |
| Labourer | 4 | 7 | 1 | 10 | 3 | 13 |
| Husbandman & farmer | | 1 | | | 4 | 3 |
| Nailor | | 1 | 2 | 1 | | 3 |
| Weaver | 1 | 3 | | 3 | | 1 |
| Carpenter & joiner | | | 2 | | | 2 |
| Blacksmith | | | 1 | | | 1 |
| Delfman | | 1 | | | | 1 |
| Tailor | | | 1 | | | |
| Pipemaker | | | 1 | | | |
| Milliner | 1 | 1 | | | | |
| Claypotter | | | | | | 1 |
| Victualler | | | | | | 1 |
| Millwright | | | | | | 1 |
| Engineer | | | | | | 1 |
| TOTAL | 8 | 20 | 42 | 14 | 7 | 28 |

* Includes changes of occupation but not multiple entries of single occupations.

an early age. They were almost certainly the recipients of daily wages of 6d or so, commonly recorded in colliery accounts alongside payments twice as big, and payments to a named person 'and son' were made regularly at Standish (41) and Skelmersdale (105) collieries.[156] These entries show that children worked as frequently and as consistently as adults. They acted as drawers for their getter fathers in Prescot (5), and John Hodson, father of the celebrated Lancashire collier girl, took both his paragon and her brother to work with him in Orrell when they were nine and seven years old. 'These little folks soon put their strength to the basket, dragging the coals from the workmen to the pit and by these efforts did the duty as it is called of one drawer.'[157] In these circumstances it would be remarkable if colliers' sons had not become miners in adulthood.

It is this pattern of recruitment which was probably responsible for the abundance of particular surnames in particular collieries (see tables 24 and 25). Such strong kinship bonds, sometimes between three generations in the same pit,[158] were probably important, too, in the recruitment of fresh labour from abundant uncles, cousins and nephews as mining expanded in one place and contracted or remained static elsewhere.[159] Whatever the importance of kinship links in promoting it, the mobility of colliers generally remained high. True, as figure 48 shows, the Adamsons and Cartwrights moved infrequently, even between baptism and adulthood. The Greenhoughs, however, were much more peripatetic and the characteristic behaviour pattern probably lay somewhere between theirs and that of the other two families. Eight out of the eighteen colliers for whom the necessary data were registered in Up Holland worked in a township different from that of their baptism and the proportion reached over sixty per cent in St Helens (see table 18). In Up Holland, thirty-two per cent moved between townships at least once during adulthood and in St Helens forty-six per cent. Predictably, given the nature of the source, a vast majority of the moves revealed by reconstitution from the registers of these two chapels (and from a more cursory study of those of Wigan and Coppull) are over very short distances. Again, as for the early eighteenth century, the registers contain some evidence of longer migrations, but it does not increase in a quantity commensurate with the increased rate of labour recruitment.[160] There are, too, by the end of the century, signs of links between Wigan and St Helens.[161]

Perhaps this pattern of highly localised labour recruitment is surprising given the amount of new labour required. However, it is probably not wholly due to the characteristics of parish registers as a source of information of this kind because the Poor Law documents of Parr township present a picture which is almost exactly similar. A high proportion of those enmeshed in this

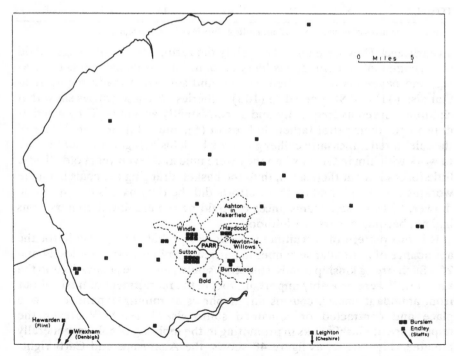

49.  Pauper migrants to and from Parr, 1740-99. Includes those bringing certificates and being removed to and from Parr.

legislation came from or went to townships adjacent to Parr: very few of these contacts stretched further than a couple of miles, even to other mining districts (figure 49).[162] Colliery agents, clerks and engineers were exceptions to this norm and even more migratory than ordinary colliers. Of the thirty-four colliery engineers whose names appear in the registers of the whole coalfield, twenty-eight were entered only once. This is itself testimony to their high rate of interparish movement. Three were registered twice, at different places over three miles apart in each case, and the three others seem to have moved between the births of each of their children almost as a matter of course (table 20).

TABLE 20. *The residential ascriptions of three colliery engineers in parish registers, 1758-96*

| Thomas Tyther (Tether, Titter) | | John Whittaker (Whitacre) | | George Nuttall | |
|---|---|---|---|---|---|
| Starford (?) | 1758 | Norley, Cheshire | 1763 | Aspull | 1790 |
| ? | 1760 | Orrell | 1764 | Haigh | 1793 |
| Haydock | 1764 | Whiston | 1767 | Wigan | 1795 |
| Whiston | 1770 | Prescot | 1771 | Leigh | 1796 |
| Sutton | 1774 | Ince | 1783 | | |
| Sutton | 1780 | | | | |

Those men who came from outside the fecund local mining community moved out of the same kinds of menial occupations as before 1740, though the list is longer because numbers were greater (see table 19). Most of these occupations were followed before the man found his way into the pits and, contrary to Challinor's surmise,[163] colliers stuck to their calling just as tenaciously as they had in the earlier part of the century. Few left the pits even in St Helens, where the rapid growth of heavy industry provided alternative jobs after the early 1770s, although some punctuated their stints with spells as a 'poper' or 'collier, poper' (table 21).

This association of occupational rigidity with residential mobility suggests that mining was attractive amongst the options open to illiterate workers.[164] Physically, the job remained at least as dangerous as it had always been. Colliers succumbed to roof falls (four deaths), falls down shafts (four

TABLE 21. *Occupational mobility of sometime colliers registered in Up Holland and St Helens, 1740-99*

| | No. of men* | No. of entries | Occupations entered | | | | | |
|---|---|---|---|---|---|---|---|---|
| | | | None | Collier | Other mining | Other | Pauper | % mining† |
| Up Holland | 30 | 134 | 1 | 129 | 4 | 0 | 0 | 100 |
| St Helens | 103 | 630 | 136 | 430 | 6 | 50 | 8 | 87 |

\* With more than one entry of occupation.
† % of occupations entered, not of entries.

deaths), fire and choke damp (four deaths) and flooding (three deaths) in Prescot between 1746 and 1789.[165] Dr Hair has demonstrated that the mortality of colliers was exceptionally high in Shropshire at this time, mainly due to accidents at work, and that the rate increased as the industry expanded.[166] There is no reason to suppose that this was not also true of South West Lancashire. The number of colliers for whom it is possible to calculate an interval between baptism and burial is too small to allow safe generalisation (table 22): if John Cartwright I, who died aged eighty-eight years in 1809, is removed from the St Helens group, the average age at burial of the remainder falls to 41.7 years. It is also apparent from Parr removal orders for colliers' widows with a few closely spaced young children that colliers died young. Almost exactly as before 1740, twelve years was all that a miner could expect to live after the baptism of his last child.[167]

TABLE 22. *Some demographic characteristics of collier families registered in Up Holland and St Helens, 1740-99*

| | Interval between marriage and first baptisms | | Age at first child's baptism | | Interval between baptisms | | Baptisms per collier | | Interval between last baptism and burial | | Age at burial | |
|---|---|---|---|---|---|---|---|---|---|---|---|---|
| | N | av. yrs | N | av. yrs | N | av. yrs | N | av. | N | av. yrs | N | av. yrs |
| Up Holland | 1 | 1.0 | 20 | 30.7 | 29 | 2.52 | 41 | 3.4 | 4 | 12.8 | 2 | 37.0 |
| St Helens (i) | 19 | 2.8 | 49 | 25.9 | 100 | 2.27 | 120 | 4.4 | 25 | 12.0 | 7 | 48.3 |
| (ii) | 16 | 1.65 | | | | | | | | | | |

N = number of colliers
(i)  Including three anomalous intervals
(ii) Excluding three anomalous intervals

The high probability of early death must have been coupled with a great likelihood of loss of time, or even complete loss of work, through injury. This was especially so at the wet gassy cannel collieries of Kirkless (101, 102, 104, 135), Aspull (12) and Haigh (2), and it may have been the reason for the St Helens colliers' spells on parish relief. Perhaps the effects of such misfortunes were ameliorated somewhat in the late eighteenth century; at Haigh (2), at least, the usual day rate was paid to men absent because of injuries from fire-damp.[168] Lay-offs probably became less too, as the relatively constant demand of canals and the drainage capacity of steam engines removed the two most important causes of stoppages. Moreover, some proprietors obviously made efforts to provide other work for their men when getting was interrupted.[169]

These partial nullifications of some of the job's disadvantages cannot have been as potent an incentive as high wages for men to take or retain it. Wage data are both more abundant and marginally easier to render into comparable statistics than for earlier periods. It is clear from table 23 that variations in the size of baskets and the measures into which they were aggregated could markedly affect rates of payment. Indeed, there is evidence that measure sizes were deliberately changed by proprietors in order to reduce the amounts paid to coalowners and colliers alike.[170] Whatever the reason, basket and score sizes varied at and between collieries in the Centre (see notes d, e, f, g, h, and i of table 23). It is obviously difficult to draw firm conclusions in such circumstances, but it does seem abundantly clear that both piece and day rates were higher than earlier and that they rose substantially in the last ten or twenty years of the century.

In Orrell, getting rates increased from 0.38d per basket in 1745 to 0.46d in 1764 and 1769-70 to about 1d in 1792.[171] This high rate was also paid at Skelmersdale (105) in the West, a remote location where one would not expect wages to be higher than elsewhere. Although getting rates were sometimes explicitly linked to prices[172] they do not seem to have risen quite as quickly, so that there was a slight fall in getting rates expressed as percentages of prices. This is not surprising when the vast increases in capital investment are considered, but the proportions of selling prices which went to the getters fell far less than the proportions which were left as profit. A combination of rising investment, falling rates of profit and relatively stable shares for the getters does not suggest that their deal was worsening. Day wages seem to have risen in parallel from 1/- to 2/6d between 1749 and 1787 at Haigh (2) and from between 8d and 1/- at Winstanley (6) in 1766 to between 1/- and 2/6d (paid to young females) at nearby Orrell in c.1795. The top day rates at remote Skelmersdale (105) in 1798-9 were even higher[173] (see table 25).

There was a geographical as well as a chronological relationship between changes in selling prices, levels of output and getting and day rates (see table 23). There was a marked gradient from the South West to the Centre before 1740. Unfortunately, little evidence of wages has survived for collieries outside the Centre for this period, but what there is suggests that this gradient still existed before the Leeds and Liverpool Canal was opened. Apart from the high rates for cutting cannel at Kirkless (104)[174] the highest recorded before 1792 was at Eccleston (99) in the South West in 1755 and 1766.[175] The Tarbock (60) rate of 0.57d per basket was also higher than any paid in the Centre before 1792: its niggardliness compared with that at Eccleston may well have been because the colliery was opened by a coalmaster from the North after advice had been given to the coalowner by the steward at Winstanley (6) in the Centre.[176] Although rates were generally low in that area until the end of the period they varied considerably from colliery to colliery. They were higher at Standish (41) in the 1740s than at Orrell (1, 74 and 132) and Winstanley (6) in the 1740s, 1760s and 1770s, and cannel getting rates were higher at Kirkless (104) than at Haigh (2) in

TABLE 23. *Wage payments, 1746-98*

| Colliery | Day wage | Getting per basket[a] | Winding or banking per basket | Other payments | Selling price per basket | Getting as % of price |
|---|---|---|---|---|---|---|
| Orrell (74) 1746-8 | | 0.38d[a] | | 12d earnest and 2/6d | 1.5d | 25 |
| Standish (41) 1746-7 | 4d–1/3d[a] | 0.50d | | | 1.5d rising to 1.75d | 33 29 |
| Haigh (2) 1749 | 1/- | | | | | |
| Tarbock (60) 1750 | | 0.57[b] | | 52 baskets of coal p.a. and earnest | 3.0d | 19 |
| Eccleston (99) 1755-6 | | 0.83d[c] | 0.28d | | 2.0d | 42 |
| Kirkless (104) 1758 | | 1.20d[d] 0.75d | | | 3.5d 1.5d | 34 50 |
| Haigh (2) 1758 | | 1.00d[e] | | 6d bonus for 12 loads a week | | 29 |
| Orrell (1) 1764 | | 0.37d[f] or 0.46d or 0.50d | 0.15d 0.17d | | 1.5d | 33 31 |
| Winstanley (6) 1766 | 8d–1/4d | 0.46d[g] 0.54d 0.58d | | 120 baskets of coal p.a. and earnest | 1.5d 2.0d 2.0d | 31 27 29 |
| Orrell (74) 1769-70 | | 0.50d[h] | 0.17d | | 2.0d | 25 |
| Kirkless (104) 1775 | | 1.00d[i] | | | 5.0d | 20 |
| Orrell (132) 1776 | | 0.40d[j] or 0.46d | | | 2.5d | 16 18 |
| Haigh (2) c.1787 | 2/6d[k] | | | | | |
| Orrell (93) 1792 | | c.1.00–1.50d[l] | | | 3.5d | 28-43 |
| Orrell (?) c.1795 | 1/-–2/6d[m] | | | | | |
| Skelmersdale (105) 1796-8 | 6d–3/-[n] | 1.00d[o] | | | 3.5d | 28 |

a. See notes to table 10.
b. Getters were paid 2/10d per saleable work of sixty baskets hauled to the pit eye by their drawers.
c. 'To allowed James Bellian for Scorage . . . Seven work of Coals at 6s. 6d p Dᵒ pays for getting & Winding . . . £2.5.6.' Coal was sold at 2d per basket or 10/- per work. Thus sixty baskets cost 66d or 1.1d each to hew, draw and wind. The last seems usually to have cost about twenty-five per cent of the total getting and winding cost.
d. 'Acct of Work done by the piece a load of the baskets at 12d and 7½d for Peney Basketts pʳ load of ten Basketts. Note the Peney Basketts sold at 1½d each.' The value of thirty-four loads five baskets and 310 penny baskets was £6.19.4, or 3.5d per basket of round cannel.
e. Ashton and Sykes, *Coal Industry* p.136. 10d a load (of ten baskets).
f. An estimate of coal got in one week

|  | £. s. d |
|---|---|
| To getting 50 scores at 1s.0d | 2. 10. 0 |
| To banking, Winding &c at 4d pSc. | 0. 16. 8 . . . |
| Value of 50 Sc. at 3s 3d p. or 1½p. | 8. 2. 6 [i.e. twenty-six baskets per score] |

If the twenty-six basket score was used in getting, the cost would be 0.46d, but 0.50d at the more usual twenty-four basket score. In the account of this colliery for 1771, getting baskets weighed 150 lb and were converted to sale baskets of 122 lb for valuation. If the scores contained twenty-six baskets of 150 lb, the hewers would have earned 0.37d per 120 lb, the normal basket size.
g. During the first few weeks of the account, coal was got at 4d per load which sold at 18d. This suggests that coal was then measured at Winstanley (6) in loads of twelve baskets, each selling at 1.5d, rather than of eight baskets as in the late seventeenth century. 'Wheeling' (i.e. drawing?) costs were variable as were 'Driving' (i.e. driving horse for winding?). The Jan. wheeling and getting costs were £1.15.8 for seventy-seven loads or 5.5d per load or 0.46d per basket.
    Feb. 17-21 was 'the first week they got by the score at 10d pʳ score and 1d pʳ for way drawing' (i.e. drawing to the underground roads along which the coal was wheeled to the pit bottom?) 'Getting at 11d pʳ.' was reckoned, plus payments for wheeling and driving, as earlier. Getting and wheeling fifty scores

cost £2.17.10d during this week, and '50 score of coals being 24 Baskets each' were valued at £10.0.0. Thus, coal now sold at 2d per basket and hewing, way drawing and wheeling cost 0.58d per basket.

    Occasionally, later in the account, 'Getting at 10d' was recorded as well as 'Getting at 11d'. With wheeling this equalled c. 12/9d per score or c. 0.54d per basket.

h.  'Getting 45 score at 12d pʳ winding and laying out 4d . . . £3. 0. 0.' There were '24 Bs to yᵉ Score at 2d pʳ Bs' reckoned in the credit account. Thus, hewing cost 0.5d per basket if the hewing and sale scores were of the same size.

i.  Coal was now reckoned by the score at Kirkless (104) (cf. note d) and manipulation of the sums in various columns of the account indicates that a score contained twenty-four baskets and normally cost 2/- to get, although odd baskets did not always cost exactly 1d. Very occasionally, '@ 2/2 a Sc' or '@ 2/3' was entered against the weekly getting totals and used to convert them to costs. In Nov. 1774, three scores twelve baskets of round cannel and one score four baskets of small cannel were valued at £1.19.8, or 5d per basket of round and 2d per basket of small, twenty-four baskets to the score.

j.  One shilling per score of twenty-six baskets of 140lb. The lower figure represents the cost in units of 120lb, the usual weight of a basket.

k.  For driving a level. Ashton and Sykes, *Coal Industry*, p.138

l.  A rough estimate based on a statement that a hewer and his drawer could earn from 5/- to 8/- a day. The average output per getter at the colliery in 1792 was calculated from the Land Tax Assessment Returns and a working year of 300 days was assumed.

m.  Earnings of a girl (drawer?) given in the tract, *The Lancashire Collier Girl*, discussed by Hair (1968).

n.  1/4d, 1/8d and 1/10d were the most common rates.

o.  Calculated by dividing the annual getting cost into the annual output.

1758.[177] Rates had risen with output and prices in Orrell by the 1790s,[178] but it is impossible to say whether they were *relatively* high or low then because the only other evidence for that decade concerns Skelmersdale (105) in 1798: this was not a canalside location and the intervening years were generally one of inflation.

Of course, the potential recruit or migrant miner would be much more likely to respond to levels of take-home pay than to these piece rates. Variations in the first were not necessarily governed by differences in the second. Hewing in thin coal was paid at a higher rate than hewing in thick coal, presumably to compensate for lower daily gets,[179] and the amounts produced by hewers could differ from colliery to colliery according to other factors such as the method of hewing used, the number and duration of lay-offs and the proportion of a man's time spent getting rather than on other generally lower paid work. *A priori,* one would expect more productive hewing techniques, fewer lay-offs and more continuous getting at the larger collieries in the more developed areas, so that actual wages may have risen faster than piece rates. Differences in wage levels between the more and less developed areas at any one time may, too, have been greater than differences in rates (although the proportions of earnings taken by fines might possibly have varied inversely).[180]

Evidence to substantiate this hypothesis is not abundant and it is, in fact, very difficult to discover what miners actually earned because of the nature of work schedules and the conventions (or lack of them) in colliery book keeping. Names appear and disappear from hewing records, day tallies and accounts of other kinds of work in those few ledgers in which they are present at all. Whether men were moving from one kind of work to another or were out of work altogether when their names were omitted cannot usually be discovered because names are given in each kind of record only in the Skelmersdale (105) accounts of the late 1790s. It does seem, however, that some degree of regularity of work routine was achieved by the late eighteenth century. The names in the weekly lists of getters at Winstanley (6) remained virtually unchanged at both pits through 1766[181] and so did those in the fortnightly accounts of Kirkless (135) and Skelmersdale (105) in the late 1790s.[182] Furthermore, although the day wages paid to different people in the same colliery varied quite widely, those due to any particular individual remained reasonably stable from week to week at both Standish (41) and Skelmersdale (105) (see tables 24 and 25). Even so, few of the men on day wages at Standish (41) in 1746-7 remained so for more than a few of the weeks for which detailed data were abstracted. Less violent flux occurred at Skelmersdale (105) in 1798 and there, at least, the intermittency was not usually due to men moving from one kind of work to another in the colliery. Both these accounts show that a nub of regular workers was supplemented by men who worked only for a short time in the years concerned. The mobility of the workforce as a whole was complemented by fluidity of employment at particular collieries.

The normal working week comprised six days according to the reckoning used in all the accounts examined, but a full week was not always worked and absenteeism reduced wages below the levels which were theoretically possible. At Prescot (5) in 1750, 'taking one week with another,' 'tis suppos'd that the colliers lose one Day in each Week at the Alehouses'[183] and in 1804 a diatribe was launched against the colliers of this coalfield, who 'idle half the week in the ale houses, which is to [them] enjoyment'.[184] However, the more prosaic evidence yielded up by account books seems to present a somewhat different picture. Colliers lost about one-quarter of the total working days through absence from Kirkless (135) in 1792-3, 1795-6 and 1799-1800. But on most of the days when absenteeism occurred the whole getting force was off together, presumably because the pits were stopped for holidays, slack sales or flooding, and the proportion of time lost to residual 'voluntary absenteeism' accounted for only some five to ten per cent of the working year. Most of this was usually due to one or two men who consistently worked only a few days a week at getting and this may have been through instruction rather than inclination.[185] It was rare for the hewers to work a full six day week at Winstanley (6) in 1766, but almost invariably they all had the same days off in any one week and this again suggests that their absence from getting was forced rather than voluntary.[186]

Only five out of the twenty men who were paid day rates at some time during the period analysed for Standish (41) in 1746 were named for more than twelve of the sixteen weeks; no one worked for all of them and even the group of most frequent attenders, who were presumably specialist day tally men, contained only one man, the auditor, who averaged more than five days a week (table 24).[187] But the impression gained from most accounts in which names appear, and this is particularly true of the later ones, is of

TABLE 24. *Day wage payments at Standish colliery (41) during sixteen weeks of 1745-6*

| Worker's name[a] | No. of weeks worked on day rates | Av. days worked per week | Modal day wage(s) | No. of weeks on modal day wage(s) | Av. day wage | Av. weekly day wage earnings |
|---|---|---|---|---|---|---|
| John Baxindell | 1 | 2.00 | 12d | 1 | 12.0d | 2/-d |
| John Berry | 7 | 4.86 | 6d | 6 | 5.7d | 2/3½d |
| John Brown | 1 | 1.50 | 7½d | 1 | 7.5d | 11d |
| Richard Brown | 13 | 3.94 | 6d | 8 | 5.6d | 1/10d |
| Robert Brown[b] | 0 | | | | | |
| William Brown[c] | 0 | | | | | |
| William Einscow[d] | 0 | | | | | |
| Henry Fisher[e] | 15 | 5.53 | 10d | 13 | 10.3d | 4/9d |
| Austin Fox | 7 | 3.88 | 14d | 4 | 12.9d | 4/2d |
| James Heaton | 1 | 2.00 | 10d | 1 | 10.0d | 1/8d |
| Edward Hunt[f] | 1 | 1.00 | 12d | 1 | 12.0d | 1/- |
| James Hunt[g] | 1 | 2.50 | 6d | 1 | 6.0d | 1/3d |
| Thomas Hunt | 1 | 3.00 | 12d | 1 | 12.0d | 3/- |
| Thomas Hollywell[h] | 1 | 1.50 | 12d | 1 | 12.0d | 1/6d |
| John Lea[i] | 16 | | | | | 13/6d |
| Samuel Lea | 14 | 4.87 | 10d | 8½ | 11.1d | 4/6d |
| Hugh Mason[j] | 4 | 1.50 | 10d & 12d | 4 | 11.0d | 1/4½d |
| 'Hus son' | 1 | 1.00 | 12d | 1 | 12.0d | 1/- |
| Robert Mason[k] | 0 | | | | | |
| John Slaytor | 13 | 4.00 | 10d | 6 | 11.0d | 3/8d |

a. Some of the Christian names have been expanded from abbreviations and the surnames are spelled as they most commonly appear in the accounts.
b. Paid with day wage men in one week for 'chifing sleck' but number of days worked not entered.
c. Paid with day wage men in five weeks for 'leading water' and 'dressing sleck' but number of days worked not entered.
d. Paid with day wage men in two week for 'dressing sough' and sinking but number of days worked not entered.
e. Colliery auditor.
f. Paid with day wage men in one week for 'dressing sleck'; number of days worked not entered.
g. Paid on day wages one week and for 'dressing lodge'; number of days not entered in another.
h. Paid for '1½d chifing' in one week.
i. Always paid the same amount for winding water and listed at the end of the day wage payments each week.
j. Paid for boring as well as day wages in one week.
k. Paid with day wage men in three weeks for 'leading water', 'drawing sleck' and 'chifing sleck' but number of days worked not entered.

almost unremitting activity by a large proportion of the men. For example, three of the twenty-six Skelmersdale colliers worked for fifty or more weeks in 1798 (table 25). Five of the fourteen day wage men worked all six days in the weeks they attended and only two averaged less than five days a week. Some of the getters must have worked long and furiously. Henry Topping was Skelmersdale's 'Big Hewer': he worked for all fifty-two weeks and earned an average weekly wage of £2. 8. 0½d, nearly £1 a week more than the next best paid man and between two and three times the average wage at the colliery.[188]

TABLE 25. *Wage payments at Skelmersdale colliery (105) in 1798*

| Worker's Name[a] | Getting | | | Day Work | | | | Other Work | | | | |
|---|---|---|---|---|---|---|---|---|---|---|---|---|
| | Weeks[b] | Wages £.s.d | Av. weekly wage £.s.d | Weeks[3c] | Av. days per week | Wages £.s.d | Av. weekly wage £.s.d | Extra weeks | Wages £.s.d | Total weeks | Total annual wages £.s.d | Av. wage per week worked £.s.d |
| Edw. Anderton | 44 | 33.18. 1¾ | 0.15. 5 | | | | | | | 44 | 33.18. 1¾ | 0.15. 5 |
| Jas. Anderton | 4 | 3. 2. 5½ | 0.15. 7 | | | | | 1 | 0. 6. 0 | 5 | 3. 8. 5½ | 0.13. 8 |
| Thos. Anderton | 22 | 16. 5. 9¾ | 0.14. 9¾ | 1 | 3.00 | 0. 6. 8 | 0. 6. 8 | 4 | 1. 1. 8 | 27 | 17.14. 1¾ | 0.13. 0½ |
| Thos. Bolding | 2 | 0. 8. 0 | 0. 4. 0 | | | | | | | 2 | 0. 8. 0 | 0. 4. 0 |
| Rich. Bradshaw | 44 | 44. 2. 0½ | 1. 0. 0½ | 2 | 6.00 | 1.10. 0 | 0.15. 0 | 4 | 11. 0. 0½ | 50 | 56.12. 1 | 1. 2. 7¾ |
| Wm. Farust | | | | 41 | 5.34 | 19.14. 0 | 0. 9. 7¼ | | | 41 | 19.14. 0 | 0. 9. 7¼ |
| Rich. Harding[d] | 34 | 51. 3.11½ | 1.10. 1½ | 3 | 4.80 | 2. 3. 0 | 0.14. 6 | 2 | 8. 8.11½ | 39 | 61.15.11 | 1.11. 8¼ |
| Rich. Hardman | | | | 2 | 6.50[e] | 2. 5. 0 | 1. 2. 6 | | | 2 | 2. 5. 0 | 1. 2. 6 |
| John Highton[f] | | | | 50 | | 22.13. 7½ | 0. 9. 1 | | | 50 | 22.13. 7½ | 0. 9. 1 |
| Jos. Highton | 36 | 27. 1. 6 | 0.15. 0½ | | | | | | | 36 | 27. 1. 6 | 0.15. 0½ |
| Luke Highton[f] | | | | 50 | | 22.13. 7½ | 0. 9. 1 | | | 50 | 22.13. 7½ | 0. 9. 1 |
| Nick Highton | 4 | 1.16. 6 | 0. 9. 0½ | 13 | 5.60 | 4.11. 0 | 0. 7. 0 | | | 17 | 6. 7. 6 | 0. 7. 4½ |
| Thos. Holland | 28 | 24. 9.10¾ | 0.17. 5 | | | | | 0 | 0. 7. 6 | 28 | 24.17. 4¾ | 0.17. 8¼ |

| Name | n | earnings | wage | weeks | rate | earnings | wage | days | amount | weeks | earnings | wage |
|---|---|---|---|---|---|---|---|---|---|---|---|---|
| Rich. Mollenex | | | | 13 | 6.00 | 2. 2. 0 | 0. 3. 2¾ | | | 13 | 2. 2. 0 | 0. 3. 2¾ |
| Thos. Mordaunt | 44 | 47. 2. 6½ | 1. 1. 5 | 48[g] | 5.72 | 21.11. 8 | 0. 9. 0 | | | 48 | 21.11. 8 | 0. 9. 0 |
| Thos. Pey | | | | | | | | 2 | 1. 0. 0 | 46 | 48. 2. 6½ | 1. 0.11 |
| Gorge Pey | | | | | | | | 4 | 1. 5. 2½ | 4 | 1. 5. 2½ | 0. 6. 3½ |
| Jas Rarstorn | | | | 33 | 5.91 | 11.18. 6 | 0. 7. 2¾ | | | 33 | 11.18. 6 | 0. 7. 2¾ |
| Rich. Rarstorn[h] | 34 | 44.16. 8½ | 1. 6. 4½ | 52 | 5.82 | 21. 5. 0 | 0. 8. 2 | | | 52 | 21. 5. 0 | 0. 8. 2 |
| and son William | | | | | | 21. 5. 0 | 0. 8. 2 | | | | 21. 5. 0 | 0. 8. 2 |
| Jas. Rylance | | | | 2 | 6.00 | 0.15. 0 | 0. 7. 6 | | | 2 | 0.15. 0 | 0. 7. 6 |
| John Rylance[i] | 24 | 13.19.10¾ | 0.11. 7 | 5 | 6.20[e] | 4.13. 0 | 0.18. 7 | 2 | 9.10. 3½ | 41 | 59. 0. 0 | 1. 8. 9½ |
| Rich. Rylance | | | | 52 | 5.78 | 23. 7.10 | 0. 9. 0 | | | 52 | 23. 7.10 | 0. 9. 0 |
| Jas. Talor | 38 | 41.17. 0½ | 1. 2. 0¼ | | | | | 2 | 1. 5.11 | 38 | 20.13. 2¾ | 0.10.10½ |
| Rich. Talor | 52 | 122.10. 2½ | 2. 7. 1½ | 12 | 5.73 | 5. 7. 5 | 0. 8.11¼ | 2 | 1. 0. 0 | 40 | 42.17. 0½ | 1. 1. 5 |
| Hen. Topping | | | | | | | | 0 | 1.13. 4 | 52 | 124. 3. 6½ | 2. 8. 0½ |
| TOTAL | | | 1. 3. 3 | | | | 0. 9.11 | | | | | 0.17. 4 |

a. Spelled as they most commonly appear in the account.
b. The account was drawn up fortnightly. The totals represent double the number of fortnights worked.
c. If less than seven days were entered for one fortnight it was assumed that only one week was worked.
d. Paid jointly with John Rylance for nine fortnights, for which half their earnings was ascribed to each.
e. Paid for thirteen days in his first fortnight.
f. John and Luke Highton were always paid jointly. Half their earnings have been ascribed to each.
g. Paid for thirteen days in his first fortnight and for '12 days and over hours' in another.
h. Sometimes they were paid jointly and sometimes only one of the pair worked. They accumulated fifty-two weeks between them.
i. See note d.

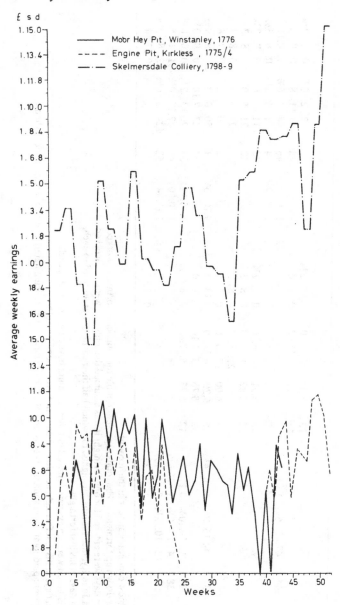

50. Variations in colliers' weekly earnings at three pits in the late eighteenth century. The first week in January has been used as week 1 in each case. The Skelmersdale figures represent fortnightly earnings divided by 2, each plotted twice. Abstracted from LRO, DDBa/Colliery Accounts, 1766, JRL, Haigh MSS/Kirkless Colliery Account Book, 1750-6 and 1774-6 and CRO, EDC/5/1816-17, Rylance decd., 1816.

Wages earned from hewing and day tally work clearly varied from week to week according to how many days were worked, how much was raised on those days and the rate paid per basket or day, which could change during the course of a year. Randomly timed fines would, of course, add to these fluctuations. Figure 50 shows the size of the fluctuations in the average weekly wages of hewers over a year at Kirkless (104), Winstanley (6) and Skelmersdale (105). It also demonstrates that the high piece rate paid to cannel hewers was compensated by their productivity: they had carefully to cut out large lumps of 'round cannel', whilst coal getters produced shattered pieces. Although this evidence is lean, it can, too, perhaps be taken to illustrate the rise in wages over the last quarter of the eighteenth century due to rising rates per basket and, even more, to increased productivity. Whatever the reason, getters' average weekly wages were 6/8½d at Winstanley (6) in 1766, 6/3½d at Kirkless (104) in 1774 and £1. 3. 3 at Skelmersdale (105) in 1798, when the day tally man's average weekly earnings were 9/11d. In Orrell (93), adjacent to Winstanley, in 1792 the getter's average weekly wage was between £1. 5. 0 and £2. 8. 0. It seems clear that average weekly wages varied more through time and perhaps through space than the piece work and day tally rates upon which they were based.

These wages were probably high in comparison with those of labourers and craftsmen. Admittedly, Gilboy's figures for the North West were paid in a less industrialised part of Lancashire and are not ideal for estimating the relative size of miners' pay.[189] Be that as it may, farmworkers in North Lancashire earned 10d a day in the hay and 1/- in the corn harvest in 1768, no more than the day wage at Haigh (2) in 1749 and less than the highest day rate at Winstanley (6) in 1766. The Lancaster labourers' day wage of 1/8d to 1/9d in 1794 was only about two-thirds of the day rate for a girl collier at Orrell in c.1795 and for men at Skelmersdale (105) in 1798. Hewers' wages were, of course, higher still. There is no evidence of colliers' wages in St Helens, but one would expect them to be at least as high there as elsewhere. If so, a good hewer would have earned as much as a skilled copper refiner there in 1780.[190] More surprising, a hewer of the calibre of Henry Topping or one on the top Orrell (93) wages of 1792 could earn more than the flatt masters on the Weaver in the same year, men whose pay was considered extravagant.[191] Even Lancashire craftsmen were earning less than the top colliery day rates in the 1790s and considerably less than hewers,[192] and so were the workers at the St Helens copper smelter in 1805.[193] Moreover, on top of the father's wage there was that of any children besides the one or two who drew for him, concessionary coal worth a couple of pounds a year and earnest, less any fines he owed. Already in the eighteenth century there was ample fuel for the outraged astonishment fired by the size of colliers' wages in this coalfield in the next century.[194]

There seems to have been an obvious causal connection between the rate at which wages increased and the way in which they varied from more to less developed areas, and the apparent ease with which colliers were recruited and the facility with which they moved from place to place. 'The labouring collier [found] his services sought after, and instead of soliciting employment, that employment courting him'.[195] However, this relationship between wages and recruitment was more complex than may appear at first sight. The standardisation into comparable forms of getting rates as expressed at different collieries takes considerable calculation and it is by no means certain that a collier of the time would have fully understood the implications of differences in rates expressed in different measures. Moreover, the actual weekly earnings of hewers and day tally men varied enormously between colliers in the same pit (see tables 24 and 25), much more than the averages between pits could have differed. No one would know what these averages were at the time, either, so that even *Carbonarius economicus* would have found it difficult to scan the opportunities and rationally select the most remunerative place of work.

Further obstacles to rational decisions about movement existed, too. Whether or not housing was provided or easily available; the frequency with which pay was received;[196] methods of getting and drawing, and even the times of day when work was done,[197] all varied considerably over short distances. There were many features of wages and working conditions suggesting that each small area fixed its rates and practices in isolation and according to local prices and narrowly parochial customs. In circumstances such as these the mechanisms of the labour market must have been rudimentary. Miners probably moved to anywhere they could get a job if they were out of work and preferred to stay put if they were not. Behaviour more studied than this was unlikely. The fact that the proprietor of Ravenhead (59) publicised the houses with gardens that he offered colliers who came to work there but did not bother to mention wage rates also indicates that there was not yet a labour market in which coalmasters bid cash for workers and colliers responded irrefragibly to differences in levels of wages. But if a miner was for some reason out of work there were irresistible inducements to remain in the industry by moving elsewhere to practise his lucrative craft.

The increase of wages seems to have accelerated at the end of the period so that one would not, perhaps, expect the relative affluence of miners to be reflected in demographic statistics aggregated for the whole period 1740-99. However, it does seem that colliers had their first child at a younger age in St Helens than in Up Holland, where the industry was not nearly so well developed when the miners for whom information was collected were alive (see table 22). The average interval between baptisms was lower in St Helens, the number of baptisms per collier was higher, despite greater mobility, and the average age at marriage of the thirteen whose baptisms and marriages were both recorded was quite low at twenty-six. It is possible to conclude tentatively, therefore, that the rapid development of mining and the high wages it brought had some influence in reducing the age of marriage, increasing the number of births and, consequently, raising the size of families. But the colliers' relative affluence seems not to have been reflected in their living standards or life-styles. True, some accumulated enough property to merit drawing up a will,[198] but generally colliers were firmly categorised with 'the lower people', as drunken ruffians rather than the aristocrats of labour. The high fluctuations in their weekly earnings and the sometimes low frequency of payment or high incidence of fines meant, of course, that the money which could be committed in a guaranteed weekly household budget must be pitched far below their average weekly wage. Thus a necessarily meagre general level of livelihood would be punctuated by irregular but frequent large cash surpluses. Moreover, a collier could not expect to be in work, or even alive, until his children were independent. The threat of parish relief for his family was ever present and this cannot have been conducive to the systematic accumulation of material possessions or short-term financial prudence. The whole domestic routine of the mining family must have been geared to the long hours spent by most of its members in the pits: on six days every week all the children over eight and the father would be absent for twelve or more hours, returning with heavily soiled bodies and filthy clothes, perhaps drenched as well. It is not at all difficult to understand the coexistence of high earnings, drunkenness and brutal and mean living conditions.

These traits were both exacerbated and set into sharp focus by the frequent isolation and specialisation of the communities in which the colliers lived. Houses must have been thrown up rapidly in the vicinity of the groups of large collieries along the canals. One can conjecture that few of these settlements matched the garden suburb image conjured up by the Ravenhead (59) advertisement. The scattered clusters of cottages amongst the pits would be inhabited by few except miners and those who served their rudimentary needs. The percentage of fathers who were colliers rarely fell below forty and sometimes rose above sixty in the whole township of Parr between 1758 and 1800. It must have reached similar levels in Orrell and the Scholes area of Wigan. There were at least as many collieries but fewer colliers in Haydock, Sutton and Windle as in Parr, some of whose denizens must thus have travelled to work in adjacent townships, as they must too, from Orrell to Up Holland, Pemberton and Winstanley and from Scholes to Haigh, Aspull and Kirkless.[199] Colliers seem to have been significantly more spatially concentrated than were their places of work.

This may, of course, have been fortuitous, due to the exigencies of the housing market rather than choice. Given their widely fluctuating incomes, cheap houses in Scholes and Parr, which later became notorious as the worst parts of Wigan and St Helens, would hold obvious attractions and it is almost inevitable that such housing will be crammed in dense uniform masses, large or small. There may have been other reasons for the residential concentration of colliers. Owing to the nature of their work, informal mutual aid was more necessary than for most and their generally high wages would ensure that some neighbour was usually able to give it. The constant threat of lay-offs or periods of low pay as even the best current technology was overwhelmed by water or fire, or rendered useless by the vagaries of coalfield geology, and the wide variations of wages from place to place, must have made it advantageous to live in a large community from which men went out to work in many different directions every day. At work they would see men from other similar communities which sent out colliers even further afield. Daily, information would be brought together about job opportunities over a wide area: even the intricacies of variations in pay rates might have been synthesised in networks of this kind. The fact that colliers lived in scattered clusters might, thus, have been an important stimulus to the residential mobility that characterised their life-style.

The apparent ease with which a labour force of the required size was recruited emanated from a complex tangle of inter-related factors. High wages, child labour and kin links meant that a high proportion of the additional workers could be recruited from families already associated with mining. Spatial variations in wages, fluid colliery labour forces and the patterning of colliery settlements encouraged the mobility that was also necessary due to spatial differences in the rate of development.

### Capital and entrepreneurship

It is apparent from table 26 that the rate of profit fell considerably in the second half of the eighteenth century. Except for the Kirkless (104) figures, which exaggerate gross profits by at least twenty per cent and possibly much more, rates of more than thirty per cent of receipts became rare. At Orrell Hall (1) profits were only eight per cent of receipts after the payment of rent[200] and elsewhere in Orrell in 1789 'the Great Colliery of Berry's [142] makes of profit no more than 3½d per ton, and the other collieries are little better'.[201] Rates were higher on the Sankey at Garswood (29), but they fell progressively from about forty-eight per cent to less than twenty per cent between 1768 and 1777.[202] However, even though the rate of profit fell in this way the incentive to run collieries probably did not because aggregate profits rose as output and prices went up. Even at Berry's rate of profit, the large Orrell collieries must have made over £1000 a year on receipts of between £15,000 and £20,000 a year in the 1790s. At small collieries the rates of profit were generally higher (see table 26), yielding a few hundred pounds a year as they had before 1750.

Not only were large outputs associated with low rates of profit, but they could only be achieved after heavy fixed capital investment. Plant requirements were considerable. All but one of the Orrell collieries worked

TABLE 26. *Colliery profits, 1750-99*

| Colliery | Area | Year | Annual profit £. s. d | Annual output tons | Profit as a percentage of receipts Before rent | After rent |
|---|---|---|---|---|---|---|
| Kirkless (104)[a] | Centre | 1752 | 617. 3. 0 | 3,472 | 85 | 71 |
| | | 1753 | 618. 1. 8 | 3,346 | 88 | 74 |
| | | 1754 | 390.16. 4 | 2,251 | 83 | 69 |
| | | 1755 | 333.16. 5 | 1,977 | 81 | 67 |
| | | 1756 | 45. 6. 3½ | 264 | 83 | 69 |
| Gillars Green (99)[b] | South | 1754 | *15. 8. 3* | 2,067 | | |
| | West | 1755/6 | 417. 1. 1 | 6,660 | 38 | |
| Orrell Hall (1) | Centre | 1764 | 36.15. 0½ | 2,714 | 28 | 8 |
| Winstanley (6) | Centre | 1766 | 322. 5. 9¼ | 3,766 | 53 | |
| Garswood (29)[c] | South | 1768 | 2375. 0. 0 | 18,121 | 48 | |
| | | 1769 | 1778. 0. 0 | 11,924 | 36 | |
| | | 1770 | 898. 0. 0 | 13,991 | 23 | |
| | | 1771 | 1260. 0. 0 | 14,938 | 33 | |
| | | 1772 | 644. 0. 0 | 14,487 | 16 | |
| | | 1773 | 725. 0. 0 | 9,531 | 28 | |
| | | 1774 | 871. 0. 0 | 8,978 | 35 | |
| | | 1775 | 541. 0. 0 | 9,984 | 20 | |
| | | 1776 | 194. 0. 0 | 6,524 | 11 | |
| | | 1777 | 304. 0. 0 | 5,982 | 18 | |
| Kirkless (104) | Centre | 1774 | 529. 1.11½ | 3,691 | 43 | 33 |
| Orrell (142) | Centre | 1789 | 584? | 40,000? | | 6 |
| Skelmersdale (105) | West | 1793 | 6. 1. 9 | 1,636 | 23 | 14 |
| | | 1794 | 155.16. 9½ | 1,023 | 51 | 45 |
| | | 1795 | 178. 2. 5½ | 2,251 | 44 | 35 |
| | | 1796 | 231.13. 1½ | 2,807 | 49 | 39 |
| | | 1797 | 86.10. 9½ | 3,318 | 23 | 13 |
| | | 1798 | 295.14. 9½ | 9,459 | 35 | 21 |
| | | 1799 | 283. 2.11 | 8,643 | 32 | 25 |

a. Excludes costs of sinking and drainage, which usually comprised twenty per cent of income and were probably higher in the wet Kirkless pits.
b. Losses in italics.
c. Output estimated from receipts.

through at least seven pits in 1799 and one probably had ten.[203] Prescot Hall (5) had nine pits as early as 1750 and Haigh (2) had six cannel and six coal pits in 1788.[204] Without the increase in productivity per pit that occurred at the larger collieries, pit numbers would, of course, have been even greater. At some of the smaller collieries where pit productivity was lower than it had been in the early eighteenth century the cost of pits must have increased its share of total costs. The saving on pits at the larger collieries was significant because the cost of sinking individual pits generally increased as the depth of workings increased. It is difficult to generalise because sinking costs varied considerably according to the firmness of the strata and their wetness as well as with depth. In a detailed estimate prepared for Viscount Molyneux at some time between 1770 and 1780 sinking costs of between 14/- and 18/- a yard were quoted, rising as depth increased, for pits in 'bass or rock'.[205] The estimate was made by a Northumberland coalmaster and seems to have been high for South West Lancashire. James Stock paid only 5/6d a yard at Park Lane (40) in the middle 1740s and William Bankes only 8/6d at Winstanley (6) in 1766.[206]

Besides digging, the shaft must be walled and equipped with a whimsey, ropes, chains and a cabin. According to the Molyneux estimate it cost £57 'to fix a Pitt after sunk with a cover for horses, Gin wheels, ropes, Tackle and to be ready to begin winding up Coals'. This may again have been high for South West Lancashire, however. Cabins and a whimsey cost only £9. 15. 5 at Haigh (2) in the late 1740s[207] and even if the ropes and 'tackle' cost the full £21 of the Molyneux estimate the equipment of the pit cannot have exceeded £30 or so. Thus, although the cost of a pit was variable it was nowhere negligible. The Bradshaighs paid £42. 2. 6 for a fully equipped pit at Haigh (2) in the late 1740s; a pit without buildings but with a sough connecting it to the main drainage channel costs £38. 19. 9 at Winstanley (6) in 1766 and James Stock paid a mere £8. 4. 6 for sinking alone at Park Lane (40) in the 1740s.[208] Other pits cost much more. In years of 'extraordinary trouble' the Holmes paid over £200 for sinking in Orrell (74) in 1769, 1771-2 and 1775-6.[209] Given a working depth of about forty yards in Orrell and a sinking cost of about 8/- a yard, the cost of a single fully equipped pit there would have been about £46. Thus, the pits alone of a large Orrell colliery would have cost not much less than £500. On the Sankey, depths seem generally to have been greater and so, too, probably, was the cost of pits. Ten pits at the ninety yards worked in the South West must have cost between £650 and £700 to sink and equip.

Unlike sinking costs, the expense of drainage increased more than proportionately with output. Indeed, productivity per pit was only high at collieries where expensive drainage equipment had been installed (see table 15). The most primitive drainage machine was an endless chain of buckets turned by a horse whimsey. It was quite cheap, 'a Horse Engine and sundry reprs.' costing £63. 2. 0 at Pemberton (30) in about 1760. But the running costs were high and the 'Expense of working the Horse Engine with four Horses and two Drivers' was £12. 2. 0 a month at Pemberton (30).[210] Sough drainage involved a heavier initial outlay but negligible running costs. James Stock paid £140. 2. 1¾ for his Park Lane (40) sough in the 1740s.[211] Both these collieries were small and soughing or engine expenses would be considerably greater at larger works. Indeed, it was probably impossible to build soughs to accommodate the amounts of water produced at the depths worked by the larger or older collieries. Sometimes supplementary power was leased from a more propitiously situated neighbour, as at Haigh (2) where the Great Sough and water engines of the Bradshaighs' own workings were aided by the (probably water powered) engines of the adjacent Halliwell colliery (101) at a cost of £50 or £100 a year.[212] But everywhere the steam engine was the only remedy for the water problem at large and at many small collieries (see above).

There are no existent records of the cost of purchasing and erecting colliery steam engines in South West Lancashire in the late eighteenth century, but it almost certainly fell from the £1500 paid by the Cases to build the first steam engine on the coalfield at Whiston (34) in the 1720s. Again, of course, the cost would vary from colliery to colliery according to the depth of the workings and the amount of water to be shifted. Two estimates were made for the installation of a steam engine in the wet Tarbock (60) mines in the 1770s, one by the steward at Winstanley (6) and one by Sir J. H. Delaval of Northumberland. The Winstanley (6) estimate gives only the cost of engine parts by weight at the works at Bersham, amounting to about £250 at £1. 10. 0 per cwt for a ten foot cylinder of fifty inches bore plus other parts. It is possible that haulage and assembly plus the construction of reservoirs, connecting channels and housing would have brought the cost of a functional engine up to the £600 to £1000 estimated by Delaval.[213] At least one colliery (34) required two simultaneously running engines to keep it dry and the collieries at the end of the Leeds and Liverpool Canal were all 'oppressed with heavy fire Engines'.[214] Somewhere around £1000 would seem a reasonable figure for a steam engine; it could be less at shallow relatively dry works but it could also be considerably more in the more common circumstances of heavy subterranean flooding.

Most of the large collieries which sold to the canals and turnpikes required a reliable all-weather connection, either a paved road or a wagonway.[215] These could be between half a mile and a mile long and they were obviously considerable undertakings. In 1753 the proprietors of Prescot Hall (5) claimed that they had spent £2000 on a steam engine and paved road to the turnpike and it is improbable that the engine absorbed much more than half the total.[216] On the other hand, the paving of a road to the small Holme colliery (74) in Orrell cost only £105. 4. 6 in 1799.[217] Besides the cost of the line itself extra ground rents had to be paid to the owners of the surface property it crossed (at £10 per acre in Orrell in the late 1790s)[218] and vehicles were required. The Clarkes had thirteen wagons on their wagonway in 1780 and most collieries of any size seem also to have had their own canal flatts plying between their wagonway staithes and Liverpool. A large flatt cost about £200 in 1763 and more than one would be required by a large colliery.[219]

Small collieries could still be opened for the two or three hundred pounds that was normal in the early eighteenth century. James Stock's (40) cost only £150 or so in the 1740s, but it was exceptionally cheap because it worked through only one pit whereas collieries of about the same size in St Helens (100) and Orrell (1) required three and six pits because of bad drainage.[220] A relatively small colliery with a few pits, tied flatts, a wagonway and whimseys must have cost about £1000 on the basis of the above figures. The Tarbucks borrowed £900 before opening a colliery of this kind in Windle (109) in 1756.[221] A large colliery with ten pits, a steam engine, wagonway, flatts and other equipment would cost at least £2000 and depending upon the size of the engine and its distance from the canal it could have cost double this. The sheer size of the investment involved must have been an important factor in the slowness of the build-up of output capacity after the opening of canals, both in terms of the number of collieries at work, and, more important, in their size. An obvious method of ameliorating the problem of heavy fixed costs was gradually to expand and enterprise to full capacity over a number of years. This kind of investment strategy is consistent with the pattern of development on the Leeds and Liverpool Canal in the 1770s and 1780s.

But this argument cannot be pushed too far. It would probably have been more difficult in mining than in many other industries to begin in a small way with little capital and gradually expand to rival the largest competitors on the basis of reinvested profits. Although the cost of pits and flatts could be spread, those of steam engines and wagonways were 'lumpy' and could not. Moreover, large capital reserves were necessary for all kinds of contingencies caused by wide oscillations in variable costs. Profits could be completely eroded in particular years even without any investment in capital expansion.[222] The running costs of large collieries were high, as the low rates of profit indicate. Pits were continually exhausted and sinking was almost continuous just to maintain capacity at large collieries. At Prescot Hall (5), for example, an average of four and a half pits were exhausted each year in raising 21,000 tons in the early 1750s.[223] Soughs and roads must be continuously extended to keep new pits within range of drainage engines and transport lines. Years of exceptional difficulties due to water, gas or geological problems led to annual losses and inadequate finances for normal annual expenditure, let alone expansion.

These problems were exacerbated by the common practice of selling 'on trust'. Credit relationships were, of course, the stuff of industrial finance at the time. But coalmasters were forced to act as 'longstops' in the system of circulating debt and credit, very much as the merchant houses did in the textile industry. This was simply because it was inevitable that as primary producers they would be owed more by customers than they could credit elsewhere. Some colliery proprietors refused to allow credit sales. The Bradshaighs, as near monopoly producers of cannel, could insist on ready cash.[224] But few were in this fortunate position and the pervasiveness of credit sales is demonstrated by the practice of allowing discounts for ready cash. In 1766 over 350 customers bought on trust at Winstanley (6). Most of them owed for small amounts and settled within the year: only £23. 9. 8½ was outstanding out of total sales of £454. 15. 4½ when the annual account was closed.[225] Even this small amount was almost equal to the cost of a new pit, but it was as nothing when compared with the sums which must have been owed by large industrial consumers to the proprietors of collieries on the canals and turnpikes. Hundreds of pounds were outstanding every year at the small Holme colliery (74) in Orrell, owed by coal merchants on the

Leeds and Liverpool Canal.[226] The trust debts at Tarbock (60) in 1757 amounted to £323. 12. 6 and nearly all of it was owed by two debtors, Wm Ashton and Wm Quick and Co.[227] The Haydock (28) steward was much exercised by the problem of settling up after credit transactions. His struggles illustrate what was probably a general problem on the Sankey owing to the preponderance of industrial consumers. In 1756 'most of my Ballance for last year lies at Northwich [saltworks] and at Copper House and Sugar House'. In February 1760

our paym.[ts] I doubt not but will be Certain but some of'em More Backward than I could wish . . . Mr Staytham & Co. Ows Near £200. I do not know what those Men are but you Order'd they sh'd have 12 Months Credit which will be up y[e] 22[d] of y[e] Next Month. In short this long Credit is Heavy – there is 3 of y[e] Northwich men has not p[d] off what sh'd have been p[d] at Mickelmast Last.

The following May the steward was 'now out of pocket Considerably Occation'd by a Northwich Bill upon Liverpool being return'd. In short Betwixt Liverpool & Northwich & some others yt. require Long Credit I have my Worke cut out'.[228] In circumstances such as these it was difficult to operate a strategy of expansion through ploughed back profits. It is unlikely that large-scale mining could be financed without large capital resources to pay for the initial lumpy investment and to tide the business over large upswings in variable costs and heavy long debts.

Landowners continued to be numerically important as colliery proprietors after 1740 (figure 51). At some time during this period, twenty landowners worked their own reserves through thirty collieries. However, these bald figures greatly exaggerate their economic significance. By 1799 there were only six gentry collieries left in the total population of thirty and families such as the Standishes of Standish (41 and 171), the Bankeses of Winstanley (6 and 122), the Heskeths of Rufford at Shevington (49), the Ecclestons of Eccleston (52, 62 and 99) and the Gerards of Ince (12 and 135), important in the industry for a century or more, had all pulled out of active mining. Moreover, all their collieries had been small right up to the time they were abandoned, leased or sold. The largest output recorded amongst them was the c.6500 tons a year at Winstanley (6) in 1766. The beginning of large-scale mining usually signalled the exit of the gentry from the industry. They remained numerically important in the Centre only until the Leeds and Liverpool Canal was built: the small workings that fed the Douglas Navigation were within their financial competence but it seems that the canalside leviathans were not.

This impression that the gentry withdrew as the investment required to work at an effective size increased is supported by an examination of the six who remained in the industry in 1799. The Holme's colliery in Orrell (74) was the only one run by gentry in that township. It was much smaller than the rest, employing only three hewers compared with the thirty-two of the next

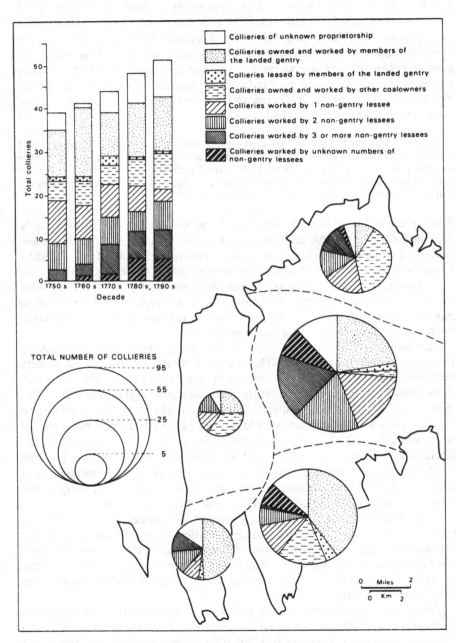

Colicries of unknown proprietorship

Collieries owned and worked by members of the landed gentry

Collieries leased by members of the landed gentry

Collieries owned and worked by other coalowners

Collieries worked by 1 non-gentry lessee

Collieries worked by 2 non-gentry lessees

Collieries worked by 3 or more non-gentry lessees

Collieries worked by unknown numbers of non-gentry lessees

Total collieries

50
40
30
20
10
0

1750 s   1760 s   1770 s   1780 s   1790 s

Decade

TOTAL NUMBER OF COLLIERIES

95
55
25
5

Miles   2
Km   2

51.  Sources of capital and entrepreneurship, 1740-99. See figure 13.

smallest Orrell colliery in 1799.[229] On the evidence of the Land Tax Assessments, the Dicconson's Shevington colliery (47) was only one-sixth and one-twelfth as big as the other two collieries in that township.[230] The other four gentry colliers were quite different, completely atypical of their class. Legh's numerous workings in the Haydock area (28, 113, 118, 124, 125, 153 and 154) contribute much to the gentry totals but most of them were very short lived. None the less, the Wood (28) and then the Florida (125) works were amongst the biggest on the Sankey, as was the Garswood (29) colliery of the Gerards. The Earl of Derby's colliery in Whiston (89) was, on the evidence of the Land Tax Assessments, quite large and so was that of the Earl of Balcarres who succeeded the Bradshaighs at Haigh (2). These four exceptions prove the rule of general gentry financial incompetence. All were large landowners. The Leghs and Gerards were soon to get the earldom and barony suitable to their lineage, wealth and power but not to their Jacobite Catholicism; the Earl of Derby was one of the wealthiest landowners in England and the Earl of Balcarres was one of the dynamos of the industrial revolution, building a group of collieries and ironworks on the Lancaster Canal at Haigh with money earned as Governor of Jamaica and from mortgages on his Balcarres estates.[231] It seems an inescapable conclusion that the gentry generally pulled out when the scale of mining increased because they did not have the reserves of capital (or perhaps the speculative acumen) of the four more illustrious survivors.

Before 1740 gentry finance had been supplemented by the money of local lessees, yeomen, small manufacturers and persons who styled themselves 'gentleman' but were not of the manorial gentry. This flow increased after 1740 and became more important than gentry finance in the Northern and Western areas. Local non-gentry capital was most important in the Centre after the construction of the Douglas Navigation (see figure 51). Most of these proprietors originated in the Central area itself and sometimes they banded together in partnerships even to work such small collieries as those on the Douglas.[232] Only four came from outside the area. Thomas Whitehead and William Briggs of Preston, at the market end of the waterway, partnered local men in Shevington (49) and Orrell (132), and John Chadwick and John Halliwell were from Wrightington and Coppull in the Northern area. Thus, finance flowed only over short distances towards the collieries on the Douglas. Its economic origins were varied. Only three of these proprietors were manufacturers; more important were those in agriculture or with shipping interests on the river.[233] Most interesting were the specialist coalmasters. Halliwell and Chadwick transferred their interests from the North as greater opportunities presented themselves in the Centre. The Halliwells were at some time involved in five collieries in the Centre (1, 49, 83, 101 and 128) and John Chadwick in three (79, 101 and 137) as well as his mining and ironworking interests in the North.

Although numerous, the collieries of such proprietors were, like those of the gentry, generally small. The largest of them was Halliwell's Orrell Hall (1), with an output of about 5000 tons a year in the 1760s. Others were much smaller. Those at Haigh (82, 90, 97, 98, 103 and 123) raised only between 500 and 1500 tons a year. Some of these proprietors remained in the industry after the Leeds and Liverpool Canal was built and men with similar backgrounds entered the industry in the South after the Sankey Navigation was opened. They were almost invariably unsuccessful in making the transition to large-scale mining. In the South, the works of the Tarbucks (109), Charles Dagnall (100), Peter Berry (80), Thomas Leigh (120), Legh Master (117) and 'Mr Mills' (119) were all small, struggling or short lived. Thomas West (157), Joseph Speakman (167) and James Orrell (43 and 134) entered the industry later in a much more quiescent phase on the Sankey and their collieries all survived until the end of the century, those of West and Orrell linked with glassworks which presumably provided their main markets. On the Leeds and Liverpool Canal, non-gentry local lessees survived until the early 1780s. But then, when the industry began to grow rapidly, they all disappeared: the longest survivor was the widow of John Chadwick who relinquished in 1788 what was by then only a partial interest in Dean colliery (137) in Orrell.[234]

That they left because they found it impossible to finance collieries of the required scale is suggested by other evidence besides inference from these chronological trends. Most of them probably had to borrow the capital necessary to begin mining and they cannot have had sufficient security for large loans. James Stock borrowed £100 of the £148. 6. 7¾ needed to open his colliery at Park Lane (40) and the Tarbucks mortgaged their Sutton and Windle holdings for the £900 they spent in opening Wiredale colliery (109) on the Sankey, even though it did not have a steam engine.[235] Charles Dagnall, a combmaker, was bankrupted by his efforts to work his flood-plagued colliery near St Helens (100) and so were Peter Berry (80) and John and Michael Jackson (93) when they tried to expand their collieries after the opening of nearby canals.[236] The difficulties of people without massive capital reserves are illustrated by the progress of Cobham and Makin in the South. They had worked the successful Prescot Hall colliery (5) for many decades and possessed 'a large capital such as few coalmasters can command' in 1763.[237] But it was not enough to prevent Jonathan Case, landowner and Liverpool merchant, from gaining control of the Prescot Hall (5) reserves in 1769. When they assembled a new concession in Hillock Street (136), to be built up to a large steam drained capacity from scratch, they formed a partnership of six which included John Williamson Esq. of Liverpool.[238] It was almost certainly this link which allowed the proprietors to 'declare that they are worth one hundred thousand pounds which they are determined to expend in support of this undertaking'.[239]

This was symbolic of the situation throughout the coalfield. Local capital, from both gentry and non-gentry sources, was inadequate to finance the large collieries which produced the upward surges of output after areas had been linked to Liverpool. Six of the eleven collieries worked in the South West after 1740 had entrepreneurial connections with Liverpool. All those which survived, except for that of the Earl of Derby, were to some extent financed from the port. Besides Williamson, E. Roberts, a Liverpool upholsterer, was a partner in Windle Ashes (116); James Gildart, who ran a large Whiston colliery (133), was an alderman, coalmerchant and saltworks proprietor in Liverpool; the Cases (34 and 127) had important connections in the Liverpool slave trade, a saltworks, a marriage alliance with the equally powerful Clayton merchant family and large mortgages drawn from Liverpool as well as Chester, and Richard Willis, who operated the other large collieries in Whiston (127 and 158), was similarly from a Liverpool merchant family.

The sources of the flows which financed expansion on the Sankey were more varied. 'Foreigners' had an interest in only nine of the thirty-one collieries which operated there, but with Legh (28 and 125) and Gerard (29) they dominated the industry. Before 1778 Garswood (29), Wood and Florida (28 and 125), the Parr Hall works of Sarah Clayton (43) and the collieries of John Mackay (59, 120, 121 and 129) were completely pre-eminent on the Sankey. Sarah Clayton, kinswoman of the Cases, was for a time probably the most important merchant in Liverpool. Like the Cases, who also had collieries on the canal (126 and 134), she was bankrupted in 1778. Thenceforth, the collieries of Mackay at Parr (121) and Ravenhead (129) were the only serious rivals of Florida (125) and Garswood (29). Mackay described himself as of Bellfield, Cheshire and it is possible that he had connections with the Sankey coal consumers on the Cheshire saltfield. Be that as it may, he borrowed £4500 in London before opening his collieries and a further £12,000 from the actor Garrick in 1776.[240] During the 1770s Mackay became the most important industrialist on the coalfield, with interests in glass and copper smelting in St Helens as well as mining. Definite links were forged later between mining on the Sankey and saltboiling. The Ashtons and Blackburnes, who bought collieries in Parr (154 and 161) in 1789 and 1793, were both Liverpool merchants with Cheshire saltboiling enterprises.[241]

The sources of capital on the Leeds and Liverpool Canal were different again. Between 1774 and 1780 twelve collieries were opened in the Central area, all but one by local proprietors who were members of the gentry (2, 6, 14, 41, 74 and 135) or non-gentry (1, 47 and 93) with freehold or leasehold interests dating from before the opening of the canal, or banded together in partnerships formed in 1773 and 1774 (104 and 132). The twelfth of these early works (139) was an augury of the future. It was leased by John Longbotham of Hargreaves, Lancashire, who had surveyed the line of the canal, John Hustler, a Bradford woolstapler and the leading canal

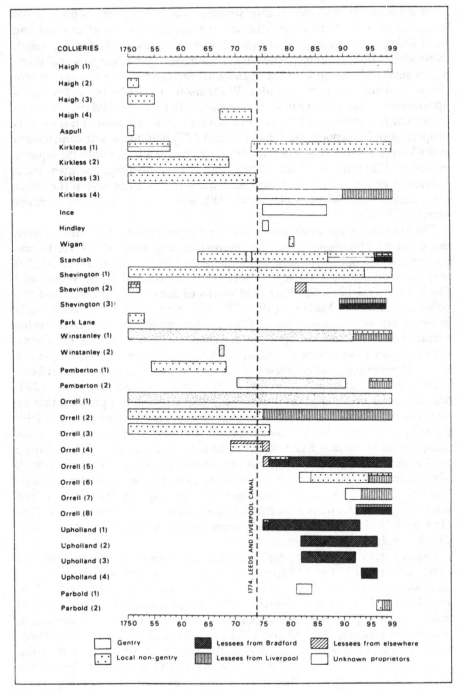

52. Sources of capital and entrepreneurship at collieries serving the Leeds and Liverpool Canal, 1774-99. Based on sources listed in Appendix 1.

proprietor, Thomas Hardcastle of Bradford, woolstapler, and John Chadwick of Burgh, an ironmaster with other mining interests in the area.[242] In 1776, the Jackson colliery in Orrell (93) was taken over by Edward Chaffers, Ambrose Lace, William Earl, Thomas Birch and Jonathan Blundell, all Liverpool merchants, and Samuel Warren, a Liverpool silversmith.[243] The pattern of subsequent development was set (figure 52). Large partnerships drawing in capital from both ends of the canal financed the expansion in the Centre and the acceleration of growth after the early 1780s only occurred after they had gained control over most of the workings. The partnerships fell into four groups. There were those headed by Hustler (137, 139, 143 and 162), those of Jonathan Blundell (93 and 135), those of John Jarratt of Bradford, 'Gent.' (143, 155 and 160), associated in two of them with John Lofthouse, a Liverpool coalmerchant, and those of William German or his heir and William and John Clarke, Liverpool bankers (6, 142, 159, 169 and 170). William Hustler joined Jarratt and Lofthouse in the lease taken out on Standish colliery (41) after it was relinquished by German, and Jonathan Blundell and John Hustler were both prominent members of the small clique which controlled the canal. Thus, there were close entrepreneurial links between all these collieries, some of which had sleeping partners from as far afield as Wisbech, drawn in through the Quaker connections of the Bradford woolstapling community.[244] The only large colliery in the Centre not run by one of these partnerships was that of the Earl of Balcarres at Haigh (2). The coal of the lesser gentry and the leases of local non-gentry came inexorably within their grasp and by 1799 only the small collieries of the Dicconsons at Shevington (47) and the Holmes at Orrell (74) had, with Haigh (2), not succumbed to them.

It seems quite clear that there was insufficient capital to finance the mining expansion of the late eighteenth century either already accumulated within the industry or within the grasp of established or putative local colliery proprietors. Local sources financed expansion of colliery numbers in the North and West, on the Douglas and in the very early years on the Sankey Navigation and Leeds and Liverpool Canal. But, except for those of Legh, Gerard, the Earl of Derby and the Earl of Balcarres, these collieries were all small and mostly short lived. Sustained production and expansion required a large jump in the scale of mining to accommodate steam pumping, wagonways and flatts. There seems to have been little chance to achieve this shift by a ratchet-like process. Sudden large infusions of capital were necessary and established coalmasters and other local people seem to have been incapable of making them. Expansion in the South West, where there had been large-scale mining since before 1740, was financed from Liverpool in the second half of the century. Liverpool and London were at least as important as the more illustrious members of the local gentry on the Sankey in the 1760s and 1770s, and in the last two decades of the century a triangular link was forged between Liverpool money, Cheshire salt and Sankey coal. Local finance was dominant in the 1770s on the Leeds and Liverpool Canal when the industry grew slowly, but large injections of capital from Liverpool and Bradford were associated with the more rapid growth of the last two decades of the century. Money had to move up the canals and turnpikes before coal could move down.

## Competition between collieries

The competition in Liverpool between the areas of the coalfield, among themselves and against the Duke of Bridgewater, has already been described. What remains is to deal with the ways in which the distribution of mining was affected by rivalry between collieries within the coalfield areas.

Imperfections due to product differences have already been treated at some length insofar as they were evident in inter-area competition. Just as the areas producing superior coals succeeded in Liverpool, Preston and the export market, so the collieries which produced the best coals in a particular area flourished and waxed fat there at the expense of better locations in that area. In the Douglas valley, Haigh (2), Aspull (12) and Kirkless (101, 102, 104 and 135) overcame their disadvantageous locations some miles away from the river and canal termini by the provision of cannel. It only occurred there and it fetched up to double the price of ordinary coal at the pithead and only slightly less in proportion in Liverpool.[245] The excellent bituminous coals of the Arley, Smith, King and Yard seams were also available through the pits which tapped the cannel. But they were barely touched until 1799, when the Lancaster Canal gave them cheap access to Preston, except to supply the Haigh ironworks and probably the Wigan market.[246] This was because fuel from the Smith and Arley seams was the sole stock in trade of the large Orrell and Up Holland collieries, which had no shallower reserves.[247] Thus, high grade bituminous fuel was not a rare commodity on the Leeds and Liverpool Canal and its price was relatively low; too low, it seems, to make its production worthwhile at the badly located cannel collieries despite the excellence of their reserves. As in the early eighteenth century, high quality seams did not guarantee a market unless they were almost uniquely rare or advantageously situated.

Winstanley colliery (6) had grown in consequence of its supplies of coal from the Smith seam in the early eighteenth century. Before the Leeds and Liverpool Canal was built, Winstanley (6) competed on the Douglas Navigation with the small collieries of Orrell, nearer to it, and on the Sankey Navigation which was far away but lacked reserves of this coal until the Rushy Park seam was opened up at Ravenhead (129) in the early 1770s.[248] But even Winstanley (6) was shut out of the Leeds and Liverpool Canal by Orrell collieries supplying the same fuel until the very end of the century.[249] The competitive strength given to the group of collieries on the south bank of the Leeds and Liverpool by location and the quality of their reserves goes some way towards explaining why mining through large-scale works was severely confined to that small part of the Central area until the very end of the eighteenth century.

High grade reserves aided the competitive position of other collieries during this period. The coking coal of Burgh (96) in the North and Laffack (118) in the South gave them unassailable advantages in the supply of the ironworks with which they eventually became linked (see above). Gildart's colliery (133) seems to have been the only one to tap the Smith seam in the South West, and this may explain to some extent its relative size and long life.[250] On the Sankey, Florida (125) and Ravenhead (129) produced coals superior to those of other collieries. This ensured the survival of Ravenhead (129), which supplied America with smith's coal through Liverpool[251] and managed to attract industrial consumers with fastidious tastes to its pit banks at the canal's terminus. Florida colliery (125) was relatively distant from the canal but at the market end, and its high quality coal merely reinforced its locational advantage and made it the Sankey's equivalent of the Orrell works.

The conscious competitive strategies of coalmasters assumed forms which had a marked effect on the chronology and spatial pattern of development in each of the three most important areas after they became linked with Liverpool. When the Sankey Navigation was built in 1757 the colliery owners of the South West seem to have cut their prices in unison.[252] But if there was collusion, there was certainly no love lost between them. The partial loss of the growing Liverpool market and the seventeen per cent price cut did not cause the immediate closure of collieries, but the difficulties they brought probably contributed to the festering rancour which eventually erupted to close Prescot Hall (5) after over two centuries of exploitation had failed by a long way to exhaust it. The spate of closures in the South West between 1769 and 1771 may generally have been due to the vigorous expansionary policies of the proprietor of Whiston (34 and 127) which accompanied the diminution of markets.[253] Jonathan Case had already accumulated immense reserves in Whiston and the neighbouring townships of Huyton, Eccleston and Prescot by the late 1760s.[254] In 1769 the six year 'fine' of Cobham and Makin on the Prescot Hall reserves expired. Robert Roper, the lessee from King's College, required £1971. 16. 7 from Cobham and Makin for the next six years but they refused to pay, presumably in the hope of extracting more advantageous terms.[255] This stratagem had worked in earlier release negotiations, and the familiar sparring commenced.[256] In January 1769 'Prescott Hall Colliery was quite at a stand; the Engine had been stopt for two months or more . . . [as the proprietors] would not be at the Expence Requisite for supporting the Colliery untill they see into whose Hands it will come'.[257] This time, the bout was ended abruptly as a triumphant Jonathan Case swept in and snatched the colliery from the partnership which had worked it since the early eighteenth century. Moreover, Roper refused to lease Cobham and John Chorley, Henry Makin's executor, any of his own coals in Whiston and Prescot, and in 1770, after they had worked out 'some remains of the Flaggy & Sir John Delfs wch. remain dry', Cobham and Chorley were dispossessed by Case who, 'with the Spirit of a Rival and Competitor', carried out his threat to 'turn them out of possession of both the Mines & Estate'.[258]

For three years or so, Case reaped the benefit of having removed his most important competitor from the South West, and worked two collieries in Whiston (34 and 127) simultaneously.[259] Chorley sought desperately for more reserves. He formed the Prescot Colliery Company in about 1772 with Thomas, son of the late Henry Makin, Robert Eden, a Prescot yeoman and John Williamson, Esq., of Liverpool.[260] The Company leased coal from the Earl of Derby to supplement their own and negotiated a concession from King's College on the Moss Delf, a deep coal which had been excluded from Case's lease at Chorley's instigation.[261] The College's permission was also required to allow a necessary tunnel to be dug beneath Hillock Street in Prescot. Case, much to his annoyance, could do nothing directly to prevent the Prescot Colliery Company beginning its operations (136) because he could not 'command a foot of the Ground contiguous to that part of the Coll. lands'.[262] But when the Company proposed building a steam engine to drain the Moss Delf in 1774 Case got up a petition to prevent it, claiming that it would jeopardise the town's water supplies.[263] The College was in a controlling position: without the tunnel the engine would be useless and without a dry Moss Delf the colliery was uneconomic. They vacillated, bombarded with claims and counterclaims, torn between the prospect of receipts from the new Company's lease of the Moss Delf and the possibility of Case ceasing to mine College coal and switching wholly to his own reserves if the College offended him. The intervention of the Earl of Derby, who also leased coal to the new Company, settled the issue. On January 16, 1775, the College informed Case that they had received a letter from the Earl's steward who 'recommends the undertaking as useful to the Public, the College and to Lord Derby, and sollicits our consent and favour'.[264] Permission was immediately granted. Case again had competition from a large steam drained colliery in Prescot: his machinations had stopped it completely from 1769-72 and crippled the new colliery until 1773, up to which time it was run with horse gins.[265]

Thus, the rash of openings and closures in the South West between 1769 and 1773 was at least somewhat due to vicious competition between coalmasters. The subsequent stability until the end of the century coincided with its abatement. The Cases were bankrupted in 1778 and their Whiston colliery (34) was thenceforth run by trustees to earn money needed to wind up the estate and pay off debts.[266] The five large collieries in Whiston (34, 89, 127 and 133) and Prescot (136) then worked on in apparent amity.[267]

Traumatic as they may have been to those concerned, the competitive blows thrown in the South West were slight in comparison with those dealt out on the Sankey. The 'string of beads' location pattern which initially existed and the subsequent concentration at the market end in the late 1760s owed much to the competition which also jerked pithead prices up and down. It began almost as soon as the waterway was opened. In 1760 Peter Berry lowered his price to 2d at St Helens colliery (80), disadvantageously situated at the terminus of the canal. Peter Legh immediately followed suit at Haydock (28), two miles nearer to Liverpool. He did not cut his price even further because his works were 'so Much Nearer ye Market than ye rest, if they Can have 'em on ye Same Terms they'l not go past us'. His intention to corner the market from his superior location was clear: 'if they at St. Hellen fall again I'll follow 'em untill Youve entirely Knok't 'em up . . . its Got to Live yt Can Live Amongst us – We do All we Can to be uppermost (Which we are by Much) . . . But I am in Hopes of Seeing Better times When some of our Adversairies are Silenc'd'.[268] The yeoman Berry was silenced when he sold his steam drained colliery (80) in 1761.[269] It was never subsequently mentioned. But Berry was relatively small beer. The collieries of Sarah Clayton at Parr (43) and Sir Thomas Gerard at Garswood (29) were different propositions. In 1760, 'Mrs Claytons Works had very little to do, it is said she will not Gate another Works . . . they Expect a Summer Sale . . . But perhaps she'l meet with a disappointment for we expect a part of yt kind of Sale at Laffock [118] and H[aydock] Stocks [113].'[270]

Legh did not have the sole initiative, however, and the scramble to corner the market intensified. A few months later Legh's steward reported that 'Sr. Thos. Gerard [29] has fallen his coal to five farthings p. Bask. for abt. 3 Weeks past'.[271] Meanwhile Legh had forced the colliery of his kinsman Legh Master (117) to close by forbidding it a wayleave.[272] By early 1762, when the price was again up to 2½d a basket, there were only six collieries on the Sankey. All of them were owned by Legh (28, 113 and 118), Gerard (29) and Clayton (43) on the lower reaches of the canal except the Wiredale Colliery (109) of the Tarbucks in Windle.[273] Two new collieries were opened in 1762, perhaps tempted by the restoration of higher prices, but neither lasted long; Mills' colliery (119) was shut by Legh, who refused a wayleave, and Thomas Leigh surrendered his lease of a new St Helens colliery (120) shortly after he took it out.[274] Clearly, the closure of the works on the higher stretches of the canal was largely due to the heavy price cutting activities of the owners of the large collieries lower down. Only the largest and best situated survived through the first five years of competition on the Sankey. Of course, this competition and price paring also explains why the Sankey did not completely take the Liverpool and Cheshire markets despite its cheaper supplies. Whilst the coalmasters of the Sankey were reducing the total production of the South in their search for dominance through cutting prices and profits, room was left in their markets for coal from the South West despite its higher price.

A new bout of fierce competition began with the entry of John Mackay into the Sankey coal trade in 1763, when he took out a new lease of St Helens colliery (120) with Thomas Leigh and opened a large works in Parr (121) on his own account.[275] Immediately, Sir Thomas Gerard 'alarmed some folks on the Navigation without any reason by Advertizg. the delivery of coals aboard of ships at Liverpool for 7/- *their ton*. The calculation will just give him the price we [Mackay and Legh] now sell for'.[276] Mackay proposed a combination with Gerard, Clayton and Legh in a venture to sell Sankey coals in Dublin, bringing extra output of about 24,000 tons a year between them and 'everlastingly keeping all new adventurers from the Navigation'.[277] Legh turned down the offer and competition continued.[278] In January 1765, Thomas Leigh, Mackay's partner at the badly situated St Helens colliery (120), proposed that prices and the quantities sold by the four dominant coalmasters shoud be regulated by agreement because 'The Gentile Men Colliars was Injuering them selves Very Much by selling Coal as they did'. Sir Thomas Gerard was agreeable 'provided Quantities Could be fixd' and a metting was arranged between the colliery agents to report to the 'Gentile Men Colliars' on proposals to 'Make the Concerns in Collearys More Agreeable & the Sooner the Better'.[279] But the coalmasters nearest Liverpool again demurred: Gerard was 'pretty well Satisfied With the Last Years Bussle' and in 1765 and 1766 he greatly increased his output at Garswood (29) and again brought pithead prices down below 2½d per basket.[280]

But conditions then changed. The collieries at the market end of the canal began to hit problems. Legh suffered attacks of flooding and damp at Haydock (28) and his original workings were reaching exhaustion.[281] Gerard's output at Garswood began to plummet, and in the early months of 1768 the 'Gentile Men Colliars' began to make overtures to Mackay. Meetings about the formation of a price fixing cartel were arranged and Peter Legh, by now subdued and morose, wrote from Bath to his steward that 'Whatever are the Resolutions of the Committee on Our Coal Affair pursue it and Hit or Miss can't prove worse than is quite familiar to Me. Ive tasted deeply on ye side of Affliction'.[282] It was now Mackay who refused to raise prices: Legh received a letter 'full of Invective' from him and wrote to his steward that

the price Coals are now sold on Sankey has been sufficiently tried to convince Every coal proprietor They must Work to ye bad . . . its highly necessary we shu'd come into an Agreement to make ye Country pay a fair price . . . if Mackay persists in obstructing our designs we must find a way to square him into terms, and Sr. Thos. and I agreeing youl bring it about.[283]

Agreement was threatened by Mackay's 'utter hatred' of Sarah Clayton, but finally a satisfactory arrangement was made in May 1768.[284]

During the next three years, Legh opened his high grade coals at Florida (125) near the market end of the canal and Mackay began to work the Rushy

Park seam on his Ravenhead (129) concession at the terminus. Other proprietors were again forced into difficulties. Clayton raised her price to a remarkable 4d per basket at Parr Hall (43) in August 1770;[285] her kinsman Jonathan Case opened a colliery in Ravenhead (126) in 1769, but was forced to close in 1773 after failing to mine the guaranteed output despite lowering his price to 2d per basket;[286] and the Tarbucks closed Wiredale colliery (109) in 1768.[287] Even before the Leeds and Liverpool Canal was opened to Liverpool, the Sankey coalmasters were in almost perpetual difficulties as each fought the other for control of supplies to the canal. Sir Thomas Gerard, whose output at Garswood (29) had continued to slump, was the first to devise an alternative strategy for comfortable survival. In 1772 he tempted the Warrington Copper Company to build a smelter on his estate with an offer of coal at substantially less than Sankey prices in return for a monopoly of their coal supply.[288] Coal-consuming industry thus came to the banks of the Sankey before the Leeds and Liverpool Canal began to bring competing supplies into Liverpool.

When this occurred in 1774 it focused the minds of the Sankey coal proprietors. A price fixing cartel was immediately formed by Legh, Gerard, Mackay and Thomas Case, acting for Sarah Clayton. The effect was to fix prices on bulk sales just below the 7/6d a ton charged by the Orrell collieries on the Leeds and Liverpool. Peter Legh emerged from the agreement in the strongest position. His high grade coal sold at 7/2d a ton to housekeepers and 6/6d a ton to shipping despite being produced nearest the market and therefore incurring the lowest transport costs. Mackay's prices were fixed at 7/0d and 6/4d, Clayton's at 6/10d and 6/2d and Gerard's at 6/6d and 5/10d.[289] Although Mackay's were the second highest prices, his Ravenhead colliery (129) was about two and a half miles further from the market than that of Legh and this must have eroded his profit margin on Liverpool sales. In about 1773 he had granted a cheap coal agreement to the Ravenhead plate glass company, in which he was a partner, in return for a monopoly of its supply, but the works were not immediately successful and the arrangement could not have ended his problems.[290] In January 1778, he entered into an agreement with Case and Gerard to limit supplies on the canal to keep up prices.[291] The Clayton–Case colliery (43) crashed with their bankruptcy in the same year.[292] Two years later, Mackay entered into a further cheap coal for monopoly supply agreement, this time with the huge Ravenhead copper works which required the high grade fuel available only in his Rushy Park seam.[293]

By 1780, Clayton had gone, Legh was supreme at the market end of the canal and Gerard and Mackay had two tied industrial markets each.[294] The long years of price warfare had finally ended. During the quarter century when it prevailed, Sankey prices and supplies had been kept down, numerous collieries had been forced to close and industry had been tempted in to assure particular coalmasters of large markets. After 1780 as these markets grew and demand rose in Cheshire, a balanced location pattern of workings strung out along the canal again developed as Cheshire and Liverpool saltboilers bought concessions and opened their own collieries to the east of St Helens. The competitive storm abated as coalmasters each monopolised their own large markets, and a marked decline in the entry and exit ratio was a consequence.

The pattern on the Leeds and Liverpool was one of quietude compared with that on the Sankey: long-lived collieries grew smoothly, closures seem to have been few, prices were stable and moved in unison and the distribution pattern remained stable until the end of the century. This calm was, like its opposite on the Sankey, due in a large measure to the competitive strategies followed by coalmasters. The overlapping partnerships of Bradford and Liverpool merchants that ran the collieries in Orrell and Up Holland which dominated the canal, and one of those in Shevington on the opposite bank, had coal yards in Liverpool and, presumably, flatts on the canal. John Hustler and Jonathan Blundell, the two dominant coalmasters of the area, were leading canal proprietors and important members of the policy making committee.[295] The unison achieved must have been to a large extent a function of the ability of this clique to control the vast majority of the best reserves at the market end of the canal, distribution and canal company policy.

As long as these coalmasters could keep prices down to 7/6d per ton in Liverpool they not only guaranteed themselves a market there but also effectively excluded most other coalowners from competing from less advantageous locations in relation to the canal. Collieries in Standish (41), Pemberton (14), Winstanley (6) and Lathom (71), at some distance from the canal, were unable to get down to these prices:[296] the small profits per ton earned by the Orrell coalmasters at these prices were compensated by the near monopoly which they achieved through them. The only real danger to the Bradford and Liverpool cartel was the Earl of Balcarres, who could supply the cannel that the Orrell coalmasters lacked but who, on the other hand, had even larger reserves of equally high grade bituminous coal and an equivalent financial and speculative acumen and ambition. He made a number of valiant attemps to bring his Haigh (2) coal into competition with that of the Liverpool and Bradford cartel.

Balcarres had two problems – getting his coal from Haigh (2) to the canal and the shortage of water in the canal between its Wigan terminus and Orrell: 'towards the Wigan End . . . the water is very short and thereby the Trade is interrupted which is always severe'.[297] In 1778 he made ambitious proposals to link his works to the canal by a branch cut, underground canal, or wagonway. He was willing to guarantee the canal proprietors dues on 40,000 tons of coal and iron a year, amounting to over £3000, in return for help in constructing his link and an improvement in the supply of water at the Wigan end of the canal.[298] The proposals were initially well received but eventually rejected under pressure from the Orrell coalmasters' lobby led by John Hustler. In 1788 James Hoare, who was then acting on behalf of Balcarres in Liverpool, wrote that 'Mr Hustler assures me that the Colliers took no part in deciding upon [the] proposals' and that the reason for rejecting them was that if they were accepted 'then Lord Derby, Mr Banks & Mr Standish . . . would expect the like indulgence and that it was their intention to make a Cut to our Canal without any Charge to Us'.[299] This suggestion was more than faintly ludicrous: none of these other coalowners had the reserves of Balcarres nor his heavy involvement in large-scale mining.[300] Even more preposterous misrepresentation was to follow. At another meeting to discuss Balcarres' proposals in 1789 it was maintained that he had claimed he would supply 30,000 tons a year to Paris if his scheme came to fruition. Inflamed by this farrago, he wrote to a canal proprietor,

> I have not a doubt that the Coalmasters at Liverpool are playing a dirty underhand game; . . . I gave in specific proposals which were approved by you, examined and found properly stated by a committee . . . I have been met (yourself alone excepted) by Chicanerie and an Interest destructive to the Interests of your . . . canal; . . . Timidity, Irresolution and Indecision counteracted [the proprietors'] Principle of Self Interest and as the Reverse of all these Qualities mark the Conduct of the Coal Proprietors, no wonder that they have been triumphant.[301]

The price of 7/6d a ton in Liverpool could not be maintained indefinitely in the face of rising costs and inflation. It was raised by 2/- in 1792, and though this increase 'occasioned a great decrease in consumption' and had to be halved,[302] it allowed Winstanley (6), Pemberton (169), Standish (41), Shevington (155) and Skelmersdale (105) collieries to compete on the canal. Those in the first three townships, each with huge reserves near the canal, posed the greatest threat to the cartel and were leased by members of the Bradford and Liverpool caucus.[303] Meanwhile, the hapless Balcarres was persecuted even further. After expanding cannel production and with £3000 worth on his pit banks awaiting shipment, the canal proprietors refused to give him adequate storage facilities at the Liverpool terminus. They allowed him only half a yard, which he could not fence and which was grossly insufficient for the needs of his colliery and ironworks. 'During the Time that I sent cannel to the half yard that I procured, I lost a great deal of money by the circumstances of it being ill fenced and not all my own . . . Cannel being of much greater value at Liverpool than coal, is much more liable to theft and Depredation than Coal, consequently it requires much better enclosures'.[304]

Balcarres' correspondence was thenceforth not concerned with the Leeds and Liverpool and in 1792 he was instrumental in pushing the enabling Act of the Lancaster Canal through Parliament. After eight years of almost continuous argument with the proprietors he got a line and policy which suited him. In the first few years of the canal's operation Balcarres cornered about three-fifths of the whole coal and cannel trade on the Lancaster Canal's southern end[305] and eventually wrenched the whole of the Preston and North Lancashire market from the Leeds and Liverpool Canal and the landsale pits: 'Being first in the field I got early possession of the coal markets of Preston and those to the north of Preston in which I have ever kept the lead notwithstanding the most powerful competition to deprive me of it'.[306] At the turn of the century he was busily engaged in buying up even further reserves, asking competitors whether it was 'worth some consideration whether we ought to remain Rivals or Quite the Reverse!', and swamping the market with coal from the Arley seam, of which he had vast reserves, at cheap prices to drive off competitors.[307] Perhaps the Leeds and Liverpool Canal caucus showed great prescience and powers of self-preservation in their dealings with Balcarres; whilst his energies were engaged elsewhere they smoothly expanded their interests as prices rose and continued to dominate the Douglas valley coal trade.

Competition became much more widespread during the late eighteenth century, integrating the collieries of the coalfield into a single economic system epitomised by a remarkable and novel uniformity of pithead prices. But the turbulent flow of this process had some profoundly variable effects along the way to its destination, or rather its destinations, because the uniformity of prices masked some fundamental differences between the competitive structures of the major producing areas of the coalfield. It is not surprising that the competitive relationships between coalmasters became the subject of intense and calculating concern. As transport cheapened and markets became larger, as collieries grew and fixed costs rose, the stakes and potential prizes as well as the power of the individual players increased. Little wonder that competition was rife. But in the mining industry there was a further and unusual twist to the familiar free market story. Because mining is an extractive and exhaustive industry, competitive success could not be permanent. Short-term victory inevitably brought more rapid extraction and involved buying up surplus reserves and/or cutting prices, profits and therefore the capital necessary to move on smoothly into deeper or more distant seams. The more complete the short-term victory the more inevitable was a renewed struggle as the victor became imbroiled in the difficulties his success brought. Case won then lost on the Prescot turnpike; Clayton, Legh and Gerard won then emerged from the succeeding bout in various states of lesser dominance on the Sankey, and all the while Liverpool's demands remained unsatisfied.

Perfect Löschian equilibrium would have produced a far different process. It would have produced numerous collieries, each as small as it was possible to be and remain profitable, each in its own spatial monopoly market area selling at a price fixed by demand and minimised costs. In such conditions there can be no short supply and as demand rises new collieries will automatically enter the industry to meet it. In this way, the rate of growth of total output would be maximised for any given rate of potential market expansion. It was the steam engine which prevented the emergence of anything approaching this situation in South West Lancashire in the late eighteenth century. It was the symbol and instrument of growth which caused the turmoil of price cutting, closures and short supply. The new technology allowed the industry to grow at an unprecedented rate, but it prevented it from growing as fast as the full surge of potential market expansion. The steam engine made it possible to produce outputs of up to 60,000 tons a year at individual collieries: that is, at least twenty per cent of the total capacity of any of the canal or turnpike markets. At first, in virgin shallow coal, it also yielded cheap unit costs at such large outputs. The larger the annual output, the sooner the heavy fixed costs would be paid off and the sooner a large annual rate of net profit would be achieved. Thus, coalmasters cut prices and hoped to make up in volume what they lost in unit profits. If competitors could be driven out of business so that windfall gains could be made in a subsequent period of raised monopoly prices, so much the better. The prospects were tantalising and there was always a potential victim working older or more distant reserves. The chances of success were obviously greater if large areas of reserves in good locations could be amassed and denied to competitors and this, presumably, was the motive behind the large purchases of Case in the South West and the Liverpool–Bradford group of partnerships in the Centre. Free competition with such large scale economies as the steam engine brought to mining seemed inevitably to lead to cutthroat competition, wasted resources, the under-supply of markets and a turmoil of leap-frogging success and failure as collieries were pillaged and exhausted without the accumulation of capital reserves.

The developments which occurred in different areas at different times in the second half of the eighteenth century bear ample witness to this inevitability. On the Douglas Navigation collieries were small and numerous and perfect Löschian competition was approached. Output and prices rose with demand and the entry and exit ratio was low. On the Sankey, in contrast, equally free competition between large collieries produced the highest entry and exit ratio recorded, price cutting, under-supply and almost perpetual crisis. The only ways out of the morass were competitive victory, which was unlikely because of the mutability of reserves, their abundance and their fragmented holding, or the cessation of free competition. This was achieved in different ways. In the South West and South coalmasters retreated from long and exhausting competition and sought tied markets in smelting and saltboiling. On the Leeds and Liverpool Canal, Balcarres had a near monopoly of the upper end of the domestic market with his cannel and a tied market for his coal in his ironworks. Nearly all the rest of the Liverpool market was held in the tight grip of the group of overlapping partnerships on the western end of the canal who moved prices in unison and kept out competitors through their possession of large areas of high grade reserves, their control of distribution in Liverpool and of canal company policy. The development of St Helens as a smelting and manufacturing town and of Wigan and its environs as areas of coal export owed much to the different competitive strategies adopted by the coalmasters on the two canals; Wigan had by far the better smelting fuels, but the cartel kept prices high and obviated the need to grab tied markets through cheap coal supply agreements. Human behaviour countervailed physical resource endowments.

### The geography of the South West Lancashire mining industry, 1740-99

What had been two mining regions became one in the late eighteenth century. Liverpool was the focus of this integration, drawing collieries from the South West, South, West and Centre into competition by the last quarter of the century. Much less important nodes in Cheshire and Preston brought the South and South West and the North and Centre, respectively, into similar rivalry. This unity was symbolised by a much less variegated pattern of prices and by the almost complete adoption of the cwt and ton system of coal measurement instead of the local and sometimes incompatible systems which had prevailed earlier (see Appendix 1).

It was the opening up of markets which were large, and which subsequently expanded further and faster as cheap and abundant coal was combined with their other economic assets, which injected the impetus to expansion into the industry. All these markets were outside the coalfield. All required the construction of a cheap transport route to catalyse their reciprocal interaction with coal. It was through competition in common markets along different transport lines that the mining industry was boosted and integrated. This process had been jolted into action when the Prescot to Liverpool turnpike and Weaver Navigation were built in 1733; it was given further small impetus by the opening of the Douglas Navigation in 1742 and the extension of the Prescot turnpike to St Helens in 1750; it reached its apotheosis in the Sankey Navigation of 1757 and the Leeds and Liverpool Canal of 1774. Both these canals, and most especially the Sankey, allowed the import of bulk raw materials into the coalfield as well as the export of coal. They therefore stimulated the growth of local as well as external markets. After about 1770 a process which could be described as an industrial revolution was in motion in the region. Mining expanded to fuel industrial production expanded on the back of increased coal output. The units of coal production and consumption increased enormously in scale and both mining and industry became imbued with an ever increasing capacity for even further expansion in consequence. Coal production and industrial consumption thus developed both an endogenous and a reciprocating dynamic.

The increases in total output and in the size of individual collieries were completely unprecedented and marched in step. In the Douglas valley the largest colliery produced about 5000 tons a year when the area's total output was about 36,000 tons a year before the Leeds and Liverpool Canal was built; in 1799 the largest colliery there produced about 60,000 tons a year and the total output of the area was about 380,000 tons a year. Thus, maximum colliery size increased at a faster rate than total output and it moved up with every new line of cheap transport. This was only possible because of the innovation of the steam engine. This, in turn, brought a large increase in capital requirements and, with the longwall mining system, a substantial increase in labour productivity. Capital could flow in from outside the coalfield but labour was less mobile. In consequence, piece rates did not fall as productivity increased and the high wages of colliers reflected their relative scarcity as demand for their skills mushroomed.

But the increased markets, new techniques and larger-scale production do not seem to have brought increased unit profits: the central pinion of industrialisation seems to have been missing. Indeed, the profit per unit of output, or as a percentage of receipts, seems, on all the evidence available, to have fallen quite sharply through the late eighteenth century. It must have been the higher aggregate profits brought by larger-scale operations which attracted the necessary investment. Furthermore, large-scale production allowed the coalmaster to influence the market price, and therefore his level of profit. But this freedom was not unconstrained: the Duke of Bridgewater was always waiting in the wings for the Liverpool price to rise above 7/6d per ton. Indeed, large collieries seem invariably to have sold at lower prices than smaller works. Perhaps this was meant as a prelude to later monopoly prices when 'adversaries' had been 'silenced' by the lower one. But the finale did not come before 1800. None the less, the strategy was always a tempting one because there were always variations in costs according to the depth of working and the age of collieries, always potential victims for the well situated works in virgin steam drained coal.

Thus, larger markets and canals brought collieries which were individually large enough to influence market prices. This brought price cutting which, in turn, reduced the capacity of the industry below its maximum potential output and cut the rate of growth. Free competition in mining reduced the rate of industrialisation in the region by preventing the full transmission of the stimulus of growing industrial demand and therefore the full reciprocation of mining and industry. This checking mechanism could only be released by erecting barriers to free competition. A number of methods were tried. The crudest was simply to acquire and hoard large areas of reserves so that competitors could gain no access. Case tried this in the South West and it formed part of the more complex composite strategy of coalmasters on the Leeds and Liverpool Canal and of the Earl of Balcarres on the Lancaster Canal later. Or tied markets could be acquired, as on the

Sankey after about 1770. Or coalmasters could act in unison in a cartel, which was tried and failed on the Sankey but which succeeded on the Leeds and Liverpool.

Of course, one would imagine that the objective of all such strategies was to push up prices after initially low ones had reduced competition. But the ceilings which could be reached were apparently always low and even then difficult to reach because of the Duke of Bridgewater and because the abundance of shallow coal posed an ever present threat from local competition. It was only in the late 1790s that Liverpool prices began to escalate steeply after two decades of absolute stability. This only occurred after members of the various intertwined partnerships on the Leeds and Liverpool Canal had managed, between themselves and numerous other sleeping partners, to assemble vast reserves in Standish, Shevington, Parbold, Pemberton and Winstanley as well as expanding their original concessions in Orrell and Up Holland. Perhaps, too, the Liverpool market, supplemented by rapidly rising exports, was too big by this time for the Duke of Bridgewater to make a punishing impression upon it. Whatever the exact circumstances, it was only at the very end of the eighteenth century that the restricted competition culminated in high prices. For twenty years before that there was a curious balance in which the prevailing form of competition allowed sustained expansion of production without the achievement of monopoly prices. These may have been difficult times for the coalmaster, forced to push up output to compensate low and possibly falling unit profits: but they were as benign as possible for the industrial consumer. It is worth reiterating in this context that as much coal was sent from the Wigan end of the Leeds and Liverpool Canal to Liverpool at the end of this period of low stable prices as went down over a quarter of a century later. On the Sankey, conditions were propitious for the industrial consumer in a different but equally potent way. Cheap coal agreements and alliances between particular collieries and particular industrial consumers provided the hole through which the coalmasters there bolted from the torments of free competition.

Thus, the concatenation of relationships within the mining industry of South West Lancashire in the late eighteenth century were such as to produce an expansion of output, the growth of coal-consuming industries on parts of the coalfield and on the Mersey, low unit profits and high wages. But the situation was inherently unstable. It could only last in full bloom as long as barriers to the entry of competitors were complete enough to prevent cutthroat competition but too imperfect to produce the high prices whose expectation brought them into being.

# CHAPTER 8

# CONCLUSION

The progress of coal production between 1590 and 1799 is shown on figure 53, which includes extrapolations based on the national rates of growth postulated by Nef and Ashton and Sykes.[1] The South West Lancashire figures do not correspond with their suggestions. In the seventeenth century expansion was much less than Nef estimated, in the eighteenth century much greater than Ashton and Sykes estimated. Growth occurred at different times and at different paces in different areas of the coalfield in the eighteenth century. The South led off in the first two decades then faded; the South West surged ahead in the 1730s; the South re-emerged to prominence in the 1750s and 1760s and the Centre, after hesitantly spluttering for three

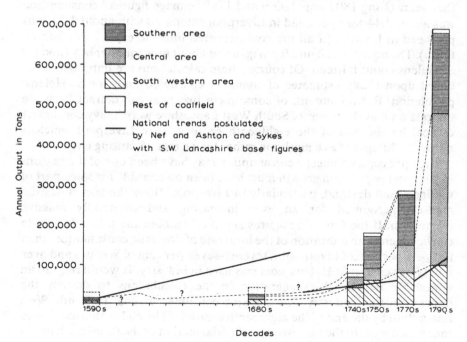

53. Output of the South West Lancashire coalfield, 1590-1799. Based on estimates given in text.

237

decades, moved at first slowly forward in the 1770s then accelerated away through the 1780s and 1790s. By the end of the eighteenth century these three areas were growing in unison and the edge of the West had just joined them. In the North and most of the West the industry was of a much smaller scale and growing considerably more slowly, although it was by no means stagnant.

In 1799 this coalfield (which, it should be remembered, comprises less than half of the total area of coalmeasures in South Lancashire) produced about 680,000 tons. It had lagged behind the national rate of growth in the seventeenth century, but by 1799 it was producing half as much coal as was exported from the Tyne and nearly as much as was exported from the Wear.[2] A very high proportion of this output was consumed within South West Lancashire and on the Cheshire saltfield. The population of this area cannot have been a quarter of that of London, yet half as much coal was consumed there as was imported into the capital by 1799.[3] Liverpool took 222,728 tons from the Leeds and Liverpool Canal alone, as well as most of the 100,000 tons produced in the South West and an unknown but considerably smaller proportion of the output of the Sankey valley collieries. If its supply in 1799 is roughly estimated at 350,000 tons and its exports as 150,000 tons (they were 102,645 tons in 1792) then Liverpool consumed about 200,000 tons in that year. Using 1801 population and 1799 tonnage figures,[4] consumption was about 2.44 tons per head in Liverpool compared with about 1.31 tons per head in London (if all the coal imported into the capital was burned there). The equivalent figure for Wigan was about nine tons per head and for St Helens about thirteen. Of course, these calculations are impressionistic, based upon shaky estimates of tonnages consumed and, for St Helens, population. But no amount of consequent fuzziness can detract from the impression that the towns of South West Lancashire were heavy consumers of coal by the end of the eighteenth century, even Liverpool which is normally thought of as a maritime rather than a manufacturing city.

High per capita domestic consumption may have been one of the reasons for this and population growth must have been responsible for some part of the increased demand, particularly in Liverpool. None the less, industrial markets accounted for an ever increasing and eventually massive proportion. If the foregoing figures are at all reliable and if London's rate can be taken as an indication of the local rate of domestic consumption, then forty-six per cent of Liverpool's, seventy-seven per cent of Wigan's and over ninety per cent of St Helens' coal was used in industry. It would require an impossible degree of exaggeration in the calculations to destroy the impression that industry was a major consumer of coal in South West Lancashire by the end of the eighteenth century. (The eighty thousand or so tons which went to the exclusively industrial market of the Cheshire saltfield were excluded from the calculations.)

Most of the coal in the region had gone to domestic hearths in the seventeenth century, but even then it was probably the consumption of the iron and non-ferrous metal industries of Wigan which picked out the Centre as the most developed mining area. Local glass and pottery and the Cheshire salt industry stimulated the development of the South in the early eighteenth century and the growth of mining in the South West owed much to Cheshire salt and Liverpool industry after the early 1730s. This long-standing inter-relationship was transformed in scale by canals, which allowed larger tonnages of coal to be shifted, bulk raw materials to be brought into the coalfield and large quantities of manufactured goods to be moved out. In the process 'factory' production of manufactures and a tenfold increase in the size of collieries were precipitated. The collieries, glass, iron and copper works of Wigan and St Helens, the saltworks of Cheshire and the Mersey estuary and the copper, iron, salt and sugar works of Liverpool, its breweries, distilleries and potteries, were all large plant, often employing more than one hundred men and sometimes several hundreds by the late eighteenth century. The coal-based manufacturers of the region served international markets through the canals, navigations and seaways articulated around Liverpool.

The cost and capacity of transport defined, in reciprocation with markets, the range of possible states of development in the mining industry. Other variables acted to determine which particular state obtained. The physical endowment became less important as technology progressed. Geomorphological constraints on the scale and duration of mining were important in the seventeenth century but not after the innovation of the steam engine in the eighteenth. The lavish abundance of shallow coal also encouraged the development of small ephemeral workings in the seventeenth century and it allowed a smooth and rapid acceleration of production in the second half of the eighteenth century, whilst the variety of coals present in the Lancashire seams permitted the subtle responses that were often necessary when manufacturing technology was rudimentary. The stimuli of transport and markets thus had rich materials on which to operated.

But three further agents were necessary for development to be produced from this interaction. The rates at which technological solutions to drainage problems, capital and labour could be supplied were important determinants of the state of the industry at any one time and of its pattern of change through time. Or perhaps one should say, more accurately, that the potentially potent constraints which might have been imposed by serious lacks or lags in any of these variables did not, in fact, materialise. Steam drainage pumps were available before rapid development began and the increased capital requirements which their installation brought flowed smoothly in. Although the scarcity of capital may have combined with other variables operating to the same end of small ephemeral collieries in the seventeenth century and over much of the coalfield in the early eighteenth, and although the correlation of slow initial growth on the Leeds and

Liverpool Canal and local capital supplies suggests a causal connection, generally the required money surged in, sometimes over considerable distances and through intricate channels. Indeed, the abundance of funds which this betokens and the complexity of the partnership and loan mechanisms which existed to marshal them is an impressive reflection of the state of economic development at the time. A labour force was slowly and gradually created in the seventeenth century in those districts where mining was most prominent. The skills, discipline and life-style of the collier were already immured in the communities of the coalfield before rapid growth began, although it is probably significant that miners from other coalfields were brought into the South West and Centre in the early eighteenth century when hitherto unexperienced difficulties occured. Thenceforth, labour recruitment seems to have presented no problems except, perhaps, during the most explosive phase of early growth along the banks of the canal in the sparsely populated Sankey valley. Increased productivity due to the more regular and specialised work patterns brought by the steam engine and the innovation of longwall and Lancashire methods of getting helped to ease the strain on labour supplies. But high wages, kin links and a high level of short distance mobility amongst the collier population were at least equally important.

If demand and transport provided the stimulus; reserves and geomorphology the matrix upon which it operated; and technology, capital and labour the enabling agents in the process of growth, it was competition which locked them together into a functioning economic system[5] – or rather, into functioning economic systems. There were two 'colliery regions' on the coalfield in the seventeenth century, one around Prescot in the South West and one around Wigan in the Centre, but over most of the coalfield the collieries were economic fragments which did not interact through competition. In the early eighteenth century, the two seventeeth-century regions expanded to include most of the areas in which they lay as well as adjacent parts of the South and North. As markets grew, so did the amount of cohesion and interaction between mining districts, but the two regions were still independent of each other. By the late eighteenth century the whole coalfield, except for the remoter parts of the West and North, was articulated into a single economic system with integrated patterns of production, prices and competition around the hub of Liverpool. It became a subsystem of the regional economy organised around the port and it was as a part of this regional economy that it grew to national significance.

The stress laid upon the spatial complexity of the economic structure of the coalfield, the cleavages between different parts of it, the variety of measure systems, markets, economic growth rates and prices, may seem misplaced. After all, the area measures only about twelve by twelve miles. A motorist can cross it today from north to south or east to west in about a quarter of an hour and, traffic in the city permitting, he can drive from central Liverpool to Chorley on the extreme northern edge of the coalfield in not much more than an hour. To stress the diversity of the area and to subdivide it into two 'regions' in the seventeenth and early eighteenth centuries may at first sight seem to be an idiosyncratic product of geographical method rather than an objective appraisal of historical reality. But this smallness is a product of contemporary perceptions, perceptions grounded in rapid transit, cheap haulage and the consequent high degree of economic cohesion over geographical space. In the seventeenth and

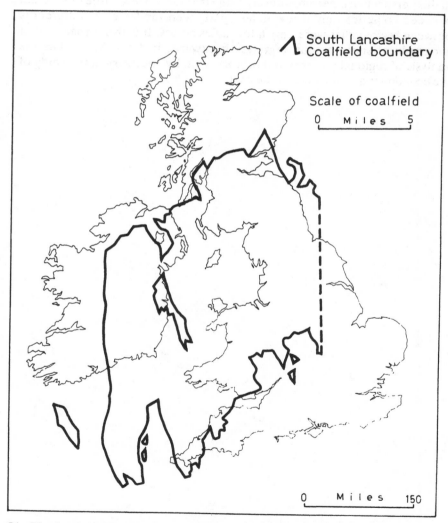

54. The South West Lancashire coalfield transformed in area according to the time taken to travel a mile in the eighteenth century.

eighteenth centuries transport was slow and expensive and the economy was much more spatially disintegrated. In a real but perhaps somewhat elusive sense the area was much 'bigger' then. If the time taken to travel a mile is used as a measure of 'economic distance', or the frictional effect of distance on economic interaction, rather than linear distance itself, then the coalfield was twenty-five times 'bigger' in the eighteenth-century than it is today.[6] The 'size' of the coalfield in terms of this eighteenth-century economic reality compared to England and Wales today is shown on figure 54. When viewed in this perspective it becomes unsurprising that there was so little interaction between collieries only a few miles apart, such fundamental differences between regions that were only a few miles across. It is this 'expansion' of geographical space as one regresses backwards in time that makes the analysis of regional patterns and processes vitally necessary in the study of the economies and societies of the past.[7]

# APPENDIX 1

## MEASURES OF WEIGHT AND VOLUME USED IN THE SOUTH WEST LANCASHIRE COAL INDUSTRY DURING THE SEVENTEENTH AND EIGHTEENTH CENTURIES

---

The weights and measures used in the mining and selling of coal at different collieries must be directly comparable for the collation of outputs, profits, rents, wages and prices. In Britain as a whole a great variety of measures was used, each coalfield employing a system that was to some extent peculiar.[1] Variation was not confined to the inter-coalfield level; within the relatively small area of South West Lancashire many types of measure, distinct in both name and content, were used. Indeed, at certain collieries the measures used in raising coal up the shaft differed from those in which it was meted to customers from the bank. In such circumstances as these, and in the absence of exact knowledge of the system of measurement used at many collieries, perfect standardisation of outputs, rents, wages and prices in terms of a consistent set of coal measurement units is impossible. None the less, some attempt must be made to eradicate as far as possible false impressions due to this particular cause. Fortunately, all large measures comprised aggregations of either baskets or cwts. To establish inter-colliery comparability, therefore, the content of the basket and cwt and the number of constituent baskets or cwts in each aggregative measure must be determined.

### The basket

The basket was the basic unit of measurement at every colliery during the seventeenth century and at a majority in the eighteenth. Like the corf of the Northumberland coalfield, it was the receptacle used to raise coal up the shaft. Basically, therefore, the basket was the smallest unit used in the measurement of a collier's production so that his piece work wages could be fixed. Not unnaturally, it also served in the sale of coal from the pit bank. The basket measure was initially one of volume rather than weight and occasionally in the seventeenth century its capacity was expressed in terms of cubic dimensions.[2] At Haigh (2) in 1664 the drawers were ordered to provide 'good and sufficient measure upon every Baskitt of Cannell that is to say lay it over the pins and higher than the Bewle and as highe at eyther end as in the middle'.[3] Unfortunately, no expression of the exact cubic dimensions of the seventeenth-century basket seems to be extant but it is clear that in certain parts of the coalfield this depended upon local custom.[4]

In the eighteenth century the basket came to be regarded primarily as a measure of weight, although the importance of tradition suggests that the capacity may have remained constant and that only the mode of expressing it altered. With the advent of description in the simpler absolute terms of weight a comparison of basket sizes in various parts of the coalfield becomes possible, revealing a remarkable almost coalfield-wide similarity. In about 1730 the contents weighed 120 lb at Shevington colliery (49) and at Porter's coalpits[5] and 1 cwt or 120 lb at Prescot Hall colliery (5) in 1735.[6] This capacity seems to have been standard over most of the coalfield later in

243

the eighteenth century, stipulated in documents relating to collieries in Haigh, Orrell, Sutton and Windle.[7] Given the prevalence of the 120 lb basket and the regard accorded to custom it seems a logical conclusion that this capacity was common to the greater part of the coalfield throughout the seventeenth and eighteenth centuries.

But exceptions to this generalisation did occur: certainly in the case of baskets used to raise coal up the shaft and perhaps, too, in the case of those used to vend it. Baskets of various sizes were used at Kirkless, Orrell and Shevington in the Wigan area. The first quantification of basket size in Orrell was made in an unexecuted agreement between the Douglas Navigation proprietors and an unknown colliery owner, where basket weight was fixed at 72 lb.[8] However, in 1756 the lessor of Orrell Hall colliery (1) fixed the size of the baskets to be used to measure output for royalty evaluation at 122 lb[9] and this measure was retained in a further lease of the same colliery in 1771.[10] This capacity varies negligibly from that which prevailed over the greater part of the coalfield, but baskets of 150 lb were used to raise coal up the shaft of this colliery and output was initially reckoned in this measure and later converted arithmetically to an expression in 122 lb baskets.[11] At adjacent collieries baskets of 140 lb ('seven score weight')[12] and 150 lb[13] were used, and Anderson cites references to Orrell baskets used in wage evaluation of 140 and 150 lb in the late eighteenth century.[14] In adjacent Shevington (47), where 120 lb baskets had been used in c.1730 and where due regard had been paid to custom in the seventeenth century, it was stipulated in a note appended to a 1745 lease of the Dicconson colliery in 1779 that baskets were to contain 140 lb of 16 oz each.[15] In neighbouring Standish (41) the accounts of 1746-9 can only be rendered consistent and comprehensible by converting the quantities reckoned as got by the colliers in the ratio of 100: 120 to obtain the tabulated figures of sales. It is a moot point whether it was the baskets of the colliery or the scores into which they were aggregated that differed in this ratio between the pit brow and the pit bank.[16]

This diversity of basket size is puzzling because the measure of 120 lb was used in selling the produce of these collieries off the Leeds and Liverpool Canal in the 1770s[17] and remarkably similar selling prices per basket prevailed throughout the area during the late eighteenth century.

Similarly exceptional baskets were used for a time at Kirkless cannel collieries (101, 102, 104 and 135) to the north west of Wigan. Fuel was sold in measures of 100 and 120 lb in 1751;[18] 124 lb baskets were used in 1776,[19] and Kirkless cannel was sold in tons of 2400 lb (i.e. twenty units of 120 lb) on the Leeds and Liverpool Canal in the same year.[20] However, various sizes of basket were used later at Kirkless. In a lease of 1779, 72 lb baskets were stipulated,[21] whilst in a lease of 1781 different basket sizes were fixed for the measurement of different types of cannel. The output of round cannel was to be assessed for rent in baskets of 132 lb but small cannel was to be measured in baskets of 80 lb.[22] Only twenty years later cannel duties were assessed in units of 120 lb on the Lancaster Canal,[23] which terminated near the Kirkless collieries. The only logical suggestion seems to be that baskets of diverse size were used in raising coal in the Wigan area and in assessing royalties but that a common size of 120 lb was used in vending.

### The cwt

The basket was superseded by this measure in certain areas of the coalfield in the late eighteenth century, particularly those supplying canals. A number of statements exist which imply that the basket and cwt were measures of equal capacity[24] and every quantification of the cwt measure quotes a content of 120 lb.[25] The eighteenth-century Lancashire cwt therefore differed from the modern one of 112 lb and the cwt measures quoted in this work are of the larger 120 lb unit.

## Conclusions regarding the basket and cwt

Over most of the coalfield in the eighteenth century the basket contained 120 lb of coal, and in the latter part of that century it was apparently considered to be equal in capacity to the cwt of 120 lb. Local custom definitely governed the size of the baskets used at collieries in Shevington, Haigh and St Helens, and it therefore seems probable that the basic unit of measurement contained 120 lb in most parts of the coalfield in the seventeenth as well as the eighteenth century. At certain collieries in the Wigan area, however, baskets of a different, generally larger capacity were used in raising coal and in assessing output for wage and rent evaluation, but even at these collieries the unit of sale seems generally to have contained 120 lb. The larger baskets used in getting were probably to allow for short weight, stone, slack and other unsaleable materials. The use of larger baskets to assess royalties was perhaps due to the measurement of output at the brow rather than the bank; insofar as the extra weight represented merchantable coal, it would serve the additional purpose of allowing rent-free coal for use about the colliery.

It can therefore be asserted with reasonable confidence that prices expressed in pence per basket or cwt were comparable over the whole coalfield in the seventeenth and eighteenth centuries. Outputs expressed in baskets or cwts are comparable for almost all areas and rents expressed as the rent payment per basket as a percentage of the selling price per basket are also comparable over most of the coalfield. Nevertheless, because of the different basket sizes used at certain collieries in the Central area it is possible that outputs and rents may be underestimated by twenty-five per cent and wages overestimated by the same amount if baskets of unstipulated size are assumed to have contained 120 lb. Where different basket sizes are known to have existed outputs, wages and royalties per basket have been converted to conform to the prevalent 120 lb measure. If, in a rare case, outputs, wages or royalties per basket have been calculated for Orrell, Shevington or Kirkless collieries without an exact knowledge of the basket size used, this is indicated in the text.

With these limitations in mind, it seems that reasonably accurate comparisons can be made between the vast majority of collieries in most parts of the coalfield by expressing prices, wages, royalties and outputs in units of baskets or cwts of 120 lb. This provides a firm standard if the commonly used larger measures can be accurately reduced to their basket or cwt contents.

## The Load

This was the least definitive term used in the measurement of coal. The term was often used in isolation, but equally often it was prefixed by either cart, wain, horse or pit and the content of some of these (fortunately rarely used) measures is difficult to ascertain.

The *cart-load* and *wain-load* seem to have been of roughly equal capacity. These terms were used very early in the period at Smithils[26] near Bolton and at Prescot,[27] where they were quickly superseded as aggregative measures by the ton and the work. Both Bailey[28] and Nef[29] state that these measures contained about one ton, but documentary evidence of their exact content in South West Lancashire has not been discovered.

The *horse-load* was used on only two known occasions[30] and on one of these it was utilised only because bad weather prevented the employment of the usual vehicle. Nef suggests that the horse-load contained half a wain-loan, or about ten baskets, but according to Aikin its capacity was, at eight to the ton, only two and a half baskets.[31]

Although these measures were used in widely separate areas the load generally fell from currency during the seventeenth century.[32] The load continued to be used as a measure of output and sales in the eighteenth century only in the town of Wigan and at the collieries of Aspull, Haigh, Kirkless and Winstanley which supplied its fuel.[33] In this area the term was usually unprefixed, although occasionally the word pit-load was used. At Little Hulton near Bolton in 1637[34] and at either Aspull or Ashton-in-Makerfield in about 1650[35] the pit-load contained twenty-four baskets. However, in the Wigan area the pit-load and load seem to have been equal measures of much smaller dimensions. A pit-load of twenty-four baskets would yield impossibly low prices of much less than one farthing per basket in Wigan in the early seventeenth century. Every quantification of the load and pit-load[36] but one at the Aspull, Kirkless and Haigh collieries gives a capacity of ten baskets. The exception is contained in a lease of coal in Haigh Park, made out in 1642, where the royalty was to be assessed per load of thirteen baskets, as it was at Ashton in 1654.[37] The allowance of measures larger than the norm for the purpose of royalty assessment was not uncommon (see *basket, score* and *work*) and probably served the purpose of allowing a proportion of output free of rent to compensate for coal mined for other purposes than sale.[38]

At Winstanley a load of exceptional dimensions was used to assess output.[39] In the seventeenth century the Winstanley (6) load contained eight baskets,[40] but nine in the 1760s.[41] This measure was replaced by the more common score at this colliery in 1766,[42] although again the Winstanley (6) measure differed in capacity from that in general use (see *score* below).

Loads of unspecified dimensions have been used in only four calculations, namely those of Wigan and Orrell coal prices in the seventeenth century, where a content of ten baskets was assumed.

**The score**

This measure was apparently never used in the South West and was mentioned only once in connection with collieries in the South.[43] Like the load it was most commonly used in the Centre, but it seems also to have served as the usual aggregative measure in the North and West.[44] In all these parts the score was the general large measure until the canal era, when it was partially ousted by the ton.

Predictably, 'one score basket of coals' often comprised twenty baskets,[45] but this was by no means always the case. At Winstanley (6) in 1766[46] and at the Holme (74) and Blundell (93) collieries in Orrell in the late eighteenth century[47] the output of the hewers was reckoned in scores of twenty-four baskets, but this was definitely not the size of sale scores at the Winstanley (6) and Holme (74) collieries.[48] In the evaluation of royalties, scores of between twenty-one and twenty-six baskets were often used,[49] probably, as with other larger than normal measures used in the fixing of rents, to allow for coal consumed in the working of the colliery. At Skelmersdale (105) in the 1790s the lessee paid 'scorage' at the rate of ¼d per twenty-four baskets although coal was sold from the bank at 5/10d per ton, i.e. twenty baskets at 3½d each.[50]

Thus, an assumption that a score of unspecified size used in the measurement of wages or royalties contained twenty baskets is liable to an error in the derived figures of up to thirty per cent. It has been made only in the calculation of the outputs of the leasehold collieries at Haigh in the mid eighteenth century from quotations of royalty rates and totals.

**The ton**

This measure was widely adopted over the whole coalfield only during the canal era, although in the vicinity of Prescot it seems to have been in continuous use from the late sixteenth century.[51] The earliest statement of the content of the ton was made in 1638, when it comprised twenty baskets at a colliery in Whiston or Prescot.[52] Later statements give the same capacity in the South West.[53] Outside that area the ton was used to measure sales on the waterways, even the early Douglas Navigation,[54] though the score continued to be used at many collieries serving it and the later Leeds and Liverpool and Lancaster Canals for all purposes except canal sales.

The ton was used for all coal measurement at collieries in the catchment of the Sankey. On that canal it was defined volumetrically at about thirty sealed bushels or sixty-three cubic feet of coal.[55] Apparently, the ton in which Sankey coal was sold in Liverpool initially contained thirty baskets or cwt,[56] but the norm of twenty was quickly adopted and all statements after 1770 quote a standard content of 20 cwt[57] (although measures as short as 15½ cwt were sometimes fraudulently meted).[58] The 20 cwt ton was also used on both the Leeds and Liverpool and Lancaster Canals.[59]

This uniformity of content is not surprising in such a nationally well-used measure, but because the local basket or cwt contained 120 lb the South West Lancashire ton was of 2400 rather than 2240 lb.

## The work

This was the most commonly used aggregative measure in the South and South West, but never current elsewhere. Like the score of the Centre, it continued to be used at collieries long after the ton had become the standard measure in waterway markets. The work was the largest measure in common use, its content being variously expressed as sixty baskets,[60] three tons,[61] or a heap of coal measuring nine by six by three and a half feet.[62] Sixty baskets or cwt was by far the most common capacity but, as with the load and score, larger versions of sixty-three or sixty-six baskets were often used to assess output for royalties.[63] The fact that the work contained sixty baskets and that larger quantities were assigned to it to provide a rent-free concession was made explicit in a number of leases.[64]

## The rook

Although used at an Ashton-in-Makerfield colliery in 1665,[65] the rook only seems to have been employed for any length of time in the seventeenth century at Winstanley (6), where it contained eighty baskets or ten loads.[66]

## The rick or nick

This measure, the largest used on the coalfield, was confined to Haigh (2) where it was employed for a short period in the mid eighteenth century. It contained twenty scores, probably four hundred baskets.[67]

## Summary and standardisation procedure

Despite the variety of measures in use, changes at particular collieries and differences between the sizes of measures on the brow and on the bank at the same colliery, it seems reasonably clear that more or less standard sets of measures were used in the sale of coal. Everywhere, the basic unit, whether called basket or cwt, contained 120 lb. In the South and South West these were aggregated into tons of twenty and works of sixty baskets, elsewhere into loads of ten and scores of twenty baskets. It seems safe to assume that quantities expressed in these units in receipts accounts conformed to these standards. So, too, probably, did statements of observers or witnesses: these would be the measures with which they were familiar as customers or provided with on enquiry at the pit bank. However, hewers' gets and royalties were often measured in units which were larger than the sale norms. Wherever this is known to have occurred the aberrant measures have been reduced by the appropriate amounts. When such measurements were made in units of unspecified size it has been assumed that they contained the standard quantities. Figures based upon these assumptions are open to error over the ranges described above: when they occur in the text their nature is made clear.

# APPENDIX 2

## THE COLLIERIES OF SOUTH WEST LANCASHIRE, c.1590-1799

The reference numbers which accompany each mention of particular collieries in the text refer to a collation of all source materials relating to each colliery. As originally conceived, the book was to include this list and two others of rents and prices as documented and as standardised for variations in weights and measures, but cost prevented it. Microfiche copies are available from the author on request.

The creation of these colliery biographies was not a straightforward process and some misallocations must have been made as a result of the sparse and fragmented nature of the source materials and the very general locational descriptions that are usually given in the documents. In many cases it was impossible to decide whether a colliery mentioned in a document represented a continuation or reopening of earlier workings or a completely different works in the same vicinity. Judgement on this question was based on probabilities: if it seemed likely that a document referred to further exploitation of an earlier colliery it was tabulated under the same reference number. Assessment of these probabilities was based on information such as location, proprietorship, coal ownership, workers' names, contemporary cross-references and such miscellaneous things as details of drainage works, depth and seams.

Figure 55 gives the locations of all the collieries defined in this way and referred to in the text, together with their reference numbers.

### List of collieries

The township in which a colliery was located is always given, together with a more detailed locational reference where possible. Township names are linked when collieries were located on estates or in hamlets which overlapped township boundaries and the exact position of the colliery within the smaller unit is unknown. Eccleston near Prescot is distinguished by (S), Eccleston near Chorley by (N). An asterisk denotes a colliery producing cannel.

| | |
|---|---|
| 1. Orrell Hall, Orrell | 12.* Aspull |
| 2.* Haigh | 13. Hindley |
| 3. Sutton Heath, Sutton | 14. Goose Green, Pemberton |
| 4. Sutton Heath, Sutton | 15. Prescot |
| 5. Prescot Hall, Prescot | 16. Windle |
| 6. Winstanley | 17. Charnock Richard |
| 7. Saltersford, Orrell | 18. Whiston |
| 8. Prescot | 19. Wigan |
| 9. Stanley Gate, Bickerstaffe | 20. Eccleston (N) or Heskin |
| 10. Blackrod | 21. Prescot |
| 11. Goose Green, Pemberton | 22. Adlington |

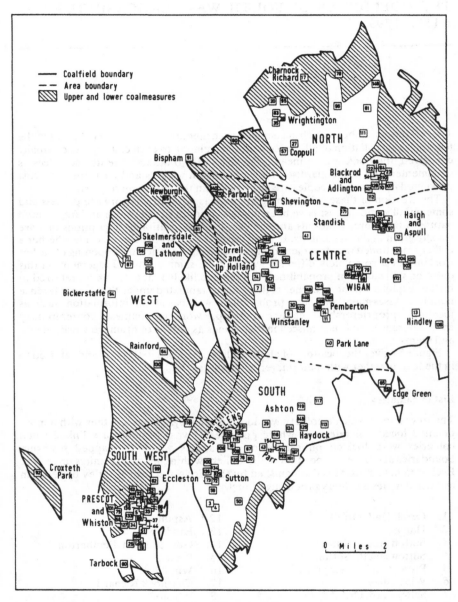

55.  Locations of all collieries worked, 1590-1799.

| | |
|---|---|
| 23. Blackrod | 76. Parr |
| 24. Whiston | 77. Shevington |
| 25. Whiston | 78. Chorley (?) |
| 26. Prescot | 79. Bottlingwood, Wigan |
| 27. Coppull Moor, Coppull | 80. St Helens, Windle |
| 28. Haydock Wood, Haydock | 81. Duxbury |
| 29. Garswood, Ashton-in-Makerfield | 82. Haigh |
| 30. Pemberton | 83. Wrightington |
| 31. Prescot | 84. Adlington |
| 32. Prescot | 85. Newburgh, Lathom |
| 33. Prescot | 86. Harrock Hill, Wrightington |
| 34. Whiston | 87. Westhead, Lathom |
| 35. Haigh | 88. Eccleston (S) |
| 36. Whiston | 89. Whiston |
| 37. Whiston | 90. Haigh |
| 38. Wigan (?) | 91. Bispham |
| 39. Wrightington | 92. Croxteth |
| 40. Park Lane, Ashton-in-Makerfield | 93. Orrell |
| 41. Standish | 94. Bickerstaffe |
| 42. Wigan | 95. Park Hall, Charnock Richard |
| 43. Parr Hall, Parr | 96. Burgh–Birkacre, Duxbury–Coppull |
| 44. Whiston | 97. Haigh |
| 45. Knowsley | 98. Haigh |
| 46. Wigan | 99. Gillars Green, Eccleston (S) |
| 47. Shevington | 100. St Helens, Eccleston (S) |
| 48. Shevington | 101.* Kirkless, Aspull |
| 49. Shevington | 102.* Kirkless, Ince |
| 50. Sutton | 103. Haigh |
| 51. Whiston | 104.* Kirkless, Aspull |
| 52. Hackley Moss, Eccleston (S) | 105. Skelmersdale |
| 53. Prescot | 106. Hardshaw, Windle |
| 54. Adlington | 107. Blackrod |
| 55. Wrightington | 108. Lathom |
| 56. Lathom | 109. Wiredale, Windle |
| 57. Chisnall, Coppull | 110. Blackrod |
| 58. Huyton | 111. Ellerbeck, Duxbury–Blackrod |
| 59. Thatto Heath–Ravenhead, Sutton | 112. Arley, Haigh–Blackrod |
| 60. Tarbock | 113. Stocks, Haydock |
| 61. Dalton | 114. Parr |
| 62. Eccleston (S) | 115. Hardshaw, Windle |
| 63. Knowsley | 116. Ashes, Windle |
| 64. Rainford | 117. New Hall, Ashton-in-Makerfield |
| 65. Edge Green, Ashton–Golborne | 118. Laffack, Parr–Haydock |
| 66. Wigan | 119. Ashton-in-Makerfield |
| 67. Parr (?) | 120. St Helens, Sutton |
| 68. Adlington | 121. Parr (?) |
| 69. Wigan | 122. Winstanley |
| 70. Wigan | 123. Haigh Park, Haigh |
| 71. Blaguegate, Lathom | 124. Guide Post, Haydock |
| 72. Sutton | 125. Florida, Haydock |
| 73. Sutton | 126. Ravenhead, Sutton |
| 74. Orrell–Up Holland | 127. Carr, Whiston |
| 75. Huyton | 128. Blackrod |

129. Ravenhead, Sutton
130. Rainford
131. Huyton
132. Naylor's, Orrell
133. Whiston
134. St Helens, Eccleston
135.* Kirkless, Ince
136. Hillock Street, Prescot
137. Dean, Orrell–Up Holland
138. Hindley
139. Ayrefield, Up Holland
140. Parbold
141. Blackrod
142. Orrell
143. Dean, Up Holland
144. Up Holland
145. Garswood, Ashton-in-Makerfield
146. Pemberton
147. Orrell
148. Chorley
149. Pemberton
150. Hardshaw, Windle
151. Hardshaw, Windle
152. Hardshaw, Windle
153. Golborne
154. Laffack, Parr–Haydock
155. Shevington
156. Skelmersdale
157. Hardshaw, Windle
158. Whiston
159. Gathurst, Orrell
160. Orrell
161. Parr
162. Dean, Up Holland
163. Broad Oak, Parr
164. Pemberton
165. Pemberton
166. Pemberton
167. Parr
168. Pemberton
169. Pemberton
170. Parbold
171. Highfield, Standish
172. Ince

# APPENDIX 3

# THE OUTPUT OF HAIGH CANNEL COLLIERY (2), 1725-73

The Haigh estate stewards' account books[1] contain a series of receipts of money paid in by the cannel pit auditors which is broken for only five years between 1725 and 1773. Exactly what these receipts represent in terms of output is not specified and the proportion of the total cannel colliery receipts that was paid in to the steward seems to have varied. It seems probable that mining expenses such as soughing, sinking, the purchase of ropes and the payment of auditors' wages were not deducted by the auditors from their takings because these were paid regularly from the stewards' accounts after the auditors' cash payments had been received. However, colliers' wages[2] were only rarely paid from the stewards' accounts[3] and it is therefore probable that these expenses were usually met by the auditors before they made their payments to the steward. It is therefore not possible to determine exactly what proportion of total annual colliery receipts was represented by the auditors' cash payments into the stewards' accounts, nor to assume that these payments represented the total cash receipts for cannel sold by the auditors.

Although the output of the colliery cannot be derived from a straight division of the annual payments to the steward by the selling price of cannel, it is possible to derive the approximate annual output of the colliery from these cash receipts by a more circuitous method. An account of the actual output of the colliery between 1748 and 1755 is extant.[4] From a comparison of these output figures with the contemporary auditors' payments to the steward (table 27) it is apparent that in every year but one the value of the cannel raised exceeded the money received by the steward from the auditors in that same year. The average difference over the whole series was in the ratio of auditors' payments = 100: output value = 133. Although this average relationship has a relatively high standard deviation it allows the estimation of an approximate annual value.

TABLE 27. *A comparison of the auditors' cash payments to the steward and the value of output at Haigh cannel colliery (2), 1748-55*

| Year | Auditors' payments £. s. d | Output of cannel tons | baskets | Value of output at 5/10d per ton £. s. d |
|------|------|------|------|------|
| 1748 | 1,171. 1. 0 | 4,157 | 0 | 1,177. 0.10 |
| 1749 | 1,121. 2. 0 | 5,664 | 14 | 1,652. 0. 4 |
| 1750 | 1,313.17. 0 | 5,755 | 10 | 1,678.13. 1 |
| 1751 | 1,152.11. 0 | 7,075 | 10 | 2,068.13. 1 |
| 1752 | 404.17. 0 | 4,339 | 0 | 1,265.10.10 |
| 1753 | 1,331. 7. 0 | 3,205 | 0 | 938.19. 2 |
| 1754 | 1,121. 1.11 | 5,485 | 10 | 1,599.15.10 |
| 1755 | 1,117. 1. 3 | 4,341 | 0 | 1,266. 2. 3 |
| TOTAL | 8,732.18. 2 | 40,023 | 4 | 11,646.15. 5 |

253

For conversion to output totals these values must be divided by the current selling price of cannel. This introduces more problems, which contribute further to the degree of inaccuracy. It was possible to produce two types of cannel, 'round' and 'small', which sold at different prices. Although round cannel apparently constituted an important proportion of output in the seventeenth century[5] it seems, on the evidence of the 1748-54 accounts,[6] that a negligible amount was produced in the mid eighteenth century. A blanket assumption that only small cannel was produced has been made, of necessity, because the exact proportion of small and round cannel output is unknown. This assumption does, of course, introduce the possibility of overestimating production by an unknown amount because round cannel sold at roughly double the price of small cannel. No great body of price information exists. Small cannel sold at 2d per basket in 1687,[7] but the price had been raised to 2.5d by c.1730.[8] It has been assumed that this rise was instigated before 1725 when the accounts begin. This price remained current until 1746, when there was an increase of unspecified amount,[9] probably to the 3.5d per basket charged in 1747.[10] This price remained operative until at least 1755,[11] but no quotation of Haigh prices was made after this date. At the Kirkless colliery 5d per basket was charged in 1774,[12] but this was probably for round cannel and one would expect a higher price at Kirkless than at Haigh in 1774 because of its closer proximity to the Leeds and Liverpool Canal, opened in that year. The assumption that 3.5d remained the current price of Haigh small cannel throughout the period 1746-73 has therefore been made, though it allows some small possibility that outputs derived from it may overestimate actual output during the later years of the series.

The selling prices detailed above have been divided into the stewards' annual receipts from the cannel pit auditors after these had been increased in the ratio of 100:130. A five year moving mean of annual output totals was calculated to counteract the auditors' occasional practice of paying money into the stewards' account in years after it was received. The resulting totals have been plotted on figure 56. Because of the possibilities of error implicit in them, these totals must be used with circumspection and none but the most general conclusions can be drawn from them. They hold some value as good approximations because they are similar in amount and trend to the accounted totals of 1748-55. However, the statistical purist might prefer to ignore them and use only the stewards' receipts, which have also been plotted on figure 56.

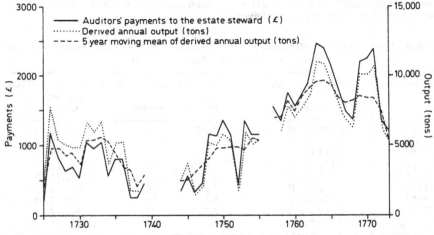

56. Haigh (2) receipts and outputs, 1725-73.

# APPENDIX 4

## THE LAND TAX ASSESSMENT RETURNS AS A SOURCE OF COLLIERY DATA

Returns have survived for nearly all the coalfield townships in almost every year after 1781 except 1798, for which there are none.[1] Those for ten townships, some in each of the three hundreds which overlap the coalfield, omit collieries which were recorded in other documents. In twenty-three other townships collieries were separately assessed, in ten of them in each year for which assessments have survived (figure 57). Once the assessors of a particular township began to record collieries they usually continued to do so, and suspicious gaps exist only in the returns of less

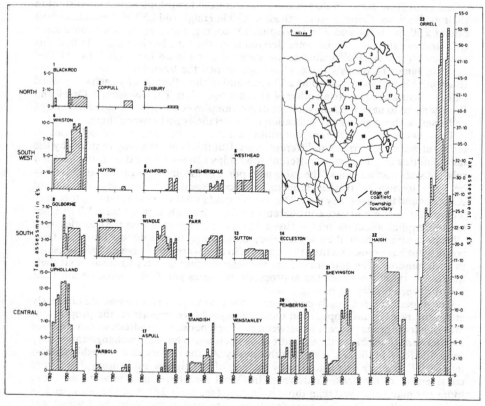

57. Colliery data from LRO, Land Tax Assessment Returns, 1782-99.

255

important mining townships. Thus, these records provide an almost comprehensive, if chronologically staggered, source of evidence for a large majority of the mining districts of the coalfield.

It is difficult to use them for any purpose beyond a simple tally of the presence of collieries. Assessments of the tax liabilities of fifty-eight collieries were entered, many for all the eighteen years, but they are of little value as indications of either the relative size and profitability of works in any one year or of the progress of individual collieries through time. There is a number of reasons for this, each bound up with the way in which the assessments were made.

According to the Act of 1693 under which the procedures were set up, money was to be levied on 'income arising from estates in ready money or debts, or in goods, wares, or any other personal estate . . . after deducting desperate debts and moneys bona fide owing'.[2] It is impossible to discover how the assessors determined the income which arose annually from most of the collieries within their jurisdiction. Usually, a tax liability was simply entered in the schedule against the occupier's name. However, further information was given at various times about the number and value of pits or the numbers of men employed in the collieries of Aspull, Standish, Orrell, Windle and Haigh (table 28). The tax assessed usually bore a direct relationship to those figures, so that in these cases it seems that the assessors gave the basis of their valuations.

It is clear from table 28 that assessors in different townships were inconsistent in their evaluations. Coalpits were valued at £20 in Haigh and £50 in Standish, cannel pits at £50 in Haigh and £76 in Aspull. Of course, these differences could have reflected variations in the incomes derived from the pits, but it is unlikely that this was completely so. Half collieries are entered in table 28 for Aspull and Haigh[3] because James Hodsons's works (101) overlapped the township boundary: in one township his pits were valued at £50 each and in the other at £76 each – identical values to those attached to all other pits in each township. The assessors in Orrell and Windle seem to have used the number of men employed as an index of the income of collieries, without giving any indication of the relationship between them.

Thus assessors in different townships estimated the incomes of collieries in different ways and arrived at different sums. But this is not the only reason why the tax liabilities of collieries in different townships cannot be used to estimate their relative sizes and incomes. The rates paid per £ of income also varied considerably between townships, further reducing any correspondence between the size of collieries and the taxes they paid. The use of a quota system in the taxation procedure meant that rates could vary considerably by the late eighteenth century. Ward is not entirely explicit about the method used, but it seems that each year the counties were allocated sums which they must raise; these were then split up between their constituent hundreds, and then these quotas in turn were split between the townships in each hundred. The assessors of the townships would then try to ensure that their quota was raised by allocating appropriate tax rates per £ of income within their jurisdiction, with Papists paying double.

The quota fixed nationally could vary from year to year and this would obviously affect the rates of tax imposed in the townships. More important, the proportions into which the quotas were split between counties, between hundreds within counties and between townships within hundreds do not seem to have been changed after they were first fixed in the late seventeenth century. As the income of a township rose so its rate fell and *vice versa,* so that by the end of the eighteenth century there were considerable differences. Coalowners in Orrell sometimes paid nearly four times more per man employed than those in Windle (see table 28), and the rates per £ in Haigh, Aspull and Standish were desultory compared to the 4/- in the £ which was supposed to apply nationally. Clearly, then, variations in the tax liabilities of

TABLE 28. Colliery assessments in the Land Tax Assessment Returns, 1783-99

| Township | Years | No. of Collieries* | Type | No. of Pits* | No. of Men* | Value per pit £. s. d | Assessment per pit £. s. d | Assessment per man £. s. d | Assessment per £ value £. s. d |
|---|---|---|---|---|---|---|---|---|---|
| Orrell | 1783 | 4 | coal | 4 | 16 | | 1. 9. 2 | 0. 7. 1½ | |
| | | | | 4 | 16 | | 1. 5. 0 | 0. 6. 3 | |
| | 1784 | 4 | coal | 3 | 12 | | 1. 7. 0 | 0. 4. 9 | |
| | | | | 7 | 28 | | 1.13. 4 | 0. 8. 4 | |
| | 1785 | 5 | coal | 6 | 27 | | 1.17. 6 | 0. 8. 4 | |
| | | | | 7 | 28 | | 1.13. 4 | 0. 8. 4 | |
| | | | | 2 | 6 | | 1. 5. 0 | 0. 8. 4 | |
| | 1786-7 | 9 | coal | 6 | 27 | | 1.16. 8 | 0. 6. 3 | |
| | 1788 | 5 | coal | | 167 | | | 0. 7. 9 | |
| | | | | | 18 | | | 0. 7.10 | |
| | | | | | 4 | | | | |
| | | | | | 51 | | | 0. 8. 4 | |
| | 1789-97 | 47 | coal | | 1111 | | | 0. 6. 3 | |
| | 1799 | 6 | coal | | 182 | | | 0. 3. 1½ | |
| Standish | 1753 | 1 | coal | 1? | | | | | |
| Windle | 1790-7 and 1799 | 26 | coal | | 245 | 50. 0. 0 | 1. 7. 1 | 0. 2. 2¼ | 0. 0. 6½ |
| Haigh | 1797 and 1799 | 2 | coal | 2 | | 20. 0. 0 | 0.12. 6 | | 0. 0. 7½ |
| Aspull | 1794 | 3 | cannel | 9 | | 50. 0. 0. | 1.11. 3 | | 0. 0. 7½ |
| | 1796, 1797 | 1½ | cannel | 4 | | 76. 0. 0 | 1. 2. 2 | | 0. 0. 3½ |
| | and 1799 | 4½ | cannel | 9 | | 76. 0. 0 | 1. 3. 9 | | 0. 0. 3¾ |

* If for more than one year or for collieries with identical rates of assessment, the figures for each year or colliery have been cumulated

collieries between townships do not reflect differences in their incomes at all accurately.

If the income of a property grew or shrank the assessors should have altered its valuation, which they did in Orrell and Windle (see table 28). However, rates also varied through time as the total income of townships changed, so that the liability of a colliery could change even if its income remained constant, or stay fixed even if its income changed. In Windle this seems not to have occurred; there, the rate per man employed was unchanged at 2/2¼d throughout the 1790s. However, in Orrell the rate was halved in 1799 and changed a number of times in the 1780s (see table 28). In Skelmersdale (table 29) the rate must have changed more frequently in the 1790s because the tax liability often altered in ways unrelated to the movements of profit, output (and therefore royalties) or receipts[4] (table 29). Changes in tax liabilities cannot, therefore, be used to chart the ways in which colliery income and output changed.

TABLE 29. *Output, receipts, profit and land tax at Skelmersdale colliery (105), 1793-9*

| Year assessed | Output in tons | Receipts £. s. d | Profit or loss £. s. d | Tax £. s. d |
|---|---|---|---|---|
| 1793 | 1634 | 350.11. 0½ | − 4. 4. 6 | 1. 3. 3 |
| 1794 | 1023 | 346. 2. 3½ | 155.16. 9½ | 0. 9. 8¼ |
| 1795 | 2250 | 515. 5. 2 | 178. 2. 5½ | 0.12. 6 |
| 1796 | 2806 | 360.11. 9½ | 231.13. 1½ | 0.18. 1½ |
| 1797 | 3318 | 678.16.10½ | 86. 4. 5 | 1.10. 2½ |
| 1798 | 9456 | 1435.12. 2½ | 295.14. 9½ | |
| 1799 | 8643 | 1415. 4. 1 | 316.15. 2 | 1. 3. 2½ |

However, on the evidence of the Windle and Orrell data it does seem that the liabilities accurately reflect the relative sizes of collieries within any particular township in any particular year. Furthermore, the pit and manpower totals recorded for Haigh, Aspull, Windle and Orrell provide surrogates which can be converted to outputs through the productivity figures given in tables 14 and 15. The average outputs per man year of two Orrell collieries (74 and 93) at the end of the century can be calculated from accounted production and the numbers of men given in the Land Tax.[5] Unfortunately, there is a considerable difference between the 704 tons of the men at the Holme colliery (74) and the 1357 tons of those at the Blundell works (93). The explanation of this discrepancy is probably that the Holme colliery (74) was very small compared with that of the Blundells (93) and, indeed, all the others in Orrell: in almost all years it had the lowest number of men recorded in the assessments of both Orrell and Windle. Its four to eight men per year compare with between sixteen and forty, usually over thirty at the Blundell colliery. Apart from its steam engine, it was typical of the collieries working in Orrell before the Leeds and Liverpool Canal was built. The Blundell works (93) was characteristic of the large speculative ventures built by Liverpool and Bradford capitalists along the canal after the early 1780s and it was one of those where the Lancashire system of getting was pioneered.[6]

Thus despite the discrepancy between them, the output per man year figures of these collieries have been used to convert the manpower data of other Orrell and Windle collieries to estimates of annual outputs (collieries 137, 142, 147, 151, 152, 157, 159 and 160). The choice of average was necessarily arbitrary. Where the number of men recorded in the Land Tax Assessments was less than ten in a particular year, the Holme (74) multiplier was used; if it was greater, that of the Blundell colliery (93) was applied. Obviously, the results are only broad approximations but they should, especially when averaged over a number of years, give a reasonable indication of the relative size of the biggest collieries on the coalfield in the 1790s.

# NOTES

For abbreviations used in notes, see bibliography

## 1. Introduction

1. See, for example, C. H. Lee, *Regional Economic Growth in the United Kingdom since the 1880s* (London, 1971) and H. W. Richardson, *Regional Economics* (London, 1969). The degree of interest shown by historical scholars in such problems is still not commensurate with Isard's contention that 'a well balanced economic history must explicitly consider spatial processes and the evolving web of interregional relationships'. W. Isard, 'Notes on the use of regional science methods in economic history', *Jour. Econ. Hist.*, 20 (1960), 597.
2. A. Thompson, *The Dynamics of the Industrial Revolution* (London, 1973).
3. For some examples of criticisms of the methodology of historical geography made by economic historians and geographers see A. R. H. Baker (ed.), *Progress in Historical Geography* (Newton Abbot, 1972); A. R. H. Baker, J. D. Hamshere and J. Langton, 'Introduction' in *Geographical Interpretations of Historical Sources* (Newton Abbot, 1970); M. J. Bowden and W. A. Koelsch, 'Reviews in historical geography', *Econ. Geog.*, 46 (1972), 214-16; H. P. R. Finberg, review in *Agri. Hist. Rev.*, 17 (1969), 77-8; M. M. Postan, 'The maps of Domesday', *Econ. Hist. Rev.*, 2nd ser., 7 (1954-5), 98-100; E. L. Jones, review in *Econ. Hist. Rev.*, 2nd ser., 16 (1963-4), 571-2; J. A. Swanson, review in *Econ. Hist. Rev.*, 2nd ser., 22 (1969), 171; and J. E. Vance, *The Merchant's World: The Geography of Wholesaling* (Englewood Cliffs, 1970), p.10.

    The target of these criticisms is the mapping or written description of spatial patterns just for the sake of describing them. Their flavour is conveyed by the following extracts. 'The historical geographer lives through the looking glass, where resources have distributions but not functions and time exists only in flashes' (Jones, Review). 'Historical geography . . . seems largely to shun any strictures other than parochial and to lack structural organization other than chronological' (Vance, *Merchant's World*).
4. D. H. Fischer, *Historians' Fallacies* (London, 1971), p.1n.
5. See A. R. H. Baker, 'Historical geography in Britain', in Baker, *Progress in Historical Geography*, pp.90-110, where the intellectual introspection of historical geographers since the Second World War is stressed.
6. G. Olsson, review in *Jour. Reg. Sci.*, 8 (1968), 253-5 and D. Harvey, *Explanation in Geography* (London, 1969), p.8.
7. For a stimulating discussion of this issue see I. Berlin, 'History and theory: the concept of scientific history', *History and Theory*, 1 (1960-1), pp.1-33.
8. M. Beckmann, *Location Theory* (New York, 1968), pp.9-10.
9. See D. M. Smith, *Industrial Location: An Economic Geographical Analysis* (London, 1971) for a full elucidation.

10. A. Weber (trans. C. J. Friedrich), *Theory of the Location of Industries* (Chicago, 1929) and A. Lösch (trans. W. H. Woglom), *The Economics of Location* (New York, 1967).
11. Lösch, *Economics of Location*, p.17. Each theoretical assumption is enumerated for the sake of clarity.
12. *Ibid.*, p.139.
13. Lösch, *Economics of Location*, pp.109-14, and Beckmann, *Location Theory*, p.28.
14. Lösch, *Economics of Location*, pp.109-14.
15. *Ibid.*, p.94.
16. *Ibid.*, p.95.
17. *Ibid.*, p.96. Beckmann's explanation of this set of circumstances is as follows. 'Under a system of mill pricing, an additional plant will reduce the demand – at a given mill price – for the product of any plant whose market area is contiguous to that of the new plant. The new plant thus shifts the demand curves at the neighbouring plant downwards.' (Because the transport cost component of the consumer prices is smaller for purchasers nearer to the new plant than to the old ones, less will be demanded from pre-existing plants in proximity to the new one.) 'As this is repeated, the demand curves will eventually just barely exceed or be tangential to the average cost curves. The industry then has no room for further plants. It has reached an equilibrium under monopolistic competition with a maximum number of plants.' Beckmann, *Location Theory*, p.35.
18. Lösch, *Economics of Location*, pp.109-14.
19. *Ibid.*, pp.27-9.
20. *Ibid.*, pp.99n-100n.
21. Anne Brontë, *The Tenant of Wildfell Hall* (London, 1967), p.274.
22. D. Harvey, personal communication.
23. W. H. Long, 'Demand in space: some neglected aspects', *Papers Reg. Sci. Assoc.*, 27 (1971), 45-60.
24. *Ibid.*, p.52
25. 'The many uncertainties, and forms of incurring losses, explain why price stability is so common, apart from the price changes involving discrimination against nearby buyers over whom a seller may have a certain degree of monopolistic control. The encroaching attempts on a rival's market area . . . probably develop only where the boundary zone contains a large cluster of buyers sensitive to price changes or whose custom is necessary for the attacking firm to break even.' Richardson, *Regional Economics*, p.36.
26. See appropriate sheets of *The Geological Survey of England*; R. C. B. Jones, C. H. Tonks and W. B. Wright, *The Geological Survey Memoir to the Wigan Sheet (no. 84)* (London, 1938); and Ministry of Fuel and Power, *Northwestern Coalfields: Regional Survey Report* (London, 1945).
27. It is difficult to analyse the repercussions of variations in royalty payments on costs because they might differ according to the quality of coal, or they could include payments for good sites or fixed capital installed before the lease was taken out. Whether or not they did so is rarely recorded, so that wrong conclusions might be drawn about the cost of leased reserves from an examination of colliery rents.
28. See H. C. Prince, 'Real, imagined and abstract worlds of the past', *Progress in Geography*, 3 (1971), 1-86.
29. See M. J. Webber, *The Impact of Uncertainty on Locations* (Cambridge, Mass., 1972).

30. W. Isard, *et al., General Theory: Social, Political, Economic and Regional* (Cambridge, Mass., 1969).
31. D. Harvey, 'Conceptual and measurement problems in the cognitive behavioral approach to location theory', in K. R. Cox and R. G. Golledge (eds.), 'Behavioral problems in geography: a symposium', *Northwestern University Studies in Geography*, 17 (1969), 35-67.
32. B. F. Duckham, *A History of the Scottish Coal Industry, 1700-1815* (Newton Abbot, 1970), pp.113-70.
33. For a discussion of the value and limitations of the systems mode of conceptualisation which is expressed by diagrams such as this one, see J. Langton, 'Potentialities and problems of adopting a systems approach to the study of change in human geography', *Progress in Geography*, 4 (1972), 127-79.
34. *Ibid.*
35. J. B. Harley, 'Changes in historical geography: a qualitative impression of quantitative methods', *Area*, 5 (1973), 69-74.

## 2. The colliery location pattern, 1590-1689

1. J. U. Nef, *The Rise of the British Coal Industry* (2 vols., London, 1932), vol. I, pp.19-20.
2. This is a more appropriate conclusion than that advanced on the basis of the upper graph alone in J. Langton, 'Coal output in south-west Lancashire, 1590-1799', *Econ. Hist. Rev.*, 2nd ser., 25 (1972), pp.29-30.
3. For the method of calculation and mathematical properties of this statistic see D. S. Neft, *Statistical Analysis for Areal Distributions* (Philadelphia, 1966), pp.19-55 and J. F. Hart, 'Central tendency in areal distributions', *Econ. Geog.*, 30 (1954), pp.48-59.
4. The standard distance statistic, expressing the amount of dispersion around the centroid in terms of grid-square units, also varied considerably, ranging from a minimum of 9.53 in the 1600s to a maximum of 11.27 in the 1680s.
5. 'Twin collieries' existed in every decade but two, e.g. (3) and (4) in the 1590s; (11) and (14) in the 1600s, 1610s and 1620s; (24) and (25) in the 1630s; (47), (48) and (49) in the 1660s, 1670s and 1680s. Collieries (2) and (12) also worked adjacent reserves on occasions between 1600 and 1640.
6. LRO, DDHe/40/74 and 15/13.
7. PRO, PL6/13/103.
8. PRO, DL1/211/G/6, DL4/75/13 and 79/26, and DL6/9/39.
9. The lease of the Prescot Hall demesne from King's College, Cambridge, to Philip Layton allowed him to mine, and he sublet the coal in 1594-7. However, later lessees were given only the right to mine coal for their own domestic use. KCC, 1/V/18 and 23, and Ledger Books no. 3, p.396 and no. 4, pp.202 and 246, and PRO, DL1/183/L/5.
10. F. A. Bailey, 'Early coalmining in Prescot, Lancashire', *Trans. Hist. Soc. of Lancs. and Chesh.*, 99 (1947), 17.
11. WPL, KP24, fol. 32.
12. WPL, Bridgeman Ledger, 1616-42, fols. 41, 62, 68 and 71.
13. In 1635 Bishop Bridgeman pronounced that 'I doe now forbid all & every of the inhabitants of . . . [Wigan] . . . to dig for Coles or to make any Soughs under any of the Streets, or any pt. of the Wast . . . as they will answer it at their perills'. King's College imposed fines against copyhold miners in Prescot in

1636. WPL, Bridgeman Ledger, fol. 213 and KCC, 1/V/26 and Mundum Book, 1637.
14. Suits were filed by the antagonistic owners of the following collieries: (1) and (7), 1598-1605 (PRO, DL1/195/S/10, 204/O/1, and 205/S/27); (2) and (12), 1601 and 1635 (PRO, DL1/211/G/6 and DL4/89/25); (3) and (4), 1588-1634 (PRO, DL4/62/5 and DL5/31/600 and St HRL, 942/73/LC); (11) and (14), 1606 and 1624 (PRO, DL4/64/22 and PL6/7/74); (22) and (54), 1687 (PRO, PL6/39/53); (29) and (43), 1656 (PRO, PL6/23/35); (36) and (51), 1685 (PRO, PL6/38/32); and (52) and (53), 1686 (WPL, Wrightington MSS/Eccleston Box/Old Deeds relating to Eccleston and Sutton from the Reign of Henry VIII/Case heard at Preston, Eccleston v. Ogle, Alcocke *et al.*, 1686).

In addition the proprietors of the following collieries were involved in suits filed against their workings by aggrieved lessors, relatives, neighbours or employees: (3), 1588-1636 (PRO, DL4/62/5, DL5/31/600 and St HRL, 942/73/LC, which lists the proceedings of four suits that were filed between 1588 and 1636); (5), 1597 and 1649 (PRO, DL1/183/L/5 and PL6/19/81); (7), 1591 (PRO, DL1/183/L/13); (12), 1626, 1627 and 1629 (PRO, DL4/75/13, PL6/9/11, and DL4/79/26); (13), 1624 (PRO, DL4/74/23); (17), 1671 (LRO, DDTa/281/Brooks v. Oldfield, Jan. 22, 1671); (24), 1636 (PRO, PL6/13/103); (25), 1636 (PRO, PL6/17/103); (34), 1642 (PRO, DL4/118/10); (39), 1649 (WPL, KP24, fol. 40); (42), 1655 (*ibid.*, fol. 32); (47), 1688 (PRO, PL6/30/108); (48), 1688 (*ibid.*); and (52), 1686 (PRO, PL6/38/108).
15. LRO, DDBa/Coal Pitt Accompts, 1676-96. As late as the 1740s a life-time of 'a little over three years' was normal for the pits of Long Benton colliery in Durham. T. S. Ashton and J. Sykes, *The Coal Industry of the Eighteenth Century*, 2nd edn (Manchester, 1964), p.11. See also L. Stone, 'An Elizabethan coalmine', *Econ. Hist. Rev.*, 2nd ser., 3 (1950-1), 97-106.
16. PRO, DL4/118/10.
17. WPL, Haigh Colliery Orders, text as beginning at end A, fol. 12 r and v.
18. LRO, DDHe/40/69 and 74; PRO, DL4/47/19 and 74/24; WPL, Account Book of Francis Sherrington, fol. 411 r and v. In two of these leases in Shevington it was covenanted that the lessees 'will keepe three getters Constantly at work if $y^e$. $s^d$. sale will Cary $y^e$. $s^d$. Coles away'.
19. 'Lancashire Royalist composition papers, 1643-60', *Rec. Soc. of Lancs. and Chesh.*, 29 (1894), 38; PRO, DL4/89/24 and WPL, Haigh Colliery Orders, end A, fols. 5r, 6r and 7r and end B, fols. 8v and 9 r and v.
20. Langton, 'Coal output'.
21. WPL, KP24, fol. 32 and Wrightington MSS/Eccleston Box/Old Deeds relating to Eccleston and Sutton . . . / Case heard at Preston . . . 1686, and PRO, PL6/7/74.
22. Two pits operated at Winstanley (6) in 1681, although usually there was only one. 'Coal pits' existed at Haigh (2) in 1595, and two pits worked there in 1636, three in 1635 and c.1675. 'Pits' were worked at Aspull (12) in 1626 and 1627, and two operated together in 1635. At Hindley (13), 1605-24 and 1685-95, and at Park Lane (40), 1665-76, 'coal pits' existed, and at least two pits were worked simultaneously at Eccleston (52) in 1686. 'Twoe eyes or shaftes for getting of Coale, and twoe Croppe pitts for the drawing of Ayre' were sunk in Sutton in c.1614. LRO, DDBa/Coal Pitt Accompts, 1676-96 and J. M. H. Bankes, 'Records of mining in Winstanley and Orrell . . .', *Trans. Hist. Soc. of Lancs. and Chesh.*, 54 (1939), 36 and 38; PRO, DL1/169/H/7, DL4/75/13 and

WPL, Haigh Colliery Orders, end A, fol. 12 r and v and end B, fols. 8v and 9r; PRO, PL6/9/11 and 15, DL4/74/24 and PL6/44/11; LRO, DDGe(M)/1184; WPL, Wrightington MSS/Eccleston Box/Old Deeds relating to Eccleston and Sutton/Case heard at Preston . . . 1686, and PRO, DL4/62/5.

23. A list of hewers is given in WPL, Haigh Colliery Orders, end A, fols. 8v and 9r. It is not clear whether this was made in 1636, 1664 or 1665, but a comparison with a list of hewers fined for transgressing the Orders given on end A, fol. 5r indicates that the date of the other list is also probably 1636.

24. WPL, Haigh Colliery Orders, end A, fol. 14. This clause was not included in the issues of the Orders dated 1636 and 1664.

25. WPL, KP24, fol. 32.

26. KCC, 1/V/16. It was reported 'yᵗ one Fletcher hath gotten Coles to the value of xx li in a pece of copyhould grounds'. Coal sold at about 1/2d per ton in Prescot at that time.

27. WPL, Bridgeman Ledger, fols. 41, 62, 68 and 71.

28. KCC, 1/V/15.

29. LRO, DDBa/Coal Pitt Accompts, 1676-96.

30. PRO, DL4/79/26.

31. PRO, DL4/118/10.

32. LRO, DDBa/Coal Pitt Accompts, 1676-96.

33. PRO, DL4/79/26.

34. LRO, DDBa/Coal Pitt Accompts, 1676-96.

35. Bailey, 'Coalmining in Prescot', p.17.

36. Output figures are quoted for summer only, or for a six month period, in PRO, PL6/30/105 and 'Lancashire Royalist composition papers', p.38.

37. See Langton, 'Coal output' for details of the derivation of these figures. The coalfield as defined here includes two of Nef's colliery districts. His estimates of a total output of 50,000 to 100,000 tons in 1700 and a fifteenfold rate of increase between the mid-sixteenth century and 1700 seem very high. See Nef, *Coal Industry*, vol. I, p.64.

38. See J. A. Bulley, 'To Mendip for coal', *Proc. of Somerset Archaeol. and Nat. Hist. Soc.*, 97 (1952), 46-78; G. G. Hopkinson, 'The development of the South Yorkshire and North Derbyshire coalfield, 1500-1775', *Trans. Hunter Archaeol. Soc.*, 7 (1951-7), 295-319; Stone, 'An Elizabethan coalmine', pp.97-106, and C. D. J. Trott, 'Coalmining in the borough of Neath in the seventeenth and early eighteenth centuries', *Morgannwg*, 13 (1969), 47-74.

39. A. W. A. White, 'Sixty years of coalmining enterprise on the north Warwickshire estate of the Newdigates of Arbury', unpubl. MA thesis (Birmingham, 1969), p.244. The receipts in 1687 suggest, at Warwickshire prices, an output of over 8000 tons.

40. R. S. Smith, 'The Willoughbys of Wollaton, 1500–1643, with special reference to early coalmining in Nottinghamshire', unpubl. PhD thesis (Nottingham, 1964), pp.99-100. Even in the sixteenth century Wollaton colliery raised between 6000 and 13,000 tons a year.

41. A. W. R. Moller, 'Coal mining in the seventeenth century', *Trans. Roy. Hist. Soc.*, 4th ser., 8 (1925), 88. Nearly 20,000 tons a year were raised at Whitehaven colliery in the later years of the seventeenth century.

42. Nef, *Coal Industry*, vol. I, p.26. Winlaton colliery raised 20,000 tons in 1581/2.

43. Bulley, 'To Mendip' (1952), pp.47-8.

## 3. Factors affecting location pattern, 1590-1689

1. Nef, *Coal Industry,* vol. I, p.103.
2. T. S. Willan, 'The navigation of the river Weaver in the eighteenth century', *Chetham Society Publications,* 3rd ser., 3 (1951); LRO,DDCr/Bundle 35E/ Money paid for the Revd. Mr. Suddell, and WPL, KP30, fol. 14v.
3. W. Harrison, 'Pre-turnpike highways of Lancashire and Cheshire', *Trans. Lancs. and Chesh. Antiq. Soc.,* 9 (1891), 113, and J. Harland (ed.), 'The house and farm accounts of the Shuttleworths . . . 1582-1621', *Chetham Society Publications,* 35, 41, 43 and 46 (1856-8).
4. Nef, *Coal Industry,* vol. I, p.381.
5. Harrison, 'Pre-turnpike highways', p.105.
6. Harland, 'Shuttleworth accounts', *passim.*
7. StHRL, 942/73/LC/Duchy of Lancaster Depositions, 30 Eliz I, 25. The Haigh tenants could only take the coal to which they had a right from their lord's pit between May 3 and Aug. 1. A. J. Hawkes, 'Sir Roger Bradshaigh of Haigh, knight and baronet, 1628-84', *Chetham Society Publications,* N.S., 109 (1945), 3.
8. LRO, DDM/1/63 and 80.
9. Sale time began in March. LRO, DDBa/Coal Pit Accompts, 1676-96.
10. Nef, *Coal Industry,* vol. I, p.196.
11. J. Files, 'Mining with special reference to Lancashire', *Trans. Inst. Min. Eng.,* 92 (1936), 292.
12. Nef, *Coal Industry*, vol. I, p.107.
13. Hawkes, 'Sir Roger Bradshaigh', p.3
14. LRO, DDBa/Coal Pit Accompts, 1676-96.
15. Harland, 'Shuttleworth Accounts'.
16. LRO, DDBa/Coal Pit Accompts, 1676-96.
17. LRO, DDM/1/63.
18. PRO, EDC/179/250/11.
19. E. C. K. Gonner, 'The population of England in the eighteenth century', *Jour. Roy. Stat. Soc.,* 76 (1913-14), 287-8.
20. PRO, EDC/179/250/111.
21. PRO, EDC/179/132/355.
22. W. G. Hoskins, *Local History in England* (London, 1959), p.239.
23. The area around Wigan, where there was much fighting in the Civil War, was badly ravaged by plague in the 1640s. There were 2000 poor, 'with no bonds to keep in the infected and hunger-starved . . . whose breaking out jeopardeth all the neighbourhood, some of them, already being at the point to perish through famine, have fetched in and eaten carrion . . . to the destroying of themselves and increasing of the infection . . .' G. T. O. Bridgeman, 'The history of the church and manor of Wigan', *Chetham Society Publications,* N.S., 15-18 (1888, 1889 and 1890), pt 2, p.462.
24. D. Defoe, *A Tour Through the Whole Island of Great Britain* (Harmondsworth, 1971), pp.540-1.
25. Lancashire itinerary of Dr. Kuerden, c.1695, in J. P. Earwaker (ed.), *Local Gleanings Relating to Lancs. and Chesh.,* vol. I (1876), pp.209 and 211.
26. At a rate of consumption of 90 cwt per household, or 45 cwt per hearth per year, there was a market for about 50,000 tons in the region mapped in figure 8, somewhat less than half of it on the coalfield itself.
27. White, 'Newdigates of Arbury', pp.8-9.
28. PRO, DL1/211/G/6.

29. PRO, DL1/195/S/10.
30. WPL, Up Holland MS/Survey of the Manor of Upholland.
31. Figure 10 has been compiled from the 'Indexes to the wills and inventories now preserved in the court probate at Chester', *Rec. Soc. of Lancs. and Chesh.*, 2 (1879), 4 (1881), 15 (1887) and 18 (1888).
32. B. W. Awty, 'Charcoal ironmasters of Lancashire and Cheshire, 1600-1785', *Trans. Hist. Soc. of Lancs. and Chesh.*, 109 (1957), 72 and 81 and WPL, KP38, fol. 21.
33. Awty, 'Charcoal ironmasters', p.82 and app. 1 and WPL, KP24, fol. 19.
34. G. H. Tupling, 'The early metal trades and the beginnings of engineering in Lancashire', *Trans. Lancs. and Chesh. Antiq. Soc.*, 61 (1949), 21.
35. Kept in the Lancashire Record Office.
36. Tupling, 'Early metal trades', p.21.
37. W. C. Sprunt, 'Old Warrington trades and occupations', *Trans. Lancs. and Chesh. Antiq. Soc.*, 61 (1949), 167-9.
38. J. Picton, *Memorials of Liverpool* (2 vols., London, 1873), vol. II, p.131.
39. C. N. Parkinson, *The Rise of the Port of Liverpool* (Liverpool, 1952), p.50.
40. Hawkes, 'Sir Roger Bradshaigh', pp.19-20.
41. H. B. Rodgers, 'The market area of Preston in the sixteenth and seventeenth centuries', *Geog. Studies*, 3 (1956), 46-55.
42. WPL, Court Leet Rolls.
43. Bridgeman, 'Church and manor of Wigan', pt 2, p.309.
44. F. H. Cheetham, 'The church bells of Lancashire', pt 4, *Trans. Lancs. and Chesh. Antiq. Soc.*, 45 (1928), 89-123.
45. WPL, Bridgeman Ledger, fols. 2, 68, 84, 87, 105, 171, 178 and 180.
46. WPL, Court Leet Rolls.
47. Inventory in the Lancashire Record Office.
48. J. Hatcher and T. C. Barker, *A History of British Pewter* (London, 1974).
49. R. J. A. Shelley, 'Wigan and Liverpool pewterers', *Trans. Hist. Soc. of Lancs. and Chesh.*, 97 (1945), 3-15.
50. WPL, Court Leet Rolls and Book of Common Council.
51. LRO, The Will and Inventory of Roger Browne of Wigan, Panmaker, viewed in 1632 and proved in 1633.
52. Hatcher and Barker, *Pewter*, p.221.
53. C. D. J. Trott, 'The historical geography of the Neath region up to the eve of the industrial revolution', unpubl. MA thesis (Wales 1964), pp.21, 22 and 27, and H. Hamilton, *The English Brass and Copper Trades* (London, 1926), p.338.
54. J. A. Twemlow, *Liverpool Town Books* (2 vols., Liverpool, 1918 and 1935), vol. I, p.264 and 400 and vol. II, p.168.
55. BMus, Add MSS/34/318, fol. 23.
56. G. Chandler, *Liverpool Under James I* (Liverpool, 1960), p.139 and R. L. Galloway, *Annals of Coalmining and the Coal Trade* (London, 1898), p.90.
57. 'The Moore rental', *Chetham Society Publications*, 12 (1847), 80 and 84.
58. A. W. R. Moller, 'The history of English coalmining, 1550-1750', unpubl. DPhil thesis (Oxford, 1923), pp.784-6.
59. Moller, 'English coalmining', pp.787-8 and 780.
60. *Liverpool Daily Post*, May 12, 1942.
61. LRO, QSP/211/25. As early as 1601 people from Ditton and Tarbock complained they were obstructed in the carriage of their coal by the state of the road in Whiston. 'Lancashire Quarter Sessions Records: Quarter Sessions Rolls, 1590-1606', *Chetham Society Publications*, N.S., 77 (1917).

62. LRO, The Wills and Inventories of Richard Case of Whiston, proved in 1646, and Henry Ashton of Whiston, proved in 1668.
63. A. F. Calvert, *Salt in Cheshire* (London, 1915), pp.88 and 138.
64. G. Owen, *The Description of Pembrokeshire* (1603; London, 1892), p.88 and Nef, *Coal Industry*, vol. I, p.157. For an interesting account of early opinions on coal fires see chapter 1 of Moller, 'English coalmining'.
65. Nef, *Coal Industry*, vol. I, p.158.
66. J. M. H. Bankes, 'James Bankes and the manor of Winstanley, 1595-1617' *Trans. Hist. Soc. of Lancs. and Chesh.*, 94 (1942), 71 and JRL, Legh MSS/Box 56/Letter from J. Brotherton, Oct. 26, 1630.
67. Nef, *Coal Industry*, vol. I, p.163
68. LRO, DDK/1601/3.
69. LRO, DDK/102/3.
70. LRO, DDK/302/20 and 1526/1.
71. PRO, St Ch Procs/8/248/1.
72. JRL, Legh MSS/Box D/Bundle A1/Summary of the evidence of Peter Legh relating to a dispute between himself and the Duke of Devonshire, 1725.
73. PRO, DL4/49/1 and 94/55.
74. H. T. Crofton, 'Lancashire and Cheshire coalmining records', *Trans. Lancs. and Chesh. Antiq. Soc.*, 7 (1889), 41.
75. See, for example, those of Robert Callon of Rainford, 1624; John Cooper of Lathom, 1639 and Thomas Sefton of Skelmersdale, 1646 and 1675, in LRO.
76. PRO, PL6/23/126.
77. These are Charles Wilson's words, used in *England's Apprenticeship* (London, 1966), pp.4 and 13.
78. WPL, KP30, fol. 14 and Bridgeman Ledger, fols. 184 and 186 illustrate the great concern over the passability of the road from Wigan to Haigh.
79. The geological information which follows has been taken from Jones, Tonks and Wright, *Geological Survey Memoir* and E. S. Moore, *Coal* (New York, 1955).
80. As pits were exhausted borings were occasionally sunk to discover whether seams extended into neighbouring land. The surveys carried out at Haigh throughout the period were probably the most thorough of their kind, and in neighbouring Aspull a search for cannel and the subsequent opening of the seam was reputed to have cost £3000 in the opening years of the seventeenth century. In both cases, however, the local presence of the seam was known before the search began. Few definitive records of *de novo* exploration exist, but there were fruitless borings off the coalfield at Kirkby in 1609, Sefton in 1662 and Liverpool in 1698. Trial shafts were sunk without profit on the Croxteth Park *Lenisulcata* outlier in 1609. See WPL, Haigh Colliery Orders, *passim;* PRO, DL4/75/13; LRO, DDM/4/1; PRO, PL6/23/12; *Liverpool Daily Post*, May 12, 1942, and LRO DDM/4/1 (Kirkby and Croxteth).
81. The pit of Sherrington's Orrell colliery (7), sunk in 1601, was only eight and a half yards deep and the shafts of nearby Orrell Hall colliery (1) which tapped the same seam must have been of about the same depth. Winstanley colliery (6) worked through shafts of fourteen and twenty-four yards in 1676 and the latter was said to be of 'more than usuall Depth'. The pit sunk by the Gerards in Parr in 1655 (43) was twenty yards deep and the shafts of Eccleston's colliery on Hackley Moss (52) were sixteen yards deep in 1686. Though the Great Sough at Haigh reached a depth of forty-eight yards at one point, there were only '32 yeards to the levell of the Mayne Cannell delf . . . so that we have 16 yards full to spare'. If the cannel was worked at this depth, the Haigh pits must

have been the deepest on the coalfield. Compared with contemporary shafts of forty-eight yards at Brandon in Durham the South West Lancashire pits were shallow, but they were similar to the East Lothian pits of twelve yards and only slightly shallower than the Pembrokeshire pits of twenty-four, twenty-eight and forty yards. The Lancashire figures, however scant, do not support Nef's claim that 'before 1700 mines at least twenty fathoms deep were common in nearly all the British coalfields'. See WPL, Account Book of Francis Sherrington, fol. 41; LRO, DDBa/Coal Pitt Accompts, 1676-96; LRO, DDGe(M)/844; WPL, Wrightington MSS/Eccleston Box/ Old Deeds relating to Eccleston and Sutton . . . /Case heard at Preston . . . 1686; WPL, Haigh Colliery Orders, end A, fol. 10; Owen, *Description*, p.90; and Nef, *Coal Industry*, vol. I, pp.350-1.

82. The eight and a half yard shaft sunk at Orrell colliery (7) in 1601 cost only £4. 10. 0 and the abortive sinkings at Kirkby and Croxteth in 1609 cost only 14/-. Although the exceptionally deep Winstanley pit (6) of 1678 cost £22. 15. 6¾ the expense was said to have been 'far more than usuall' because of depth and geological problems. Sinking costs more typically amounted to 15/- per yard at Winstanley (6) in 1678, plus 5/- towards 'boring if occasion be to drain the pit'. See WPL, Account Book of Francis Sherrington, fol. 41; LRO, DDM/4/1; LRO, DDBa/Coal Pit Accompts, 1676-96; WPL, Haigh Colliery Orders, end A, fol. 12, and PRO, PL6/23/126.

83. PRO, DL1/169/H/7. The owner of coal under a piece of land occupied by someone else in Ashton in 1654 had the right to dig and make pits anywhere he thought fit, all rights of passage over the land in question to and from the pits for carting coal, slack, rubbish, stone and timber, the right to dump slack, rubbish and water onto the land, to lay 'the said Coales Cannell or Stonnes uppon the Banke in loads pyles or other wyse uppon all or any of the said Landes which . . . may be most beneficial for the uttering, venting and selling of the coal', and liberty to make soughs or trenches to carry away the water from the mine into the hollows or soughs of any other mines in Ashton. LRO, DDGe(M)/95.

84. WPL, Bridgeman Ledger, fol. 233 and KCC, 1/V/5 and Mundum Book, 1637.

85. *Trans. Rec. Soc. of Lancs. and Chesh.*, 32 (1896), 160-71; StHRL, 942/73/LC; PRO, DL4/94/55 and St Ch Procs, 8/248/1; WPL, Wrightington MSS, Eccleston Box/Old Deeds relating to Eccleston and Sutton . . . / Case heard at Preston . . . 1686; LRO, DDK/295/7 and PRO, PL/6/38/107.

86. Before c.1600 water drained naturally from the Sherrington's coal in Orrell (7). No water problem was encountered in the Bankes' mines in Winstanley (6) before the sinking of an abnormally deep shaft, but this may have put drainage beyond the reach of a pre-existing sough rather than introduced the difficulty. In a Shevington (48) lease of 1666 'coles both wet and dry' were mentioned, though the use of a sough at this colliery suggests that most of the reserves were in the former category. PRO, DL1/195/S/10; LRO, DDBa/Coal Pitt Accompts, 1676-96, and LRO, DDHe/40/69.

87. As early as 1573 a sough was needed to drain the seams of James Worsley and Edmund Winstanley in the Brockhurst area of Winstanley. Bankes, 'Records of mining', p.33.

88. The total cost of sinking and operating an Orrell colliery (7) for an unspecified time in 1601 was £13. 16. 10, and £3. 9. 4 was paid to various workmen 'for winding water'. WPL, Account Book of Francis Sherrington, fol. 41.

89. A horse in good harness could pull roughly fifteen times as strongly as a man, but it cost about four times as much to feed and required labour to superintend

it. T. K. Derry and T. I. Williams, *A Short History of Technology* (Oxford, 1960), p.249.

90. At Sutton Heath (3 and 4) at the turn of the sixteenth century; at Hindley (13) between 1604 and 1624; at Goose Green (11 and 14) 1623-4, and possibly at two Whiston collieries (36 and 51) in the late seventeenth century. StHRL, 942/73/LC and PRO, DL4/62/5; PRO, DL4/74/24; PRO, PL6/7/74, and LRO, the Inventories of Henry Ashton and Henry Lathom, both of Whiston, 1688 and 1691.

91. Buckets and chains are mentioned in connection with the six engines that worked in Sutton, Whiston and Pemberton.

92. For a description of the different kinds of drainage machines in use at this time see G. Agricola (trans. H. C. and L. H. Hoover), *De Re Metallica* (1556; London, 1912), pp.172-96. Small stones were thrown into the Hindley (13) machine in an attempt to ruin it; this would have done considerable damage to rag and chain and suction pumps, but not to a chain and bucket engine. PRO, DL4/74/24.

93. PRO, DL4/62/5.

94. LRO, DDHe/59/84 and 40/74. By the end of the century colliery concessions in both townships comprised groups of properties in different ownership, assembled so that soughs could be dug.

95. LRO, The Inventory of Henry Ashton of Whiston, 1668 and PRO, DL4/74/24. In the Trent valley a rag and chain pump cost £40 in 1610 and an engine drawing fifteen tons of water an hour, built in the same year in South Wales, cost £100. Nef, *Coal Industry* vol. II, app.O.

96. LRO, DDGe(M)/1184 and DDBa/Coal Pitt Accompts, 1676-96.

97. WPL, Haigh Colliery Orders, end A, fol.8 and 'Lancashire Royalist composition papers', pp.34 and 44.

98. Before the sough was built wheels and pumps were employed, but large-scale flooding occurred. WPL, Haigh Colliery Orders, *passim* and PRO, DL4/89/25. This sough was exceeded in length by few others in Britain and enjoyed considerable fame in the seventeenth and eighteenth centuries. Ashton and Sykes, *Coal Industry*, pp.33-4.

99. PRO, DL4/79/26.

100. WPL, KP24, fol. 32.

101. PRO, PL6/19/81.

102. LRO, The Will and Inventory of Peter Martlu of Lathom, 1689.

103. PRO, PL6/13/103 and 19/81.

104. PRO, DL4/89/25 and WPL, Haigh Colliery Orders, end A, fols. 5, 6 and 7, end B, fols. 8 and 9.

105. Nef, *Coal Industry*, vol. I, pp.411-28.

106. LRO, The Wills and Inventories of Peter Martlu and Thomas Brighouse, both of Lathom, 1689 and 1692.

107. PRO, PL6/19/81 and DL4/118/10.

108. WPL, KP24, fol.32 and Bridgeman, 'Church and manor of Wigan', pt 2, p.397.

109. The last two entries in the Haigh Colliery Orders are bonds of 1662 and 1669. At Winstanley (6) there was an annual hiring, but no evidence that bonds were signed there has survived. LRO, DDBa/Coal Pitt Accompts, 1676-96.

110. The original draft of the Orders was signed by the whole workforce, and in the hiring bonds colliers agreed to abide by 'such Articles as are Contained in this booke'. The Haigh colliers none the less remained 'the arrantest knaves in nature' to whom other stratagems, such as claiming a full week's wages when little time was worked, remained open. R. Challinor, *The Lancashire and Cheshire Miners* (Newcastle, 1972), p.13.

111. WPL, Haigh Colliery Orders, end B, fols. 8 and 9.
112. WPL, Haigh Colliery Orders, end A, fol. 9.
113. LRO, DDBa/Coal Pitt Accompts, 1676-96.
114. PRO, DL4/89/25; WPL, Haigh Colliery Orders, end B fol. 9. and LRO, QSP/229/18.
115. WPL, Haigh Colliery Orders, end B, fols. 8 and 9.
116. LRO, DDBa/Coal Pitt Accompts, 1676-96.
117. The auditors as well as the colliers signed the Orders with crosses or abstract symbols. Not even initials were used.
118. PRO, PL6/9/15.
119. PRO, PL6/9/15; LRO, DDHe/40/69 and DDBa/Coal Pitt Accompts, 1676-96.
120. LRO, DDBa/Coal Pitt Accompts, 1676-96 and WPL, Account Book of Francis Sherrington, fol. 41.
121. These rates seem to have been about the same as those paid on other coalfields at the same times. See Nef, *Coal Industry,* vol. II, pp.182 and 190.
122. WPL, Haigh Colliery Orders, *passim.* See also Account Book of Francis Sherrington, fol. 41, and LRO, DDBa/Coal Pitt Accompts, 1676-96.
123. A woman drawer was drowned at Aspull (12) in the 1630s and another was disabled after falling down a pit in Parr (43) in 1666. PRO, DL4/89/25 and LRO, QSP/299/18. Injuries were only usually recorded if colliers were so badly hurt that they had to apply at the Quarter Sessions for poor relief.
124. JRL, Haigh MSS/Box EIII/Bundle I/Account Book of Roger Bradshaigh, 1591-1624.
125. LRO, QSP/686/3 and 4 and 710/19. See also LRO, QSP/554/17, 107/17, 179/11 and 239/19. All these injury cases relate to colliers from the South and South West. It is curious that although many colliers from the Centre petitioned for other things, none did so because injury prevented them from working.
126. LRO, QSP/203/39, 412/9 and 710/16.
127. Colliers petitioned the Quarter Sessions because they and their families of up to six children had nowhere to live in Charnock Richard in the North in 1666, in Haigh in 1669, and in Parr in the South in 1666. In the last case the collier 'applied himself to Mr. Byrom Lord of Parr who knowinge him to be a collier and to Live wthin a myle of Cole mynes in Parr wherein he hath been imployed: did consent that yo$^r$ pet$^r$ might erect a Cottage in Parr'. LRO, QSP/291/26, 335/12 and 283/23.
128. WPL, Haigh Colliery Orders, end B, fols. 9 and 10 and PRO, DL4/89/25.
129. In 1653 Thomas Green, who was born in Chorley but had worked as a collier in Adlington until he took ill with 'the hot fever', petitioned for relief to Standish parish because his children had been born whilst he was working in Adlington in that parish. In 1690 it was ordered that 'Robt. Jackson collier & his wife may be removed out of Pemberton unto Adlington the place of his last settlement he gets Coles with Raph Deane was not above a yeare with M$^r$ W$^m$ Mollyneux', the coalowner and lord of Pemberton manor. These mines were opened up by John Robinson, who 'came into the district wanting work' in 1606. LRO, QSP/75/17 and 682/17 and PRO, DL4/64/22.
130. If a collier was referred to as of a particular place whilst being employed in another place it seems reasonable to assume that he lived in the former. Most of the places were over a mile apart and Bryn in Ashton is seven miles from Hackley Moss near Prescot. PRO, PL6/38/107; LRO, DDBa/Coal Pitt Accompts, 1676-96; PRO, PL6/23/35 and WPL, Haigh Colliery Orders.

131. WPL, Haigh Colliery Orders, end B, fols. 8 and 9.
132. Two pitmen were employed in Parr colliery (43). One was from Ashton and one from Eccleston, but both were called Robinson. It is unfortunate that parish registers are not sufficiently detailed in the seventeenth century to allow a search for the exact relationships between colliers with the same surname.
133. See PRO, DL1/211/G/6 and DL4/89/25. The latter is remarkably voluminous. The interrogatories alone run to seventy-one items and the written evidence is accompanied by a remarkable large-scale plan of the works which is many square feet in area. In c.1605 the hewing of pillars that were too thin caused the Bradshaigh workings to collapse over eight acres, occasioning £1000 worth of damage.
134. LRO, DL1/211/G/6 and DL4/75/13.
135. WPL, Haigh Colliery Orders, *passim*. A loose sheet in the Winstanley colliery accounts contains a list of purchases of equipment for a new pit. The total expense was £2. 14. 2.
136. After he had acquired a number of adjacent concessions to allow him to sough dry coal which he leased from the Weatherby family, so that 'the mines are now efficiently drained of water which otherwise would have prevented the getting of coals', Browne was thrown off the property by the lessors. PRO, PL6/13/103.
137. The owners of manors have been classified as 'gentry' and other freeholders as 'non-gentry coalowners'. See J. Langton, 'Landowners and the development of coalmining in South-west Lancashire, 1590-1799', in H. S. A. Fox and R. A. Butlin (eds.), *Change in the Countryside: Essays on Rural England, 1500-1900*, I. B. G. Special Publications, 10 (1979).
138. The very high rents in parts of the centre may to some extent be illusory because it seems to have been customary in Shevington, at least, for the lessor to share some of the costs of mining. LRO, DDHe/40/69 and 74 and WPL, Wrightington MSS/Shevington Box/Blackburne v. Rigby and Holme, Jan. 1, 1663.
139. Bridgeman 'Church and manor of Wigan', pt 3, p.463. When in the hands of Parliamentary Commissioners, collieries seem to have been even more starved of funds than when in Royalist hands. See 'Lancashire Royalist composition papers', pp.34 *et seq.*
140. PRO, PL6/44/81.
141. 'Lancashire Royalist composition papers', pp.34 *et seq.* and LRO, DDGe(M)/872.
142. PRO, DL4/118/10.
143. PRO, PL6/23/35 and LRO, DDGe(M)/844.
144. LRO, DDHe/50/84.
145. LRO, DDHe/15/13.
146. The complexity of the land holding pattern which emerged is made evident in the uncatalogued Willis MSS. LRO, DDWi.
147. PRO, DL4/118/10; LRO, DDWi/18/Ogle v. Fazakerley, Legh and Bankes, June 22, 1664; LRO, DDK/295/7; PRO, PL6/38/107 and WPL, Wrightington MSS/Eccleston Box/Old Deeds . . . relating to Sutton and Eccleston . . . / Case heard at Preston . . . 1686.
148. LRO, The Will and Inventory of Henry Ashton of Whiston, 1668.
149. PRO, PL6/19/81 and DL4/118/10.
150. Both Sir Roger Bradshaigh and William Bankes were Members of Parliament and Deputy Lieutenants of Lancashire. They were also kinsmen. Hawkes, 'Sir Roger Bradshaigh'.

151. The Gorsuch, Catterall and Gerard families held parts of Aspull manor and the Sherringtons, Orrells and Bisphams had rights to smith's coal in Orrell.
152. Stanley Gate lay between the Shuttleworth estates on the lower Douglas and Smithils. Bankes' coal was carted to Warrington, Walton and Liverpool on odd occasions, but it probably went as ballast in a vehicle sent to fetch something. In 1684 coal was taken to Liverpool 'when Statues were fetched'. LRO, DDBa/Coal Pitt Accompts, 1676-96.
153. PRO, DL4/89/25 and DL1/195/S/100.
154. PRO, DL4/89/25; F. J. North, *Coal and the Coalfields in Wales* (Cardiff, 1926), p.73 and Galloway, *Annals*, p.328.
155. Defoe, *Tour*, p.547.
156. In 1600 cannel cost about 0.4d per basket and in 1635 1½d a basket at Haigh (2) and 2d at Aspull (12).
157. PRO, DL4/89/25.
158. PRO, DL4/89/25 and JRL, Haigh MSS/box EXI/Letters &c . . . 1639-1713.

## 4.  The colliery location pattern, 1690-1739

1. Wrightington colliery (55) opened in 1688, closed in 1710 and reopened in 1723. Tarbock (60) worked 1692-4, reopened in 1717 and closed again before 1733. Haydock Wood (28) reopened in c.1707 after fifty years' closure, but no other record of its activity has survived for before c.1750. The pit of Baldwin and Young at Haigh (82) worked only from 1728 to 1734, and the Earl of Derby's collieries at Newburgh (85), Westhead (87) and Whiston (89) operated from 1730 to 1732, 1731 to 1737 and 1732 to 1735, respectively. LRO, DDHe/50/84, 55/15, 50/86 and 116/2; LRO, DDM/4/2, 4/21, 1/5 and 4/5 and R. S. France, 'Lancashire Papist estates in 1719', *Rec. Soc. of Lancs. and Chesh.*, 98 (1943), 108 (1960); 98, p.195. JRL, Legh MSS/Box 58/Letter, Alanson to Legh, Ascension Day, 1707; Box 43/Letter, Richardson to Legh, Sept. 22, 1716. JRL, Haigh MSS/Bradshaigh Accounts, 1731-41. LRO, DDK/2021/14, 16, 22, 23, 28, 31 and 42.
2. PRO, PL6/54/110, 57/7, 59/51, 61/11 and 65/45; LRO, DDHe/58/42; LRO, DDGe(M)/847-62, and LRO, DDK/1620/25-7.
3. The output of 10,500 tons a year plotted for Prescot Hall (5) in the 1730s on figure 17 is the average of 6000 tons raised in 1731 and c.15,000 tons raised in 1735. Copy of the MS 'Prescot Hall and coal mines in 1761', in the possession of Mrs D. Bailey, and KCC, 1/V/42/33.
4. LRO, DDK/1553/73 and 2021/16 and 22.
5. LRO, DDHe/55/15; LRO, DDK/1743 and JRL, Haigh MSS/A/5/56.
6. LRO, DDBa/Coal Pitt Accompts, 1676-96, and Bankes 'Records of mining', p.45. In 1716 this colliery made £100 clear profit, which would require an output of between 1500 and 2000 tons at the rates of profit earned there between 1676 and 1696 and in 1766.
7. See Appendix 3 for an account of the method used to calculate estimates of the output of Haigh colliery (2).
8. JRL, Haigh MSS/Bradshaigh Accounts, 1731-41.
9. An undated document, which was probably written in c.1730, records that Shevington colliery supplied 2000 tons of coal a year to Preston. A high proportion of the colliery's output is unlikely to have been absorbed by such a distant market. JRL, Haigh MSS/Collieries Folder/Early Documents relating to Colliery Affairs/Account what Coals, Slate, &c.
10. It should be noted that the collieries of the North are, on the whole, much less

well documented than those of other parts of the coalfield; accounts are extremely rare, and only tentative suggestions can be made about outputs.

11. A plan of the colliery drawn in c.1712 shows five pits, but coal was probably not raised through 'old Ginn pitt' or 'Cheain pump pitt'. LRO, DDGe(M)/861.
12. StHRL, 352/04207/LB and LRO, DDCs/3/36.
13. The collieries (72 and 73) were the subject of a court case in which an output of 4000 tons per year was claimed. The output of the Barn Hey workings in Sutton (72) can be estimated at 1800 tons a year from rent receipts, so that it is probable that the 4000 tons statement refers to both collieries. PRO, PL6/54/10 and 55/7.
14. 12,240 works were raised between May 1718 and July 1722. PRO, PL6/61/11. (A work contained about 3 tons. See Appendix 1.)
15. Thatto Heath colliery (59) was not recorded by name between 1727 and 1770, although it may have been the colliery which operated in Sutton in c.1735. Parr Hall (43) was not recorded between 1727 and 1756; Garswood (29), which worked on a well-documented estate, was not recorded between 1736 and 1761, and Haydock (28) not between 1735 and 1750. The Longworth collieries in Sutton (72 and 73) were leased in 1709 and one of them was re-leased for fourteen years in 1712, but only one subsequent record of their operation in 1717 has survived. Berry's colliery in Windle (80) was not recorded between 1726 and 1754, and neither was Edge Green colliery (65) between 1717 and 1779. It could be a coincidence of documentary survival that gaps exist in the records of all these collieries in the 1730s and 1740s, but the most likely explanation is that mining collapsed in the South in these decades.
16. LRO, DDM/4/2 and KCC, 1/V/41/11 and 42/41.
17. LRO, DDK/1553/39.
18. LRO, QSP/1339/9.
19. KCC, 1/V/42/33.
20. KCC, 1/V/74.
21. KCC, 1/V/42/35.
22. KCC, 1/V/74.
23. The occupations of fathers of baptised children do not give unbiased estimates of the number or the proportion who followed a particular kind of employment in a population. Bachelors and men outside early adulthood are omitted; adherence to the established church and its rituals could have varied systematically with occupations and so could age at marriage, family size and the age-structures of workforces. Thus, these data can only be interpreted in a tentative and generalised way.
24. For the methods of estimation used to calculate this table, see Langton, 'Coal output'.
25. There were even larger workings in the North East, where a Grand Allies colliery raised c.75,000 tons a year before the end of the seventeenth century. However, the largest collieries in Fife, Nottinghamshire, Cumberland and non-Tyneside Durham do not seem to have surpassed Prescot Hall (5) by much, and collieries as small as the South West Lancashire norm of the seventeenth century were still amongst the largest in the early eighteenth century on coalfields where there was no navigable water. Nef, *Coal Industry*, vol. I, pp.26, 30, 48-9, 57 and 71; Bulley, 'To Mendip' (1952), p.49, and G. G. Lerry, *The Collieries of Denbighshire – Past and Present* (Wrexham, 1946), p.116.

## 5. Factors affecting location pattern, 1690-1739

1. Gonner, 'The population of England', pp.261-96 and A. Toynbee, *Lectures on the Industrial Revolution of the Eighteenth Century in England* (London, 1913), pp.9-10. The most reliable recent estimates are slightly smaller than those of Gonner. P. Deane and W. A. Cole, *British Economic Growth 1688-1959* (Cambridge, 1964), p.103.
2. Defoe, *Tour*, p.548. For an indication of the distribution of the increase of population in the county based on statistics for ecclesiastical divisions see L. W. Moffitt, *England on the Eve of the Industrial Revolution*, 2nd edn (London, 1963), pp.140-2.
3. *Census of Population, 1841: Enumeration abstracts*, British Parliamentary Papers, 1841, preface, pp.36-7.
4. PRO, EDC/179/250/11 and T. C. Porteus, *History of Standish* (Wigan, 1927), p.46.
5. Moffitt, *England*, p.142.
6. Moffitt, *England*, p.140 and Parkinson, *Liverpool*.
7. Based on 'Indexes to wills'.
8. It should be stressed again that the wills' evidence greatly underrepresents the numbers of craftsmen. According to Tupling, over a hundred nailors were enumerated in the chapelry of Up Holland between 1699 and 1739, and eight out of thirteen people who took out leases from the Gerards in Ashton between 1716 and 1757 were hingemakers. Tupling, 'Early metal trades' and LRO, DDGe(M)/812.
9. The information on iron supplies is taken from Awty, 'Charcoal ironmasters'.
10. The information on Wigan crafts is based upon a count of 426 occupations entered in the borough court records and parish register between 1720 and 1729. About fourteen per cent were in both non-ferrous metals manufacturing and weaving.
11. *Victoria History of the County of Lancaster (VCH)*, (8 vols., London, 1906-14), vol. II, p.355.
12. Sprunt, 'Old Warrington trades'.
13. T. C. Barker and J. R. Harris, *A Merseyside Town in the Industrial Revolution: St Helens, 1750-1900* (Liverpool, 1954), p.12.
14. Barker and Harris, *St Helens*, pp.108 and 110; LplRO, Salisbury MSS/681/11, 12 and 21.
15. T. E. Gibson, *Blundell's Diary: Selections from 1702-1728* (Liverpool, 1895), p.177.
16. PRO, PL6/65/45.
17. Gibson, *Blundell's Diary*, p.154 and KCC, 1/V/69.
18. LRO, DDWi/Probate Box/The Will of Thomas Makin, June 2, 1739.
19. PRO, PL6/65/45.
20. LplRO, Salisbury MSS/681/11.
21. KCC, 1/V/69, p.9, 1/V/74 and 1/V/42/18.
22. As early as 1674 royalties were received for 'pipe clay & Slurry Clay' in Sutton and the Earl of Derby sold potting clay and glass sand from his Rainford estates throughout the early eighteenth century. At least two mug houses operated in Rainford in 1703, owned by the Bisphams and Tunstalls. JRL, Legh MSS/Box 61/Letter, Pennington to Bowdon, Dec. 6, 1674; LRO, DDK/2021/14, 16 and 18 and DDGe(M)/1124.
23. W. R. Scott, *Finance of English, Scottish and Irish Joint Stock Companies to 1720* (3 vols., Cambridge, 1910–12), vol. II, p.470.

24. LRO, DDK/406/90 and T. C. Barker, 'Lancashire coal, Cheshire salt and the rise of Liverpool', *Trans. Hist. Soc. of Lancs. and Chesh.*, 130 (1951), 86 and 87.
25. Barker, 'Lancashire coal', p.91.
26. LRO, DDK/406/90.
27. Moller, 'English coalmining', pp.784-90.
28. Willan, 'Navigation of the Weaver', pp.2-4
29. A number of fines were imposed in Ashton manor court between 1690 and 1710 for illegally gathering timber and digging coal and the latter offence was quite frequently punished in Wigan. Carr furnace drew its charcoal supplies over eleven miles from Hulton Park, and the Chadwick furness at Birkacre obtained theirs from Bickerstaffe over a similar distance. Awty, 'Charcoal ironmasters', p.101; LRO, DDCs/3/99 and DDK/1809/23 and WPL, Euxton MSS/Bundle 9/9 and Court Leet Rolls.
30. LRO, DDWa/Ince Box/Ince, 1690-1710/Gerard to Ince, Sept. 10, 1700 and DDHk/Hawkshead Surveys/Standish, 1800.
31. Barker and Harris, *St Helens*, p.7 and LRO, DDK/2001/16.
32. LRO, DDK/1526/2-8.
33. Blundell, *Blundell's Diary*, p.125 and LRO, DDGe(M)/1124.
34. LRO, DDK/1620/25.
35. G. H. Tupling, 'The turnpike trusts of Lancashire', *Mems. and Procs. of the Manchester Lit. and Phil. Soc.*, 4 (1952-3), 39-62.
36. WPL, Court Leet Rolls/Box I/Roll 107: there are numerous other presentments about the poor state of the roads in the town in the records of the Court Leet and Quarter Sessions.
37. Gibson, *Blundell's Diary*, p.203.
38. F. A. Bailey, 'The minutes of the trustees of the turnpike from Liverpool . . .', *Trans. Hist. Soc. of Lancs. and Chesh.* pt 1, 88 (1936), 159-201 and pt 2, 89 (1937), 31-91; pt 1, 160-4 and Gibson, *Blundell's Diary*, p.212. The coalpit road in Parr was also in continual need of repairs, which were often effected by the colliers.
39. PRO, PL6/61/11 and 65/45.
40. The maximum tolls fixed were as follows: ½d for horses carrying coal, 1d for other horses; 2d–4d for coal carts, depending upon the number of horses, and 6d for carts carrying other goods; 6d for coal wagons and 1/- for other wagons. Carts carrying earthenware were to pay 8d irrespective of the number of horses. Bailey, 'Turnpike minutes', 161.
41. *Journal of the House of Commons*, 19 (1718-21), 211, 248 and 264; LRO, DDCl/1064 and A. P. Wadsworth and J. de L. Mann, *The Cotton Trade and Industrial Lancashire, 1600-1728* (Manchester, 1931), pp.214-16.
42. JRL, Haigh MSS/Box EIII/Bundle II/Letter, Leigh to Bradshaigh, July 21, 1739.
43. JRL, Haigh MSS/Collieries Folder/Early Documents relating to Colliery Affairs/Account of what Coals, Slate, &c.
44. LRO, DDM/1/141.
45. The ordinary cost of drawing was 2½d per score, but at forty yards 'odd drawing' rates came into operation: 3d per score from forty to fifty-six yards, 3½d from fifty-six to seventy-two yards and so on to 8½d per score between 232 and 248 yards from the eye of the drift. Ashton and Sykes, *Coal Industry*, p.62n.
46. Porteus, *Standish*, p.33.
47. The reason for the closure of 1739 was not specified, but in 1746 drainage was

difficult 'by Reason of the Thick rusty water and ye abundance of Oaker that is more than ordinary Natural to that Coal . . . two old soughs in ye same Mine Being now Intirly Stop'd'. LRO, DDK/1640.

48. There were shallow pits at Edge Green (65), where the colliery was first opened at only fourteen yards, and some of the pits at Wrightington (83) in c.1730 were only twelve yards, though others were fifty-seven yards deep. In 1735 the pits producing glass-coal at Prescot Hall (5) were twelve yards at the crop and twenty yards down-dip and the shafts of the Harrington colliery at Huyton (58) were only twenty yards in 1690. There were deeper pits elsewhere and at these same collieries after long-continued working. Haydock colliery (28) worked at thirty-eight yards in 1717; a maximum depth of forty yards was agreed upon at Edge Green (65) in 1712 and this depth was worked at Orrell (74) in 1711. The reason for the pumps at Haigh (2) was probably a descent below the forty-eight yard level drained by the Great Sough, but the deepest colliery by far seems to have been that of seventy-two yards at Charnock Richard (17). LRO, DDGe(M)/856; Porteus, *Standish*, p.33; KCC, 1/V/74; LRO, QSP/710/79; JRL, Legh MSS/Box Q/C/3; LRO, DDGe(M)/855; LRO, DDBa/6/4 and WPL, Haigh Colliery Orders. Pits of less than thirty yards seem to have been uncommon in Britain in the early eighteenth century.

49. LRO, DDHe/55/15, 50/86, 116/2 and 58/42.

50. JRL, Haigh MSS/A/5/56; LRO, DDK/1620/25 and DDLm/8/13.

51. In 1728 Sir Roger Bradshaigh was told that 'Mr. Chadwick wants to have ye benefitt of your Engine, And complains much of your having Diverted the Water from its antient course to supply your Works'. It thus seems that the Haigh engine was water powered and that Bottlingwood colliery (79), where Chadwich was steward, also needed water from the brook which marked the boundary between Haigh and the Bottlingwood area of Wigan for drainage. JRL, Haigh MSS/Box EIII/Correspondence of Alexander Leigh/Letter, Leigh to Bradshaigh, Jan. 14, 1728.

52. LRO, DDBa/6/4 and DDGe(M)/861.

53. JRL, Legh MSS/Box 43/Letter, Richardson to Legh, Sept. 8, 1708 and Box 59/Letter, Ford to Legh, June 6, 1750.

54. JRL, Haigh MSS/Box EIII/Correspondence of Alexander Leigh/Leigh to Bradshaigh, April 5, 1737. The following account of the outbreak is taken from this bundle of letters, with supplementary information from JRL, Haigh MSS/Bradshaigh Accounts, 1731-41.

55. LRO, DDM/4/2, 4/21, 14/11 and France, 'Papist estates', p.195. It is not clear whether 'the three Rowley brothers of Shropshire' were employed in 1716 or at a later reopening in 1750. However, the workings of 1716 were 'more regular than any of the subsequent proceedings', which suggests that Shropshire skill was employed to dig them.

56. KCC, 1/V/74.

57. LRO, QSP/686/3 and 4 and 710/19; Gibson, *Blundell's Diary*, p.152 and MS in the possession of Mrs D. Bailey.

58. LRO, DDWi/Box 12/Abstract of the Title of H. Case to the Manor of Whiston.

59. JRL, Legh MSS/Box 46/Correspondence of J. Stafford/Letter to P. Legh, Feb. 21, 1735; LRO, DDHe/58/42 and WPL, Wrightington MSS/Box 10/Mr Standish aboute getting Coles . . . 1679, and Dicconson and Hesketh: Articles Coal in Shevington . . . 1731.

60. LRO, DDHe/55/15, 50/86 and 112/18; PRO, FEC1/C/17 and LRO, DDTa/283/34.

61. JRL, Haigh MSS/Collieries Folder/Colliery papers &c miscellaneous and

Amusing/Tho. Walker's Draught of ye Sough at Monk's Pitt; Bradshaigh Accounts, 1731-41 and Box EIII/Correspondence of Alexander Leigh/Letter, Leigh to Bradshaigh, Jan. 14, 1728.

62. LRO, DDCs/4/4; CLM, Piccope MSS/186 and JRL, Haigh MSS/Collieries Folder/Articles of Agreement between the Heirs of Ralph Dean ... and Thomas Gerard.

63. LRO, DDK/2021/14, 22-4 and DDK/1640.

64. LRO, DDHe/55/15 and JRL, Legh MSS/Box 45/Correspondence of J. Stafford/Letters to P. Leigh, Feb. 21 and Dec. 22, 1735.

65. PRO, FEC1/C/7, LRO, DDTa/283/34 and DDHe/116/2.

66. LRO, DDGe(M)/849-51 and PRO, PL6/61/11.

67. KCC, 1/V/40/1 and LRO, DDWi/Box 12/Abstract of the Title of H. Case to the Manor of Whiston.

68. Most of the colliers registered in Eccleston and Douglas chapel were from Wrightington. The registers or bishops' transcripts of all the coalfield parishes and chapelries are held by the Lancashire County Record Office. The registers of Up Holland and St Helens in the eighteenth century have been published as *Lancashire Parish Register Society*, 23 (1905) and 107 (1968).

69. Both Up Holland and St Helens were at that time chapelries in the parishes of Wigan and Prescot and there were other churches and chapels within relatively easy reach of the townships they served. It may be that the deficiencies in marriages were due to their solemnisation in more imposing parish churches. I searched the baptism and burial registers of all nearby churches and chapels for the names of St Helens and Up Holland colliers, and any relevant information was included in the data from which the tables of this chapter were compiled.

70. LRO, Prescot Baptism Registers, 1720-33 and 1734-51.

71. JRL, Haigh MSS/Bradshaigh Accounts, 1731-41 and LRO, DDBa/Up Holland Box/Hiring of Colliers, 1744-8.

72. A number of colliers who had married daughters of Wigan men, some from as far away as Parr, were refused freedom and ordered to leave the town as were other 'foreign' colliers. WPL, Court Leet Records/Box I/Rolls 68, 7? 95, 97, 123 and 138.

73. StHRL, Parr Township Papers/Removal Orders and Settlement Papers. There are thirty-six documents covering this period, which provided the information plotted on figure 27.

74. Occupations are rarely present in the Parr documents, but some of those paupers who were colliers or colliers' widows can be identified by matching names with those on the forms used for family reconstitution. Compared with removals of deceased miners' families there were few of families still headed by a collier, but see, e.g., those of John Cartwright and family in 1767, John Travers and family in 1791 and James Johnson and family in 1797. The detailed depositions of Edward Fairhurst on Sept. 8, 1795 and James Johnson on May 1, 1797 provide considerable information on the criteria used to establish the place of settlement.

75. For example, Burslem (Staffs.), Hawarden (Flints.), Clifton (Worsley), Blackburn, Bedford (Leigh), Hindley, Abram, Shevington, Standish and Aspull. Some of the names which became conspicuously common in the colliery workforce in Parr during the eighteenth century were first recorded in settlement papers brought into the township, e.g. Peter Cartwright I deposited a certificate from a Cheshire township in 1731 and Richards Sumner I deposited one from Bedford in the same year. Parr mining families also sent

members to mining townships elsewhere, e.g. Thomas Greenhough took a Parr certificate to Ince in 1694 and William Rigby took a Billinge certificate to Wrexham in 1740.

76. The wife of Robert Lancaster, collier of Yorkshire, was buried in Up Holland in 1713. Her husband was not otherwise mentioned in the Up Holland registers, though Edward Lancaster, collier, had a child baptised in 1751. Joseph Lancaster and his wife were removed in 1718 from Shevington to Crosthwaite, Cumberland, whence they were said to have come with Joseph's father Robert. Robert Patrick Lancaster of Coppull Moor was a colliery agent and coalmaster in Shevington and Wrightington during the first three decades of the century. Lancaster is an unusual name in the registers of coalfield parishes, but this is not, of course, ground for an assumption that the three Robert Lancasters were the same, highly peripatetic, person.

77. JRL, Haigh MSS/Bradshaigh Accounts, 1731-41 and LRO, DDM/4/21.

78. Only two of the seven auditors in the 1720s were named Lowe. JRL, Haigh MSS/Bradshaigh Accounts, 1731-41, which in fact begin in 1724.

79. Petitions to the Quarter Sessions by injured miners in need of relief were quite common, and so were compensation payments and expenditure on brandy and medicaments in the accounts at Haigh (2), Winstanley (6) and Tarbock (60). LRO, QSP/*passim;* JRL, Haigh MSS/Bradshaigh Accounts, 1731-41; LRO, DDBa/Coal Pitt Accompts, 1676-96 and DDM/4/2.

80. LRO, DDBa/Coal Pitt Accompts, 1676-96, DDBa/Up Holland Box/Hiring of Colliers, 1744-8, DDM/4/2 and RCSt/6, and KCC/1/V/74.

81. The 'work' measure of three tons was supposed to represent the daily output of one hewer, and a six day working week was customary. However, it is clear that 500 tons was nearer to a hewer's normal annual get than 900 tons which would be produced at a work per day for 50 six day weeks (see table 15). A get of twelve loads of ten baskets entitled a hewer to a bonus at Haigh (2) in 1758. At 120 baskets a week and fifty weeks a year, a cannel getter would produce 300 tons a year, but it seems that they usually managed less than this. See table 14 and Ashton and Sykes, *Coal Industry,* p.136.

82. Variation between the incomes of different colliers' families would, of course, be increased even further by the employment of wives and different numbers of children in the pits.

83. JRL, Haigh MSS/A/5/56.

84. WPL, Haigh Colliery Orders, end A, fol. 19r and LRO, DDBa/Coal Pitt Accompts, 1676-96. Some proprietors still had trouble with auditors; the accounts of Thatto Heath colliery in 1718 were incomprehensible and worthless scraps of paper. This experience cannot have been universal because three auditors questioned about coal sales to Preston in c.1730 had served twenty, nineteen and twenty years in their current jobs. PRO, PL6/54/110 and JRL, Haigh MSS/Collieries Folder/Early Documents relating to Colliery Affairs/Account what Coals, Slate, &c.

85. LRO, The Wills and/or Inventories of Thomas Brighous, 1692, Edward Greenhough, 1693, William Brown, 1727, Robert Brown, 1748 and Richard Turner, 1750. All except William Brown leased parcels of land or had crops and animals valued. James Pollett, collier, of Atherton had a pair of fustian looms let to a weaver when his inventory was taken in 1710 and Gilbert Allens, collier, of Abram left two cottages in Ashton.

86. KCC,1/V/41/11.

87. LRO, DDBa/Coal Pitt Accompts, 1676-96; LRO, DDM/4/2; Bankes 'Records of mining', p.44; PRO, FEC1/S/66; LRO, DDK/2021/22; KCC,

1/V/74 and LRO, DDK/2021/31, 33 and 37.
88. These figures may have been exceptionally high for the colliery. They were
estimated by an agent of King's College, who realised that their Prescot coal
was yielding them far less than it should. The agent reported that 'the way of
Work for some time past has been, to get all they can, upon the Crop (That
being to be got at 12d p$^r$ Work less Wages, as well as less Charges in Sinking)
and y$^e$ Deep Work is some hundred yards behind the other, whereas . . . the
Deep Work should be rather before than behind the Crop and This is the
reason why there appears so large a Profit this pres$^t$ y$^r$ 1735'. KCC, 1/V/74.
89. It should be stressed that the tabulated profit figures represent annual balances.
They do not, except in the case of Whiston (89), include money spent in years
before coal came into production. Sums for the amortisation of fixed capital
installations were not accounted annually except at Prescot Hall (5), where the
1/- per work cost of drainage was for 'if with Fire Engines, expenses of building
[them] included'.
90. There is no record of the cost of a whimsey, but the two gin horses, pasture rent
and annual supply of oats cost £29. 5. 0 at Westhead (87) and the harness and
gear of a gin horse cost £4. 9. 11 at Tarbock (60). In addition, the drainage pit
could cost in excess of £50. Thus, if the machine itself cost only £20 or so, the
total cost of whimsey drainage installations would be about £100.
91. LRO, DDM/4/2 and RCSt/6.
92. LRO, DDK/1620/25 and DDLm/8/13.
93. JRL, Haigh MSS/A/5/56.
94. KCC, 1/V/74.
95. LRO, DDK/2021/22-4 and LRO, DDGe(M)/1185.
96. LRO, DDM/4/2 and DDGe(M)/852 and DDBa/Coal Pitt Accompts, 1676-96,
JRL, Haigh MSS/Bradshaigh Accounts, 1731-41 and Legh MSS/Box Q/Map
of Haydock Colliery, and Porteus, *Standish*, p.33.
97. At Haydock in 1714 'Drifts are very wide: and the Coales gotten (generally
speaking) clean which causes something more Timber to be used than ordinary
for Propps . . . but I take y$^t$ to be repaid double by the Quantities gotten out of
those places supported by Propps where wee used to Leave Pillars.' This entry
does not necessarily suggest that longwall techniques had been introduced, but
perhaps only that props were used to allow smaller pillars in pillar and stall
workings. JRL, Legh MSS/Box Q/C/3.
98. PRO, PL6/61/11 and KCC, 1/V/74.
99. KCC, 1/V/74.
100. LRO, DDK/2021/22-4 and 31. These sums include all costs, from sinking and
soughing to horse harnesses and colliers' earnest on hiring. Costs were
particularly high at Newburgh (85) because of a long sough, which took three
years to build before production began.
101. CLM, Piccope MSS/vol. III, p.186 and LRO, DDCs/4/4. The costs in
Lancashire do not seem to have been exceptional. It was estimated that it
would cost between £150 and £200 to open Esclusham colliery in Flintshire in
c.1715, but within a few years of its beginning this two-pit colliery required a
more efficient source of drainage power than horses, which presumably
involved more spending. Whitehaven colliery in Cumberland cost 'upward of
half a million Sterling' before c.1730. Lerry, *Collieries of Denbighshire*, p.116
and J. E. Williams, 'Whitehaven in the eighteenth century', *Econ. Hist. Rev.*,
2nd ser., 8 (1955-6), 399.
102. The fire-damp explosions at Haigh (2) in 1737 and 1738 caused a loss of over
£1000 in receipts. Opening and closing pits and the sough, dissipating damp

and transporting, entertaining and hiring colliers from Shropshire and Newcastle cost £45.12.4. JRL, Haigh MSS/Bradshaigh Accounts, 1731-41.

103. LRO, DDM/4/2 and PRO, PL6/54/110 and 55/7.

104. JRL, Legh MSS/Box 59/Letter, Gatley to Legh, Feb. 24, 1756.

105. The Langton line died out and their lands eventually passed to the Duke of Bridgewater. The Gerard lands were held in trust, a living allowance paid out of colliery profits, until 1714. The estate and colliery were sold to a relative from Wigan at sometime between 1717 and 1720, and on their insolvency they were taken in trust by Sir William Gerard of Garswood for his son. PRO, PL6/44/81; LRO, DDGe(M)/872 and *VCH*, vol. IV, pp.118-19.

106. *VCH*, vol. III, p. 171; JRL, Legh MSS/Box 49/Letters, Byrom to Legh, 1703-5; LRO, DDGe(M)/845 and DDWi/Box 12/Abstract of the Title of Henry Case to the Manor of Whiston.

107. PRO, FEC1/C, D, E, H and S contain details of the administration of the Chorley, Dicconson, Eccleston, Hesketh and Standish estates during sequestration. Those for the Standish estates are particularly detailed.

108. LRO, DDGe(M)/853, 855 and 883.

109. PRO, FEC1/D/82-8.

110. LRO, DDBa/3/3 and 6/4.

111. JRL, Haigh MSS/Bradshaigh Accounts, 1731-41.

112. The rent of 38.3 per cent at Tarbock (60) in 1716 was charged to lessees from outside the coalfield, who were probably not conversant with local rates, for the nearest colliery to Liverpool and Dungeon. The 1732 Shevington rent of 37.5 per cent was charged for coal that was already well drained and accessible from the lessees own existing pits, as was the coal for which thirty-six per cent of selling price was charged in 1736. Rents seem to have been expressed as so many pounds per acre of coal on other coalfields and are therefore difficult to compare with the Lancashire figures. However, twenty-five per cent was charged in Eastern Scotland in 1719, when the rate was about seventeen per cent at Saltcoats in the west. Ashton and Sykes, *Coal Industry,* p.177-8 and N. M. Scott, 'Documents relating to coalmining in the Saltcoats district in the first quarter of the eighteenth century', *Scot. Hist. Rev.,* 19 (1922), p.90.

113. LRO, DDHe/50/84 and 86 and 55/15, DDTa/283/Lease, Brooke to Lancaster, 1712, DDHk/Brooke Accounts, 1726-46, and the Will of John Chadwick of Birkacre, 1755; WPL, Wrightington MSS/Box10/Dicconson and Hesketh: Articles of Coal in Shevington ... 1731; JRL, Haigh MSS/Bradshaigh Accounts, 1744-5; PRO, FEC1/C/17, H/15 and D/82, and Porteus, *Standish,* p.33.

114. JRL, Haigh MSS/Box/EIII/Letters of Alexander Leigh, and LRO, DDBa/The Case of Howarden v. Bankes *et al.*/Letter from J. Halliwell.

115. LRO, DDBa/6/4 and WPL, Wrightington MSS/Box10/Dicconson and Hesketh: Articles of Coal in Shevington . . . 1731.

116. PRO, PL6/61/11 and 65/45.

117. LRO, DDK/2021/32.

118. France, 'Papist estates', p.195.

119. Even the Cases were severely embarrassed financially in the 1720s and 1730s when they took out large mortgages from Nicholas Fazakerly of Liverpool and from the Claytons of Liverpool and Parr, with whom they intermarried. LRO, DDWi/Box 12/Abstract of the Title of Henry Case to the Manor of Whiston, and Barker and Harris, *St Helens,* pp. 24 *et seq.*

120 See, for example, the notebook compiled by the College Bursar in the early eighteenth century. KCC, 1/V/38.

121. The first mentions I have traced relate to Cobham and Makin after they had become concerned in Prescot Hall colliery (5). Cobham died sometime before 1731, Makin at an unknown date. The wills of their heirs, made in the 1740s, show a complete preoccupation with the colliery, in which the families continued their partnership until the 1760s. KCC, 1/V/42/35; LRO, The Will of Thomas Cobham of Prescot, proved 1746 and DDWi/Probate Box/The Will and Codicil of Henry Makin, 1746.
122. DDK/2002/14 and 2021/4, 14 and 16.
123. JRL, Haigh MSS/Box/EXI/Letters etc. . . . . 1639-1719/Evidence given in a Case of 1714 concerning the use of Horrocksford Bridge, and Collieries Folder/Early Documents relating to Colliery Affairs/Account what Coals, Slate, &c. £150 per year were made from sales to Preston and the Fylde at 2½d and 3d per basket of 120 lb. Thus, between 600 and 720 tons were sold there out of a total ouput of about 5000 tons a year.
124. PRO, PL6/65/45.
125. The industrial markets for smith's coal were mainly in domestic smithies, and householders could obviously not respond to coal supplies in the location of their work in the same way that the proprietors of salt and glass works did. The Smith and Cannel seams are geologically deep coals which were brought to the surface by faulting and erosion in very few places.
126. KCC, 1/V/74.
127. JRL, Haigh MSS/Collieries Folder/Early Documents relating to Colliery Affairs/Account what Coals, Slate, &c.
128. LRO, DDK/1620/25 and 26.

## 6. The colliery location pattern, 1740-99

1. A much fuller discursive account of the mining industry in this period can be found in my PhD dissertation, 'The geography of the south west Lancashire mining industry, 1590-1799' (2 vols., Wales, 1970), vol. I, pp. 187-256.
2. JRL, Haigh MSS/Bradshaigh Account Books, 1731-41, 1744-54 and 1754-65.
3. LRO, DDWi/Box 43/Plan and Letter from Mr Moss, 1746 and appended comments, and MS copy in the possession of Mrs D. Bailey, 'Prescot Hall and coal mines, 1761'. The same reference numbers have been retained for collieries unless there was a complete change of site even if there was a change of name, as at Whiston where the new works were called New Paradise colliery.
4. JRL, Haigh MSS/Bradshaigh Account Books, 1731-41, 1744-54 and 1754-65.
5. *Ibid.*; LRO, DDGe(M)/1185 and RCSt/6 and JRL, Haigh MSS/Kirkless Colliery Account Book, which comprises a jumbled account of the Kirkless cannel colliery, 1750-6 and 1774-6.
6. LRO, DDBa/Up Holland Box/Hiring of Colliers, 1744-8.
7. WPL, Wrightington MSS/Eccleston Box/Papers concerning the Eccleston Family of Eccleston/Bundle of Documents concerning Charles Dagnall, and LRO, DDCs/1/1.
8. LRO, DDK/2021/37 and DDM/4/5. In 1757 trust coal, banked reserves and fuel used by the owner's family amounted to 1653 tons at Tarbock (60), probably representing about half of total output. LRO, DDM/1/143.
9. MS copy in the possession of Mrs D. Bailey, 'Prescot Hall and coal mines, 1761' and KCC, 1/V/42/18.

10. KCC, 1/V/42/14.
11. KCC, 1/V/40/1.
12. LRO, DDK/406/90 and DDWi/Box 12/Abstract of the Title of Henry Case to the Manor of Whiston. The personal allowances alone would have required an output of about 10,000 tons a year at a rate of profit of about forty per cent of receipts.
13. JRL, Haigh MSS/Kirkless Colliery Account Book.
14. WPL, Euxton MSS/4/20 and 21.
15. PRO, DL1/503(1); WPL, Leigh-Pemberton MSS/Deeds Reserving Rents and Coalmines, Orrell/Mr Porter's Account of Coal got at Hall Orrell, and LRO, RCSt/6, which contains the first record of extraction in the Highfield area of Standish.
16. LRO, DDGe(E)/1257. Gillars Green (99) was last recorded in 1756, but in 1757 its proprietor was offered 'the water ingan which I think would be of greate sarvis to you'. LRO, DDCs/1/1 and WPL, Wrightington MSS/Eccleston Box/Eccleston Family of Eccleston/Bundle of Documents concerning Charles Dagnall.
17. KCC, 1/V/74/Letters, Chorley to King's College, Sept. 25, 1769, Palmer to King's College, Jan. 8, 1769 and Case to King's College, 1771.
18. LRO, DDM/4/27 and LplRO, *Lpl Gen Advt*, Feb. 1, 1771.
19. PRO, PL6/87/10.
20. LRO, DDWi/A Map and Survey of the Land . . . belonging to James Gildart . . . Oct. 10, 1770 and KCC,1/V/74/Petition against the Erection of an Engine, Nov. 15, 1774.
21. WPL, Wrightington MSS/Eccleston Box/Old Coal Leases, Eccleston and Sutton/Eccleston to Case, July 31, 1769, appended note and LRO, DDCs/1/1.
22. For example, (80), (109) and (120). Buckley's colliery in Hardshaw (115) was offered for sale immediately it was opened in 1759 and not subsequently recorded. LplRO, *Wms Lpl Advt*, July 20, 1759.
23. For example, (28), (118), (117) and (119). The last two were only mentioned when subjected to the acrimony of the powerful Leghs, who cut their wayleaves and shut them up. It seems reasonable to presume that they would not have been allowed to operate for long.
24. LRO, DDGe(M)/1201.
25. LplRO, *Lpl Gen Advt*, Nov. 1, 1771.
26. Barker and Harris, *St Helens*, p.35. See Appendix 4 for estimates of hewer productivity.
27. JRL, Legh MSS/Box 51/Letter from R. Leigh, Feb. 17, 1760.
28. LplRO, *Wms Lpl Advt*, Dec. 16, 1757.
29. *Ibid.*, Feb. 16, 1759 and *Lpl Gen Advt*, June 19, 1778.
30. JRL, Legh MSS/Box 59/Letter from R. Dodge, Jan. 8, 1758.
31. JRL, Legh MSS/Box 51/Letter from R. Leigh, Feb. 14, 1760.
32. JRL, Legh MSS/Box 43/Letters, Richardson to Legh, Sept. 29, 1769 and Aug. 18, 1770.
33. Barker and Harris, *St Helens*, p.43 and LplRO, *Lpl Gen Advt*, Feb. 2, 1770, May 24, 1771 and July 23, 1773.
34. See Appendix 4 for conversion factor. Only one colliery was mentioned subsequently at Thatto Heath and/or Ravenhead and one has been plotted on the relevant maps and graphs.
35. JRL, Legh MSS/Box 51/Letter, Leigh to Legh, Jan. 1765 and WPL, Wrightington MSS/Eccleston Box/Old Coal Leases, Eccleston and Sutton/Eccleston to Leigh, Feb. 17, 1760 and 1763.

36. LplRO, *Wms Lpl Advt,* June 29, 1759 and Nov. 13, 1761.
37. WPL, Wrightington MSS/Eccleston Box/Old Coal Leases, Eccleston and Sutton/Eccleston to Case, July 31, 1769 and appended note of Oct. 8, 1773.
38. The colliery was stopped in early 1769 and later in that year its lease passed into the hands of the proprietor of Whiston (34). The old partnership worked out coal not subject to the new lease until 1770, when the colliery was closed. KCC, Prescot MSS/Coal Accounts, 1763-81 and 1/V/74/Letters, Palmer to King's College, Jan. 18, 1769, Chorley to King's College, Sept. 25, 1769 and Case to King's College, 1771.
39. KCC, 1/V/74/Petition against the Erection of an Engine, Nov. 15, 1774.
40. LRO, DDGe(E)/1257 and LplRO, *Lpl Gen Advt,* Feb. 1, 1771.
41. LRO, DDCs/1/1.
42. In 1757 trust coal, banked reserves and supplies to the owner's family amounted to 1653 tons. It is unlikely that the total output was much more than twice as large. LRO, DDM/1/143.
43. LRO, DDWi/Map and Survey of the Land . . . belonging to James Gildart . . . Oct. 10, 1770.
44. Case raised between 2000 and 11,000 tons a year from the College lands and unknown amounts from the Earl of Derby's reserves besides, presumably, a majority of the output from his own coal on which he paid no royalty. An enormous cash balance of £3959.13.11¾ was recorded in the 'colliery book' at Whiston (34) in 1777. KCC, Prescot MSS/Coal Accounts, 1763-81 and 1/V/74/Letter, Chorley to King's College, Sept. 25, 1769; LRO, DDK/2021/20, 86, 88, 90 and 91 and 1816/27 and PRO, PL6/87/10.
45. JRL, Haigh MSS/Bradshaigh Accounts, 1765-74; LRO, DDCa/1/47 and DDBa/Colliery Accounts, 1766, The Case of Howarden v. Bankes, 1760, and Accounts and Rentals/20.
46. See Appendix 3, and WPL, Leigh-Pemberton MSS/Deeds Reserving Rents and Coalmines, Orrell/Mr Porter's Account of Coal got at Hall Orrell, and Title Deeds to Orrell Hall Estate/Lease, Heirs of J. and S. Prescott to Halliwell and Heskin, Mar. 25, 1756. The Orrell (1) output excludes coals which were allowed rent free, supposedly for use around the colliery and compensation for difficulties. There were twenty-five different kinds of allowance which consumed about one-third of the coal raised. This situation illustrates firmly the somewhat chimerical value of even accounted outputs and their imperfect comparability from colliery to colliery. WPL, Leigh-Pemberton MSS/Deeds Reserving Rents and Coalmines, Orrell/Allowances for Mr Halliwell by the Articles for wch. no Lord's Part are to be Paid.
47. LplRO, *Lpl Gen Advt,* Dec. 12, 1777 and LRO, DDWa/Coal, Leigh Estate, 1741-97/Lease, Leigh to Longbotham, Jan. 20, 1775.
48. James Harrison (150) had no connection with mining until he got a job at Parr Hall (43) in 1779, and the colliery of Nicholas Ashton in Parr (154) was an example of the entrepreneurial links between Cheshire salt and St Helens coal which only developed towards the end of the century. LRO, DDBa/Bispham Estate Papers/7/4, and Barker, 'Lancashire coal', p.97.
49. JRL, Haigh MSS/Kirkless Colliery Account Book, 1750-6 and 1774-6 and Bradshaigh Accounts, 1765-74; WPL, Leigh-Pemberton MSS/Orrell Coal Accounts, 1771-6 and LRO, RCSt/6. The reserves worked by Longbotham were incorporated into those of another colliery in 1776. LRO, DDWa/Coal, Leigh Estate, 1741-97/Lease, Leigh to Chadwick, Hustler and Hardcastle, Sept. 29, 1776.
50. Eight different people paid land tax on collieries in Pemberton, all but two for

only one or two years. It is unlikely that there were six collieries of such transience and it may be that pit auditors, who handled all colliery moneys and perhaps actually paid over the tax to the assessor-collectors, were entered in the returns of this township. However, no connection can be established between any of the names which succeed each other from year to year and eight separate collieries have, perforce, been counted, but it would obviously be wrong to include them in calculations of the length of colliery lives.

51. LRO, Land Tax Assessment Returns/West Derby/Whiston.
52. LRO, Land Tax Assessment Returns/Leyland/Blackrod and Coppull.
53. LRO, DDBa/Accounts and Rentals/20 and WPL, Leigh-Pemberton MSS/Orrell Coal Accounts, 1771-6.
54. WPL, Leigh-Pemberton MSS/Accounts of Dean Colliery, 1776 and LRO, RCSt/6. The Dean accounts are very scrappy and run for only a few months. Between two and eight getters were employed in the colliery, which produced 332 scores in an accounted thirteen week period. The scores there probably comprised twenty-six baskets and the baskets 140 lb, so the annual output at the average of the given weekly rates would have been c.3500 tons.
55. LRO, DDBa/Accounts and Rentals/20 and D. Anderson, 'Blundell's collieries: the progress of the business', *Trans. Hist. Soc. of Lancs. and Chesh.*, 116 (1965), 113.
56. Anderson, 'Blundell's collieries', p.113
57. CRO, EDC/5/1816-17, Rylance decd, 1816.
58. WPL, Euxton MSS/5/2, 6/24 and 14/6 and JRL, Haigh MSS/Kirkless Colliery Account Book, 1750-6 and 1774-6, Bradshaigh Accounts, 1744-54, Cannel Account, 1747-8, and Cannel, 1749.
59. WPL, Wrightington MSS/Eccleston Box/Dagnall Bundle/Accounts of Charles Dagnall's Colliery.
60. LRO, Land Tax Assessment Returns/West Derby/Orrell, 1783-5. An undated estimate made at Tarbock (60) assumed that five hewers would work in each pit of a five-pit steam drained colliery.
61. LRO, Land Tax Assessment Returns/West Derby/Orrell, 1788-99 and Anderson, 'Blundell's collieries', p.113. The Holme and Blundell colliery output per man figures were obtained by dividing the numbers of men recorded in the Land Tax Assessments into annual accounted outputs.
62. Calculated from weekly accounts which give the daily attendance and output records of hewers. WPL, Euxton MSS/5/2, 6/24 and 14/6.
63. LRO, DDGe(M)/1201, and see above p.143.
64. Barker and Harris, *St Helens*, p.82 and Galloway, *Annals*, p.330.
65. 'Two new Fire Engines' were built in about 1780. LRO, DDWi/Unsorted and Unnumbered Box/Indenture of Feb. 26, 1801.
66. LRO, DDBd/48/3/5.
67. J. Aikin, *A Description of the Country from Thirty to Forty Miles round Manchester* (London, 1795), p.292 and LRO, DDX/905.
68. In 1791 there were four collieries with a total of forty-three men, one with eleven men in 1796 and two with twenty men in 1799. LRO, Land Tax Assessment Returns/West Derby/Windle.
69. The derivation of these estimates is given in Langton, 'Coal output', pp.47-8.
70. Langton, 'Coal output,' p.51.

## 7. Factors affecting location pattern, 1740-99

1. The Bridgewater and Lancaster Canals are dealt with only incidentally and the

abortive scheme to open a canal from Wigan through St Helens to Liverpool is ignored. For details, see C. Hadfield and G. Biddle, *The Canals of North West England* (2 vols., Newton Abbot, 1970), vol. I and J. R. Harris, 'Early Liverpool canal controversies', in J. R. Harris (ed.), *Liverpool and Merseyside: Essays in the Economic and Social History of the Port and its Hinterland* (London, 1969), pp.78-97.

2. Hadfield and Biddle, *Canals,* pp.60-5. In 1768 Leigh held twenty-nine of the thirty-six shares. In 1758 he claimed that the navigation had cost £23,000. LRO, DDWa/Coal, Leigh Estate, 1721-97/Sale Agreement, A. Leigh to H. Leigh, Aug. 1, 1768.
3. JRL, Haigh MSS/Box EIII/Letters of A. Leigh/Bundle 2/Leigh to Bradshaigh, April 22, 1739.
4. JRL, Haigh MSS/Box EIII/Letters of A. Leigh/Bundle 2/Leigh to Bradshaigh, Mar. 7, 1740 and Ashurst to Leigh, Jan. 14, 1740; JRL, Haigh MSS/Collieries Folder/Early Documents relating to Colliery Affairs/Letter, Leigh to Bradshaigh, Mar. 31, 1731.
5. Hadfield and Biddle, *Canals*, p.64.
6. LRO, DDCl/1064.
7. *Ibid.*
8. JRL, Haigh MSS/Box EIII/Letters of A. Leigh/Bundle 2/Leigh to Bradshaigh, April 22, 1739. In 1758 duty was charged even on coal bound from Tarleton to the Fylde, but Leigh petitioned successfully against this. LRO, DDCl/1064.
9. JRL, Haigh MSS/Box EIII/Letters of A. Leigh/Bundle 2/Leigh to Bradshaigh, April 22, 1739, April 9, 1740 and Jan. 2, 1741.
10. JRL, Legh MSS/Box 51/Thoughts on the Dublin Trade from Sankey (n.d., 1763).
11. LRO, DDCl/1064 and DDCa/1/47.
12. The coaster *Liverpool* carried seventy-five and a half tons of cannel and thirty-eight tons of coal and the *Sincerity* fifty-seven tons of cannel and twelve tons of coal from Tarleton to Dublin in 1753. LRO, DDCa/1/47.
13. LplRO, H.380/MD/43 and JRL, Haigh MSS/Bradshaigh Accounts, 1744-54.
14. LRO, DDBa/Account of Sundry Trusts, 1766.
15. The proprietors of Haigh (2) and Pemberton (30) had their own flatts and those of Aspull (12), Winstanley (6), Park Lane (40) and Orrell (1) and (74) sold to the boats of the Tarleton Trading Company. LRO, DDHk/Hawkshead Surveys/Wigan, Survey in Part (n.d.); LRO, DDBa/7/12, Accounts and Rentals/20, 1769-73 and Account of Sundry Trusts, 1766; LRO, DDCa/1/47; JRL, Haigh MSS/Correspondence of Sir R. Bradshaigh, 3rd Bart. and his wife/Leigh to Bradshaigh, 1720-46/Leigh to Bradshaigh, Feb. 14, 1746 and LplRO, H.380/MD/43.
16. For Leigh's concerns see LRO, DDCa/1/47 and LRO, DDWa/Coal, Leigh Estate, 1721-97/Sale Agreement, A. Leigh to H. Leigh, Aug. 1, 1768.
17. LRO, DDCa/1/47 comprises the accounts of the Tarleton Trading Company, 1752-4.
18. WPL, Leigh-Pemberton MSS/loose sheet, Unsigned Articles of Agreement, 1769; LRO, DDCa/1/47 and JRL, Haigh MSS/Kirkless Colliery Account Book, 1750-6 and 1774-6.
19. The Company paid £89. 1. 2 in navigation tonnage dues between Dec. 8, 1752 and Aug. 30, 1754. Assuming that only coal was shipped, this represents 1780 tons at 1/- a ton over thirty-one months. In June and July 1754 each flatt shipped an average of 102 tons per month.
20. JRL, Haigh MSS/Bradshaigh Accounts, 1754-65.

21. LRO, DDBa/Accounts and Rentals/20.
22. LplRO, Holt and Gregson MSS/vol. X, p.231.
23. Hadfield and Biddle, *Canals,* p.64.
24. Hadfield and Biddle, *Canals,* pp.43-7 and 53-9 and T. C. Barker, 'The Sankey Navigation: the first Lancashire canal', *Trans. Hist. Soc. of Lancs. and Chesh.,* 100 (1948), 121-55.
25. Hadfield and Biddle, *Canals,* p.43.
26. University of Liverpool, Department of Geography, 'Some aspects of the human geography of Liverpool in the late eighteenth century', Second Year Students' Field Week Report (mimeo., 1976).
27. Barker, 'Lancashire coal', p.95.
28. *Ibid.* (quoting Thomas Pennant).
29. KCC, 1/V/42/12 and 15.
30. KCC, 1/V/42/13.
31. KCC, 1/V/42/14.
32. LplRO, *Wms Lpl Advt,* Feb. 12, 1762.
33. MS copy in the possession of Mrs D. Bailey, 'Prescot Hall and coal mines, 1761'.
34. LplRO, Binns MSS/vol. XXXI. The canal seems periodically to have been closed for repair over protracted periods, e.g. 'Notice is hereby given that the WATER in Sankey Brook NAVIGATION will be drawn off and the Passage will be stopped for about three weeks'.LplRO, *Wms Lpl Advt,* June 20, 1761.
35. LplRO, *Lpl Gen Advt,* Nov. 1, 1771.
36. *Ibid.* Despite its limitations, the shareholders were 'revelling in near twenty per cent for their capital annually'. Hadfield and Biddle, *Canals,* p.53.
37. For example, the saltworks in Liverpool were supplied from this area rather than via the canal. LplRO, *Lpl Gen Advt,* Aug. 24, 1772.
38. *Ibid.*
39. Barker and Harris, *St Helens,* pp. 29-30 and LplRO, *Lpl Gen Advt,* Jan. 17, 1772.
40. LplRO, *Wms Lpl Advt,* Jan. 8, 1762 and Oct. 10, 1766; *Lpl Gen Advt,* Nov. 1, and Nov. 8, 1771.
41. LplRO, *Lpl Gen Advt,* Nov. 1, 1771.
42. LplRO, Binns MSS/vol. XXXI.
43. LplRO, *Lpl Gen Advt,* Oct. 21, 1774.
44. Hadfield and Biddle, *Canals,* pp.73-4.
45. LplRO, Holt and Gregson MSS/vol. X, pp.238-9.
46. LRO, DDBd/48/8/4, 5, 6 and 7 and H. F. Killick, 'Notes on the early history of the Leeds and Liverpool Canal', *Bradford Antiquary,* N.S., 1 (1900), 169-238.
47. LRO, DDBd/48/8/6.
48. LRO, DDBd/48/8/7.
49. Leeds and Liverpool Canal coal shipments are given in summary for 1781-92 and in more detail for 1786-9 in the Holt and Gregson MSS, and also in two sets of abbreviated accounts for 1783-5 and 1792-1800 in collections held in the Lancashire Record Office. For the seven years during which they overlap, these sets of figures agree almost perfectly. The Holt and Gregson summary data are of tons of 22 cwt, so the others must also be in this measure. Furthermore, the Holt and Gregson summary figures do not include coal shipped down the Douglas Navigation; the 1786-9 figures do, and in the accounts of 1783-5 'coals to Tarleton' on the Douglas are not included in the Leeds and Liverpool coal tonnages, which merely duplicate the quantities given as 'coals to Liverpool' in the Douglas account. Thus, coal to Tarleton was

not included in the Holt and Gregson figure for 1792. This agrees with the figure for that year in the 1792-1800 account, which also, therefore, omits coal which did not travel on the Liverpool line. (Only three per cent of the coal which travelled from the Wigan end along this line was unloaded before Liverpool in 1789.) Coal to Tarleton usually made up about twenty per cent of the total tonnage shipped out of the Douglas valley in 1783-5, 1786-8 and 1791, the years in which figures for the Douglas are extant, so that the total volume taken out of the Central area by canal in 1800 must have been the accounted figure × 22 ÷ 20 + c.twenty per cent, or 303,160 tons of 20 cwt. LplRO, Holt and Gregson MSS/vol. X, pp.233 and 238-9 and vol. XIX, p.98; LRO, DDBd/48/8 and LRO, DP/175/p.65.

50. LplRO, Holt and Gregson MSS/vol. XIX, pp.77-8.

51. Coal exports were measured in chaldrons, but the content of this measure is unknown. Much of Liverpool's exports went to the colonies and coal exported to these markets was measured in Winchester chaldrons. According to a Directory of 1836 all Liverpool's coal was exported in Winchester chaldrons or tons, the dock dues being in the ratio of 1.33: 1. If all the eighteenth-century exports (which included coal sent to Ireland) and coastal shipments were meted in this measure, then the tonnages would be as in figure 40: a Winchester chaldron contained $^{11}/_{21}$ Newcastle chaldrons and a Newcastle chaldron contained 2.65 tons, so that the Winchester chaldron:ton ratio was 1.39:1, similar to the ratio for chaldron: tonnage dues in 1836. It may be, however, that a chaldron of two tons was used on the west coast, as it was until the late seventeenth century, in which case the coastal shipments, at least, should be doubled to yield even higher tonnages. BMus, Add MSS/3874/fol. 32; Nef, *Coal Industry*, vol. II, p.370; LplRO, Holt and Gregson MSS/vol. XIX, p.49 and vol. X pp.379-81; LplRO, W. Enfield, *Essay* (Graingerised by Gregson), p.80; and *Gore's Liverpool Directory*, 1836.

52. LRO, DDBd/48/8/5.

53. LplRO, Holt and Gregson MSS/vol. XIX, p.46.

54. LRO, DDBd/48/8/7 and JRL, Haigh MSS/Canals Box/Letter, Balcarres to Stanhope, 1789.

55. LplRO, Holt and Gregson MSS/vol. XIX, p.78.

56. LplRO, Holt and Gregson MSS/vol. X, pp.241-3.

57. LplRO, Holt and Gregson MSS/vol. X, pp.231-3; LRO, DDBd/48/8/5 and JRL, Haigh MSS/Canals Box/Letter, Balcarres to Stanhope, 1789.

58. LplRO, Holt and Gregson MSS/vol. X, p.231, and LRO, DDBd/48/8/5.

59. Sankey coal was henceforth to be sold only by 'sworn agents in whom the public may confide' and full measure was to be 'certified by a ticket from the keeper of the weighing machine with each cart'. LplRO, *Lpl Gen Advt,* Dec. 2, 1774.

60. It is difficult precisely to compare prices because they varied according to the quantity and type bought and the place of delivery. The most expensive Sankey coal was from Haydock colliery (125) at 7/2d per ton (4.3d per cwt) to the householder in Dec. 1774. House-coal from Warren's Orrell colliery (93) cost 7/6d per ton and 4.5d per cwt in the next month. The least expensive Sankey coal was 6/6d per ton (3.9d per cwt) from Garswood colliery (29). However, in loads of less than one ton coal from all the Sankey collieries cost 4.5d per cwt. The Sankey prices per ton were for coal 'delivered ... at the doors of housekeepers' and the 4.5d per cwt price was for coal 'delivered at the yard'. Warren's 7/6d per ton price for coal (and his 10/- per ton for smith's coal and 14/6d per ton of best and 11/- per ton of small cannel) also included free cartage to any part of Liverpool with a ticket to certify full measure. The cwt

prices of 4.5d, 6d and 8d for best house-coal, small and round cannel respectively were for coal taken from the yard by the customer. The poor, for whom the small loads were intended, had to pay a higher price and the cost of carting. LplRO, *Lpl Gen Advt*, Dec. 2, 1774 and Jan. 27, 1775.

61. Later in 1776, Warren published prices of 6/10d per ton of 20 cwt of 120 lb of house-coal, 9/4d per ton of 20 cwt of 140 lb of smith's coal and 12/6d per ton of 20 cwt of 120 lb of cannel. However, all these prices were exclusive of carting, which was 8d a ton extra and brought all prices except that for cannel up to their January levels. LplRO, *Lpl Gen Advt*, Nov. 1, 1776.

62. LRO, DDBd/48/8/5.

63. LRO, DDBd/48/8/5 and 7.

64. The Rushy Park seam was the St Helens area equivalent of the Smith, a first quality coal in both areas. The Florida seams were also first quality coal in Haydock, although not so good in St Helens. The Smith seam was also mined under this name in Prescot, but the names given to other seams there in the eighteenth century are obsolete and the quality of the coal they produced cannot be ascertained. Jones, Tonks and Wright, *Geological Survey Memoir*, and LplRO, Binns MSS/vol. XXXI, A LIST of the Several CANNEL and COAL MINES . . . 1772.

65. LplRO, Holt and Gregson MSS/vol. X, pp.231-3. In 1833 the Leeds and Liverpool Canal supplied 270,753 tons to Liverpool (not much more than at the turn of the century), the Sankey Navigation 123,928 tons and the Prescot turnpike 120,000 tons. Hadfield and Biddle, *Canals*, p.176.

66. LRO, DDBd/48/8/5.

67. Sankey coal sold at 8d a ton cheaper to shipping than to householders. It also cost 8d to cart coal from the terminus of the Leeds and Liverpool Canal to shipping and to householders and no separate price was published for coal delivered on board ship from the Leeds and Liverpool. Thus, in Dec. 1774 Sankey coal ranged from 5/10d to 6/6d a ton on board vessels and Douglas coal must have cost 7/6d a ton. Sankey prices to shipping were 6/3d in 1785 when Douglas coal still, presumably, cost 7/6d. LplRO, *Lpl Gen Advt*, Dec. 2, 1744 and LRO, DDBd/48/8/5.

68. This is clearly implied in a number of contemporary descriptions in which the export trade is equated with the shipment of Leeds and Liverpool coal. LplRO, Holt and Gregson MSS/vol. X, pp. 231-3 and LRO, DDBd/48/8/5, 6 and 7.

69. Calvert, *Salt in Cheshire*, p.123.

70. LRO, DP/258.

71. JRL, Legh MSS/Box 59/Letter, Gatley to Legh, Feb. 24, 1756 and Box 43/Letter, Richardson to Legh, Aug. 18, 1770.

72. KCC, 1/V/38 and 1/V/42/14 and WPL, Leigh-Pemberton MSS/Agreement, Mar. 2, 1776. Disbursements for paving are common in accounts of the period.

73. D. Anderson, *The Orrell Coalfield, Lancashire, 1740-1850* (Buxton, 1975), pp.108-11 and E. H. Clegg, 'Some historical notes on the Wigan coalfield', *Trans. Inst. Min. Eng.*, 117 (1957-8), 795-7.

74. Anderson, *Orrell Coalfield*, pp.108-11.

75. On different occasions the Earl of Balcarres pursued the possibility of building a wagonway and a canal from Haigh to the Leeds and Liverpool Canal but the concessions he required from the canal proprietors were never forthcoming. JRL, Haigh MSS/Canals Box/Letters, Balcarres to the Leeds and Liverpool Canal Proprietors, April 28, 1788, and Balcarres to Stanhope, 1789.

76. Clegg, 'Historical notes' and Aikin, *Description*, p.291.

77. LplRO, *Lpl Gen Advt*, July 14, 1769 and June 19, 1778; JRL, Legh MSS/Box

43/Letter, Richardson to Legh, Sept. 29, 1769 and LRO, DDGe(E)/864 and (M)/818.

78. Bailey, 'Turnpike minutes'.
79. *Ibid.,* pt 2, p.33.
80. The takings at the Ashton gate were much less than those at the other gates and by 1764 the Ashton section was 'much out of repair'. Bailey, 'Turnpike minutes', pt 2, pp.46 and 50.
81. Collieries were closed at Gillars Green (99), Windle Ashes (116) and Eccleston and Sutton (100 and 126).
82. Bailey, 'Turnpike minutes', pt 2 and Aikin, *Description,* p.363.
83. Bailey, 'Turnpike minutes', pt 2.
84. *Ibid.*
85. JRL, Legh MSS/Box 49/Letter, Worsley to Legh, Sept. 25, 1752; Box 51/Letters, Leigh to Legh, July 15, 1759 and Feb. 17, 1760; Box 59/Letter, Lilliman to Orford, July 28, 1762 and Box 56/Letter, Legh to Orford, April 30, 1768. Legh blocked competitors' wayleaves to the turnpikes and was concerned about the relative advantage of his Haydock collieries in relation to the turnpikes and to the common roads which could be used to by-pass them and avoid tolls.
86. Aikin, *Description,* p.306.
87. Harrison, 'Pre-turnpike highways', p.113.
88. JRL, Haigh MSS/Canals Box/Broadsheet . . . Directed to the Committee of the Lancaster Canal Navigation, July 2, 1799. Coal from Orrell and Haigh had a considerable quality advantage over that from the collieries of the North; in the early nineteenth century coal from Orrell cost 17/10d a ton in Preston, that from Haigh 16/- and that from Adlington in the North 12/6d in reflection of their qualities. Hadfield and Biddle, *Canals,* p.165.
89. Awty, 'Charcoal ironmasters', p.103 and LplRO, *Lpl Gen Advt,* Oct. 14, 1774.
90. Awty, 'Charcoal ironmasters', p.104; LRO, DDX/379/2; LRO, The Will of Thomas Chadwick, 1780, and LplRO, *Lpl Gen Advt,* June 4, 1773.
91. LplRO, *Lpl Gen Advt,* Nov. 28, 1766.
92. LplRO, *Lpl Gen Advt,* May 5, 1775.
93. JRL, Haigh MSS/Canals Box/Earl of Balcarres' Objections to the Proposed Line of the Leeds and Liverpool Canal, n.d. and A. Birch, 'The Haigh Ironworks . . .', *Bull. of the John Rylands Library,* 35 (1953), 316-34. Robert Daglish built *The Walking Horse,* the first Lancashire locomotive, at Haigh for use on the Orrell wagonways.
94. LRO, DDGe(E)/815 and LplRO, *Lpl Gen Advt,* Feb. 22, 1771.
95. LRO, DDGe(E)/842.
96. Birch, 'Haigh ironworks', p.327. Ironstone was leased out in Parr and Windle in 1759 and an ironstone getter was brought into Parr from Endley, Staffs., in 1743 but he became a collier. LRO, DDGe(E)/815 and StHRL, Parr Poor Law Papers/April 22, 1743.
97. J. J. Cartwright, *The Travels through England of Dr. Richard Pococke . . . during 1750, 1751 and Later Years,* Camden Society (2 vols., London, 1888), vol. II, p.8.
98. The colliery was linked to the ironworks by a short canal. JRL, Haigh MSS/Canals Box/Articles of Agreement, Lewis v. Haliburton, June 2, 1795.
99. Laffack (118) seems to have been the only colliery in the district which could produce coal that would 'sowder', but 'piking the best of the Coale out for the furnis has Don the other part of the Coale a Deale of hurt', reducing its value by 1/- a work. JRL, Legh MSS/Box 51/Letters, Leigh to Legh, Feb. 17, 1760 and Searjeant to Legh, May 4, 1760.

100. LplRO, H.380/MD/43; LRO, DDCa/1/47 and LplRO, *Lpl Gen Advt,* Oct. 14, 1774.
101. PPL, *Preston Guardian,* April 26, 1884.
102. LplRO, *Wms Lpl Advt,* Nov. 28, 1766.
103. LplRO, *Lpl Gen Advt,* June 4, 1773.
104. Birch, 'Haigh Ironworks', p.326 and JRL, Haigh MSS/Canals Box/Letter, Balcarres to Stanhope, 1789.
105. Awty, 'Charcoal ironmasters', p.109.
106. Aikin, *Description,* p.312
107. Unless otherwise indicated, information on the glass industry is taken from Barker and Harris, *St Helens,* pp. 108-19.
108. Aikin, *Description,* p.313.
109. J. R. Harris, 'The copper industry in Lancashire and North Wales, 1760-1815', unpubl. PhD thesis (Liverpool, 1952), p.93.
110. Sailcloth-making was Warrington's main industry but pins, locks, hinges, iron founding and brewing and malting were also important. Aikin, *Description,* p.302.
111. LplRO, Holt and Gregson MSS/vol. XIX, pp.77-8 and Harris, 'Copper industry', p.93.
112. LRO, DDGe(E)/459 and 864. All necessary coal was to be supplied by the Gerard's colliery (29) at 3d, 4d or 5d a ton below rates on the Sankey depending upon the prevailing price.
113. Unless otherwise indicated, information on the copper industry is taken from Harris, 'Copper industry', Barker and Harris, *St Helens,* pp.75-89 and J. R. Harris, *The Copper King* (Liverpool, 1964).
114. Barker and Harris, *St Helens,* p.84.
115. *Ibid.* and LRO, DDGe(E)/459.
116. LRO, DDK/1500/5, 11, 13 and 14 and 1526/12, 15 and 16.
117. Cartwright, *Richard Pococke,* vol. II, p.8.
118. LRO, DDK/2021/18-20.
119. Aikin, *Description,* p.21.
120. Hadfield and Biddle, *Canals,* p.188. Preston received 21,000 tons of coal along this stretch in 1799.
121. Aikin, *Description,* p.311.
122. Dagnall's coal sold at 3d per basket in 1760, ½d more than Prescot and 1¾d more than Garswood prices. He mined high grade coking coal from the Rushy Park seam.
123. Some of this variety may have been more apparent than real due to unrecorded variations in the contents of measures, a strategy which seems to have been common on the Sankey (see above p.228).
124. The Skelmersdale price was recorded later than those of the collieries on the Leeds and Liverpool, which may also have risen in unison with prices in Liverpool itself in the 1790s.
125. LRO, DDM/4/14 and 15.
126. MS copy in the possession of Mr D. Anderson/undated sheet relating to letters of 1734. These may represent the seams worked by Clarke and Co. (170) in the late 1790s.
127. LRO, DDMc/68/14.
128. LRO, DDHc/18/2.
129. LRO, DDM/4/18.
130. See, for example, WPL, Wrightington MSS/Eccleston Box/Old Coal Leases in Eccleston and Sutton/Eccleston v. Case, July 31, 1769 and Eccleston v.

Mackay, June 24, 1778.

131. LplRO, Binns MSS/vol. XXXI, A LIST of the several CANNEL and COAL MINES . . . 1772. Numerous as yet unworked seams in the area between Wigan and St Helens are listed in this document.

132. In 1789 it was claimed that 'above 50 stat. acres of coal, four feet thick, are now exhausted annually' in that area. Both the Arley (Orrell four foot) and Smith (Orrell five foot) mines were about four foot thick there, so that about twenty-five acres must have been exhausted each year around 1789.

133. LplRO, Binns MSS/vol. XXXI, 'A LIST of the several CANNEL and COAL MINES . . . 1772' and JRL, Haigh MSS/Alexander, 6th Earl of Balcarres: Management of Mining Estates, 1812-22.

134. Pits were generally deepest in the South West. Though coal thirteen and a half yards deep was advertised for sale in Huyton in 1771 (LplRO, *Lpl Gen Advt,* Dec. 20, 1771), this was exceptionally shallow. Pits of between sixty and ninety yards were worked at Whiston colliery (34) in 1746 (LRO, DDWi/Box 43/Letter, Moss to Clayton, Mar. 24, 1746); sixty yards was reached at Tarbock (60) in the 1750s (LRO, DDM/4/27); and seventy yards at Windle Ashes (116) in 1771 (LplRO, *Lpl Gen Advt,* Feb. 1, 1771). At Prescot Hall (5) the 'shallow' coal was mined at thirty yards in 1750 (cf. twelve and twenty yards in 1735) and the deep coal at sixty and sixty-six yards (KCC, 1/V/38). Various depths were later recorded at this colliery: forty to fifty yards in 1758 (MS draft in the possession of Mrs D. Bailey, 'Coroners inquests at Prescot' by F. A. Bailey; Mr Bailey published a less detailed version of this paper in *Trans. Hist. Soc. of Lancs. and Chesh.,* 86 (1934), pp.21-40); forty-eight yards in 1759 (*ibid.*); eighty yards in 1767 (KCC, 1/V/74); and the engine pits of 1744 and 1753 were seventy and ninety-three and a half yards, respectively (KCC, 1/V/38).

In the upper Sankey basin similar depths were reached after the opening of the Sankey Navigation. New reserves at four yards were offered for sale in Windle in 1759 (LplRO, *Wms Lpl Advt,* July 20, 1759) and crop coal was worked at Sutton (126) in 1763 (WPL, Wrightington MSS/Eccleston Box/Old Coal Leases, Eccleston and Sutton/Eccleston to Case, July 31, 1769). However, two of the seams worked at Ravenhead (129) were forty-five and fifty yards deep in 1778 (*ibid.,* Eccleston to Mackay, June 24, 1778); St Helens colliery (120) worked at sixty and seventy yards in 1762 and 1763 (*ibid.,* Eccleston to Leigh & Co., May 19, 1762 and LRO, DDSc/12/132); and the Haydock collieries of Peter Legh (28, 124 and 125) descended from twenty-two yards in 1708 (JRL, Legh MSS/Box 55/Letter, Gatley to Legh, Feb. 24, 1756) to forty-four yards in 1758 (JRL, Legh MSS/Box 55/Letter, Dodge to Legh, Jan. 8, 1758) to 95 and 120 yards in 1783 (LRO, DDGe(M)/873).

In the middle Douglas basin workings were generally shallower. Ninety yards of walling stone were bought for a pit at Haigh (2) in 1745 (JRL, Haigh MSS/Bradshaigh Accounts, 1744-54), but the same Cannel mine was tapped at forty-three yards two inches on lower ground at Kirkless (104) in 1752 (*ibid.,* Kirkless Colliery Account Book 1750-6 and 1774-6). The pits of Holme's Orrell colliery (74) were twenty-nine to thirty yards deep in 1747 (LRO, DDBa/Up Holland Box/ Hiring of Colliers, 1744-8) and coal lay at forty yards in the nearby Livesey holding in 1773 (LRO, DDX/233/5). James Stock's Park Lane colliery (40) was only twenty-nine yards deep in the mid 1740s (LRO, DDGe(M)/1185).

As before 1740 deep pits existed on the *Lenisulcata* measures. On one

Parbold holding the coal plunged from fourteen and a half to forty-seven to sixty-seven yards between three bore holes (MS copy in the possession of Mr D Anderson, undated note, c.1794).

135. JRL, Haigh MSS/Bradshaigh Accounts, 1744-54 and 1765-74.
136. JRL, Haigh MSS/Kirkless Colliery Account Book, 1750-6 and 1774-6.
137. Cartwright, *Richard Pococke,* vol. I, p.206.
138. Bailey, 'Coroners' inquests', p.27.
139. WPL, Wrightington MSS/Eccleston Box/Old Coal Leases Eccleston and Sutton/Eccleston to Leigh, May 19, 1762.
140. JRL, Legh MSS/Box 43/Letters, Richardson to Legh, Feb. 13 and Sept. 17, 1767.
141. This technique seems to have been used at Haigh (2) where 'a horse-hide for the Bellows, currying', cost 15/2d in 1748. JRL, Haigh MSS/Bradshaigh Accounts, 1744-54.
142. JRL, Legh MSS/Box 43/Letters, Richardson to Legh, Jan. 4 and Feb. 10, 1767 and July 31, 1768.
143. See numerous letters in WPL, Wrightington MSS/Eccleston Box/ Bundle of Papers concerning Charles Dagnall.
144. KCC, 1/V/38 and LRO, DDWi/Unsorted Box/Indenture of Feb. 26, 1801.
145. PRO, DL1/503(2)/Dec. 5, 1760.
146. WPL, Leigh-Pemberton MSS/loose papers/Letter, Jackson to Leigh, Oct. 30, 1769.
147. JRL, Haigh MSS/Kirkless Colliery Account Book, 1750-6 and 1774-6. The horse, wind or water powered engines and soughs of Haigh (2), Orrell (1 and 74), Florida (125) and Windle Ashes (116) were all overwhelmed before the eventual installation of steam engines.
148. JRL, Legh MSS/Box 51/Letter from R. Leigh, Feb. 14, 1760 and Box 59/Letter, Serjeant to Legh, Nov. 10, 1762. No later complaints about labour supplies exist in the voluminous Legh correspondence.
149. The first advertisement for 'any number of good COLLIERS or SINKERS' appeared in 1770 and the last, for 'THIRTY COLLIERS with able DRAWERS' in 1773, just before coal from 'The NEW PITTS' at Ravenhead (59) was offered for sale. LplRO, *Lpl Gen Advt,* Feb. 20, 1770, July 23, 1773 and Aug. 20, 1773.
150. JRL, Haigh MSS/Bradshaigh Accounts, Disbursements, 1750.
151. A number of Orrell (74) and Tarbock (60) bonds have survived and earnest was paid at annual hirings at Winstanley (6). LRO, DDBa/Up Holland Box/Hiring of Colliers, 1744-8 and Colliery Accounts, 1766 and DDM/4/5.
152. LRO, Land Tax Assessment Returns/West Derby/Orrell, 1782-99.
153. A. Aspinall, *The Early English Trade Unions* (London, 1949), pp.7-8. 'Colliers and cannellers from distant works' came to watch the 'progress of our [Orrell] people'. The only cannel collieries were to the east of Wigan.
154. The reconstitutions for Up Holland comprise all those colliers who were first registered as adults before 1764, including all later entries for them. All the families of St Helens colliers first mentioned as adults before 1790 were reconstituted, again including all subsequent entries for those men. Thus, the Up Holland families do not include any formed or substantially completed after the Leeds and Liverpool Canal was opened.
155. A labourer and two husbandmen were killed, on different occasions, whilst at work in the Prescot pits. Bailey, 'Coroners' inquests' (MS).
156. LRO, RCSt/6 and CRO, EDC/5/1816-17, Rylance decd, 1816.
157. Bailey, 'Coroners' inquests' (MS) and P. E. H. Hair, 'The Lancashire collier girl', *Trans. Hist. Soc. of Lancs. and Chesh.,* 120 (1968), 63-86.

158. There were adults and their fathers as well as adults and their children at work in the same colliery in Ashton and Prescot. LRO, DDCs/3/39 and Bailey, 'Coroners' inquests' (MS).

159. In an area where the industry was stable a fecund population would continually produce a surplus of juveniles and young adults who were inured to and trained in mining. This may be the reason for the occasional appearance of colliers 'of Blackrod' in the North in the registers of the churches and chapels of the South and Centre throughout the eighteenth century.

160. Between Sutton and Hindley in 1752; Newcastle and Rainford, 1756; Farnworth (Bolton) and Pemberton, 1762; Wrightington and Wigan, 1782-4; Hindley and Orrell, 1786-92; Blackrod and Wigan, 1782 (2) and 1789; Parr and West Leigh, 1792-4; Bolton and St Helens, 1797 and Clifton (Manchester) and Orrell, 1797-9.

161. It is impossible to decide the direction and type of movement from entries of non-local townships in a particular register. Movement may have been either way and involved the collier alone, leaving his wife and family behind, or the whole family.

162. A comparison with similar evidence for four Gloucestershire townships is salutary. A high proportion of the Poor Law contacts made with Stonehouse, Stroud, North Nibley and Kingswood were with places outside the county, reaching as far as London, Lancashire and Devon. See University of Liverpool, Department of Geography, 'Some aspects of the historical geography of Gloucestershire and Gloucester in the eighteenth century', second year students' field week report (mimeo., 1975), p.13.

163. Challinor, *Lancashire and Cheshire Miners*, p.17.

164. Only three of the twenty-six colliers and wives married after the Hardwick Marriage Act could sign their names.

165. Bailey, 'Coroners' inquests' (MS). The ages of thirteen of the dead were given. Only one was over sixteen and it seems likely from this and other details in the inquests that enquiries were only generally made into child deaths.

166. P. E. H. Hair, 'Accidental death and suicide in Shropshire, 1780-1809', *Trans. Shropshire Archaeol. Soc.*, 59 (1969-70), 63-75.

167. The average age at burial of the seven whose baptisms and burials were registered was 48.3 years. The average age at marriage of the thirteen whose baptisms and marriages were registered was 26.2 years and the average age at which they had their first child was 28.4 years, over two years more than the age for all forty-nine whose own and first child's baptisms were registered. The average age at marriage plus the product of the average number of baptisms and the average interval between baptisms plus the average number of years lived after the last child's baptism gives an estimated collier life-span of 48.1 years. The average age at the baptism of the first child plus the product of the average number of baptisms minus one and the average interval between baptisms plus the average number of years lived after the baptism of the last child gives an estimate of 45.6 years. These discrepancies indicate the danger of drawing conclusions from such small and incomplete samples as these but it seems reasonably clear that colliers could not expect to live beyond their middle or late forties.

168. JRL, Haigh MSS/Bradshaigh Accounts, 1744-54 and 1765-74.

169. JRL, Legh MSS/Box 43/Letter, Richardson to Orford, June 4, 1767 and Box 58/Letter, Richardson to Orford, Sept. 1, 1767 and WPL, Wrightington MSS/Eccleston Box/Dagnall Bundle/Letter, Dagnall to Eccleston, Aug. 24, 1757. Men were, however, laid off if getting stopped for long periods. JRL,

Legh MSS/Box 58/Letter, Richardson to Orford, Sept. 1, 1767.
170. LRO, DDM/4/5 and 6. The Winstanley steward advised the coalowner at Tarbock in c.1760 that 'if there is good sale you may help the getters by reducing the weight which helps the Lord more than the undertaker'.
171. LRO, DDBa/Up Holland Box/Hiring of Colliers, 1744-8; WPL, Leigh-Pemberton MSS/Deeds Reserving Rents and Coalmines, Orrell/notes on reverse of Mr Porter's Account of Coal got at Hall Orrell; LRO, DDBa/Accounts and Rentals/20 and Aspinall, *Trade Unions,* p.8.
172. As at Winstanley (6). See note f of table 23. In his advice to the owner of Tarbock the Winstanley steward said that 'if prices advance wages usually advance with them'. LRO, DDM/4/6.
173. LRO, DDBa/Colliery Accounts, 1766; Hair, 'Collier girl', and CRO, EDC/5/1816-17, Rylance decd, 1816.
174. JRL, Haigh MSS/Kirkless Colliery Account Book, 1750-6 and 1774-6. The measures used at the colliery changed between these two periods and the rate per basket fell slightly in consequence. See note h of table 23.
175. LRO, DDSc/1/1.
176. LRO, DDM/4/5 and 6.
177. LRO, RCSt/6; WPL, Leigh-Pemberton MSS/Deeds Reserving Rents and Coalmines, Orrell/notes on reverse of Mr Porter's Account of Coal got at Hall Orrell; LRO, DDBa/Accounts and Rentals/20 and Colliery Accounts 1766.
178. Anderson, *Orrell Coalfield,* p.125 and Aspinall, *Trade Unions,* p.8.
179. The rates suggested to the owner of Tarbock by an unknown correspondent were 1/10d per work in seven foot coal, 2/2d in six foot coal, 2/6d in five foot coal, 2/10d in four foot coal and 3/2½d in three foot coal. The Winstanley steward suggested rates ranging from 2/4d to 3/3d per work. LRO, DDM/4/5 and 6.
180. For a very complete list of fines for Tarbock (60) in 1750 see LRO, DDM/4/4.
181. LRO, DDBa/Colliery Accounts, 1766.
182. WPL, Euxton MSS/5/2 and CRO, EDC/5/1816-17, Rylance decd, 1816.
183. KCC, 1/V/38, p.36.
184. *Remarks on the Salt Trade . . . in reference to the Weaver Navigation the Coal Trade and the Revenue Laws* (1804), pp.23-4 (quoted in Barker and Harris, *St Helens,* p.163n).
185. WPL, Euxton MSS/6/24, 14/6 and 5/2. At Meadow pit, 1772-3, 24.8 per cent of the working days were lost through absence but only 5.8 per cent on days when all the getters were not out. At Brook pit, 1772-3, the percentages were 25.7 and 10.4; at Meadow pit, 1795-6, 23.4 and 7.9, and at Peter Hilton's pit, 1799-1800, only seventeen per cent of the total days were lost by absence of all kinds and one-third of it was due to James Wareing alone. There is no evidence of St Monday in these accounts, Tuesday being the day most commonly not worked. The pits were shut for one day at Christmas and two at New Year.
186. LRO, DDBa/Colliery Accounts, 1766.
187. LRO, RCSt/6.
188. CRO, EDC/5/1816-17, Rylance decd, 1816. It could be argued that Topping's high earnings indicate the extent to which the other getters worked less than full time. Unfortunately, the Skelmersdale (105) accounts contain no record of the days worked by hewers, but the Kirkless (135) and Winstanley (6) records do not support such an argument unless the men deliberately concerted their individual patterns of absenteeism, which seems highly unlikely.
189. E. W. Gilboy, *Wages in Eighteenth Century England* (Cambridge, Mass., 1934), pp.174 and 180-5.

190. Barker and Harris, *St Helens*, p.86. A copper refiner earned 16/- a week in 1780.
191. Barker and Harris, *St Helens*, p.73. Flatt masters were said to earn £90 a year and to have rejected 50 gns in 1792.
192. The highest craftsman's rate quoted in Gilboy, *Wages*, is 2/10d.
193. Wages at St Helens copper works varied between 1/6d and 3/- a day in 1805, no more than colliers earned ten years earlier. Barker and Harris, *St Helens*, p.88.
194. Barker and Harris, *St Helens*, p.163n. The data presented here substantiate the conclusion of Ashton and Sykes that colliers were highly paid and not the opposite contention made by both Nef and Gilboy. However, the wages offered in an advertisement in the Liverpool press in 1795 for miners in the midlands were only 2/6d to 3/- a day, much less than a good hewer could earn in South West Lancashire. Ashton and Sykes, *Coal Industry*, pp.147-8; Gilboy, *Wages*, p.185; Nef, *Coal Industry*, vol. II, pp.195-6 and Barker and Harris, *St Helens*, p.91.
195. Barker and Harris, *St Helens*, p.163n.
196. Every six weeks at Tarbock (60) in 1750. An apprehensive steward wrote from Haydock (28), where the pits were producing little in 1767, that there was 'not ¼d left and the machine, Carters, Labourers at Haydock &c are wanting and have none'. LRO, DDM/4/4 and JRL, Legh MSS/Box 43/Letter, Richardson to Orford, Jan. 4, 1767.
197. The getters at Tarbock (60) bound themselves to work at night if the coalmaster should instruct them to, whilst at Prescot (5) it was customary to work through the night. LRO, DDM/4/4; KCC, 1/V/38, p.36 and Bailey, 'Coroners' inquests' (MS), where all the times given for accidents are between midnight and noon.
198. Sixteen did so. One left £36, mostly in cash owing and bonds, another £13.6.9 plus £2 p.a. income from a piece of land.
199. Information on journeys to work is very difficult to come by. Colliers gave places of residence in Burtonwood, Newton and Rainhill, all just off the coalfield, in parish registers, and one copperhouseman of Burtonwood must have travelled two or three miles to work in St Helens. William Fowler of Coppull, 'banksman at Standish coalpits', must also have travelled over two miles to work. An attempt was made to match lists of colliers given in mining archives with names and places of residence given in parish registers. Relatively few could be coupled and there is always the possibility that there were two colliers with the same name, a different one in each source of information. For what they are worth, the results are as follows. Of the six colliers who worked in the collieries on the border of Winstanley and Pemberton (122 and 30) in 1760 and whose names can be traced in the registers of Up Holland and Billinge, four lived in Orrell, one in Up Holland and one in Winstanley. Of ten at work at Winstanley (6) in 1766, four lived in Orrell, four in Winstanley, one in Up Holland and one in Pemberton. Three workers at Haigh (2) in 1769 lived in Wigan, seven in Haigh and five in Aspull; one of them moved from Haigh to Standishgate and then to Wigan Lane in Wigan between 1787 and 1795 and another from Haigh to Wigan Lane between 1787 and 1791. Three of the Haigh (2) auditors of 1788 lived in Haigh, Aspull and Wigan Lane. Some of the workers at Kirkless (135) in 1792-3 and 1799-1800 gave places of residence in Ince, Haigh, the contiguous Scholes, Woodhouse and Whelley areas of Wigan, Winstanley and Orrell in the registers of the 1790s. Few of these distances need have been much more than a mile and those that were could have been bridged by migration between

the dates of registration in the two sources.

200. WPL, Leigh-Pemberton MSS/Deeds Reserving Rents and Coalmines, Orrell/Mr Porter's Account of Coal got at Hall Orrell.
201. JRL, Haigh MSS/Canals Box/Letter, Balcarres to Stanhope, 1789.
202. LRO, DDGe(M)/1201.
203. LRO, Land Tax Assessments/West Derby/Orrell, 1799.
204. JRL, Haigh MSS/Colliery Accounts, 1788.
205. LRO, DDM/4/7.
206. LRO, DDGe(M)/1185 and DDBa/Colliery Accounts, 1766.
207. JRL, Haigh MSS/Bradshaigh Accounts, 1744-54.
208. The Park Lane (40) pit was, at forty-nine yards, the shallowest recorded in this period. LRO, DDGe(M)/1185 and DDBa/Colliery Accounts, 1766.
209. LRO, DDBa/Accounts and Rentals/20.
210. LRO, DDBa/The Case of Howarden v. Bankes/Howarden's Acct. of claim agt. Wm. Bankes . . . 1760.
211. LRO, DDGe(M)/1185.
212. JRL, Haigh MSS/Bradshaigh Accounts, 1765-74.
213. The lower price was for an engine with a thirty-six inch cylinder to draw 3565 hogsheads per sixteen hours, the higher for one with a forty-four inch cylinder to draw 5851 hogsheads per sixteen hours (LRO, DDM/4/7). According to John Curr in 1797 the total purchase and installation costs of colliery steam engines ranged from £928 for a thirty inch cylinder machine to £2741 for one with a sixty inch cylinder. Anderson, *Orrell Coalfield,* p.88.
214. JRL, Haigh MSS/Canals Box/Letter, Balcarres to Stanhope, 1789.
215. Collieries (43), (109), (125) and (29) were definitely linked to the Sankey by wagonway and there seems to have been a network, begun at (93) and (139), which linked most of the large collieries in Orrell to the Leeds and Liverpool Canal. An underground canal linked (41) to the canal in Standish. LplRO, *Lpl Gen Advt,* June 19, 1778 and July 14, 1769; JRL, Legh MSS/Box 43/Letter, Richardson to Legh, Sept. 29, 1769; LRO, DDGe(E)/864 and (M)/818; Anderson, *Orrell Coalfield,* pp.108-11 and Clegg, 'Historical notes', pp. 795-7.
216. KCC, 1/V/42/14. W. T. Jackman, *The Development of Transportation in Modern England* (2 vols., Cambridge, 1916), vol. II, p.472n, gives a cost of £3000 per mile for wagonways in South Wales in the early nineteenth century.
217. LRO, DDBa/Accounts and Rentals/20.
218. Anderson, *Orrell Coalfield,* p.111.
219. JRL, Legh MSS/Box 51/Letters, Mackay to Legh (n.d.) and reply Leigh to Mackay, Sept. 9, 1763. This was the cost of a flatt to carry twenty works. The common size on the Sankey was forty tons, two-thirds that capacity.
220. WPL, Wrightington MSS/Eccleston Box/Dagnall Bundle/Accounts of Charles Dagnall's Colliery, 1747-65 and Leigh-Pemberton MSS/Deeds Reserving Rents and Coalmines, Orrell/Mr Porter's Account of Coal got at Hall Orrell.
221. PRO, PL6/85/35.
222. Dagnall (100) spent £200 to combat flooding in 1757 and Howarden (30) £454.10.1½ between July 1758 and Jan. 1760. Neither colliery was big enough normally to yield much more than £200 per year profit.
223. KCC, 1/V/38.
224. Alexander Leigh was castigated for not paying immediately for cannel from Haigh (2) in 1746 and the colliery's advertisement tickets had 'no Trust' printed on them. JRL, Haigh MSS/Sir Roger Bradshaigh, 3rd Bart., Correspondence, family and others/Alexander Leigh to Sir R. Bradshaigh,

1720-46/Feb. 14, 1746 and Bradshaigh Accounts, 1754-65.
225. LRO, DDBa/An Account of Sundry Trusts . . . 1766.
226. LRO, DDBa/Accounts and Rentals/20. The Blundells owed £964 in 1790.
227. LRO, DDM/1/143.
228. JRL, Legh MSS/Box 59/Letter, Gatley to Legh, Feb. 24, 1756 and Box 51/Letters, Leigh to Legh, Feb. 17, 1760 and May 2, 1760.
229. LRO, Land Tax Assessment Returns/West Derby/Orrell, 1799.
230. *Ibid.*, Shevington, 1799. Dicconson, who was double taxed as a Papist, paid only 5/1½d. Lofthouse & Co. (155) paid £1. 10. 5 and German & Co. (49?) £3. 0. 9.
231. For an autobiographical account of his career see JRL, Haigh MSS/Alexander, 6th Earl of Balcarres, Management of Mining Estates. Proud of his ancestry though he was, he urged his heir to remember that 'Colliers we are and colliers we must ever remain'.
232. For example, Samuel Bold of Wigan, yeoman, Ralph Bradshaw of Wigan, yeoman, James Bradshaw of Parbold, yeoman, Richard Culcheth of Orrell, yeoman and William Briggs of Preston, wine merchant (132). LRO, DDX/233.
233. For example, James Bradshaw (1 and 132), Alexander Leigh (30) and Winstanley, Hatton and Barton of the Tarleton Trading Co. (104).
234. LRO, DDWa/Coal, Leigh Estate, 1741-87/loose paper in Lease, H. Leigh to J. Longbotham, Jan. 20, 1775.
235. LRO, DDGe(M)/1185 and PRO, PL6/85/35. The mortgage was granted by the Rev. Richard Garnett of Middleton Cheney, Northants.
236. PRO, PL6/83/32 and LplRO, *Wms Lpl Advt*, July 20, 1759 and *Lpl Gen Advt*, Dec. 23, 1774 and Feb. 3, 1775.
237. KCC, 1/V/74/ Letter, Chorley to Smith, June 16, 1763.
238. The full partnership was Thomas Makin of Knowsley, Esq., John Williamson of Liverpool, Esq., John Chorley of Prescot, Gent., John Houghton of Prescot, Gent. and Robert Eden of Prescot, Gent. KCC, 1/V/74/Petition against the Erection of an Engine, Nov. 15, 1774.
239. KCC, 1/V/73/undated and unsigned loose sheet.
240. Barker and Harris, *St Helens*, p.113n.
241. An unexecuted lease was drawn up on coal in Parr and Ashton in 1784 between the Gerards and Thomas Marshall, Esq., a Cheshire saltboiler. In 1801 Gerard coal in Windle was leased by William Leigh of Liverpool, merchant, John Hewson of Middlewich, 'doctor in phisick', John Thompson of Northwich, salt proprietor, George Leigh of Middlewich, salt proprietor, John Leigh of London, 'Gent.', Joseph Leigh of Liverpool, merchant, George Jackson of Liverpool, merchant, William Carter of Northwich, salt proprietor, John Whiteley of Ashton-in-Makerfield, merchant and Thomas Bridge of Eavenham, Cheshire, salt proprietor. LRO, DDGe(E)/870 and (M)/469.
242. WPL, Leigh-Pemberton MSS/Lease, Leigh to Longbotham, Jan. 20, 1775.
243. LRO, DDBb/1/1.
244. For example, John Jarratt of Bradford, merchant, John Lofthouse of Liverpool, coalmerchant, Joseph Dawson of Idle, Yorks., dissenting minister and John Hardy of Bradford, attorney (160) and Edmund Peckover of Bradford, woolstapler, Jonathan Peckover of Wisbech, banker, William Hustler of Bradford, woolstapler, James Monk of Burscough, Lancs., timber merchant, John Jarratt of Bradford, 'Gent.', John Lofthouse of Liverpool, coalmerchant, Joseph Danson of Royds Hall, Bradford, ironmaster, John Hardy of Horton, Bradford, 'Gent.', Henry Ellam of Standish, 'Gent.' and

William Robinson of Up Holland, yeoman (41). LRO, DDX/233/8 and DDX/905.

245. The pithead price of a basket of cannel rose from 5d at Kirkless (104) in 1774 to 5½d and 6½d at an adjacent works (135) in 1792. The price of 6½d remained current until 1799 when it jumped to 8d.

246. It is remarkable that there is no record of coal being produced at cannel collieries except at Haigh (2) at the end of the century where it was mined in separate pits. In 1788 there were six coal and six cannel pits at Haigh (2). JRL, Haigh MSS/Colliery Accounts, 1788.

247. In 1778, Orrell coal was actually cheaper than Standish coal at the pithead, selling at 2½d compared with 3d, although it was more expensive in Liverpool where smith's coal cost 9/4d per 2800 lb and ordinary coal 6/10d per 2400 lb. In 1818, Orrell coal from the Leeds and Liverpool cost 16/- a ton in Preston and coal from Standish and Pemberton only 12/5d per ton. See LplRO, *Lpl Gen Advt,* Nov. 1, 1776 and Hadfield and Biddle, *Canals,* p.165.

248. In 1761 it was complained that 'there has been a Kind of Smiths Coals carried down the Sankey Navigation, and sold in Liverpool for Winstanley Coals, tho' much inferior in Quality ... there is now in Carting, down the Sankey Navigation, a Quantity of the right Winstanley coals ... the Goodness and Usefulness of the said Coals is so well known in Liverpool ... as also their Superiority in Quality to those lately sold in Liverpool'. LplRO, *Wms Lpl Advt,* May 15, 1761.

249. JRL, Haigh MSS/Canals Box/Letter, Balcarres to Stanhope, 1789.

250. LplRO, Binns MSS/vol. XXXI, A LIST of the CANNEL and COAL MINES ... , 1782.

251. LplRO, *Lpl Gen Advt,* Aug. 20, 1773.

252. LplRO, Holt and Gregson MSS/vol. X, p.177.

253. Case employed someone to 'solicite and increase the Sale of Coal which was then greatly decreased' in 1765. PRO, PL6/87/10.

254. KCC, 1/V/74 Letter, Chorley to King's College, Sept. 25, 1769 and LRO, DDK/1816/27 and 2021/20, 86, 88, 90 and 91.

255. KCC, 1/V/74/Letter, Chorley to King's College, Sept. 25, 1769.

256. As in the five years before the drafting of the new lease of 1763. KCC, 1/V/42/14 and 1/V/42/74 and MS in the possession of Mrs D. Bailey.

257. KCC, 1/V/74/Letter, Palmer to King's College, Jan. 18, 1769.

258. *Ibid.* and KCC, 1/V/74/Letter, Case to King's College, 1771.

259. PRO, PL6/87/10.

260. KCC, 1/V/74/ Petition against the Erection of an Engine, Nov. 15, 1774.

261. KCC, 1/V/74/ Copy of the Case for Mr Palmer's Opinion made at Prescot, Nov. 25, 1767.

262. KCC, 1/V/74/ Petition against the Erection of an Engine, Nov. 15, 1774.

263. *Ibid.*

264. KCC, 1/V/74/Letter, G. C. to Case, Jan. 16, 1775.

265. The Earl of Derby's coal was at first worked for only a few months in 1772. LRO, DDK/1816/27 and KCC, 1/V/74/ Petition against the Erection of an Engine, Nov. 15, 1774.

266. LRO, DDWi/Unsorted Box/Indenture, Feb. 26, 1801 (An Account of the Trusteeship of J. Leigh and J. Lyon, 1778-1801).

267. The Gildart colliery (133) had tied markets in the owner's coal-dealing and salt-refining businesses in Liverpool, Case's colliery (34) in the salt refinery at Dungeon Point, Willis' colliery (127) in the owner's staith on the Mersey and the Earl of Derby's in the vast markets provided by his own and his kinsmen's

many mansions. This possession of tied markets by all but one of the South Western collieries in the late eighteenth century must have reduced the necessity and urge for fierce competition.

268. JRL, Legh MSS/Box 51/Letter, Leigh to Legh, Feb. 17, 1760.
269. The colliery was first advertised for sale in 1759 and the price reduction may have been a desperate bid to keep it in business when no buyer was found. LplRO, *Wms Lpl Advt,* June 29, 1759 and Nov. 13, 1761.
270. JRL, Legh MSS/Box 51/Letter, Leigh to Legh, Feb. 17, 1760.
271. JRL, Legh MSS/Box 51/Letter, Leigh to Legh, May 2, 1760.
272. *Ibid.*
273. This colliery was highly capitalised, with a wagonway to the canal, two flatts and a Liverpool coal office. PRO, PL6/85/35 and LplRO, *Lpl Gen Advt,* July 14, 1769.
274. JRL, Legh MSS/Box 59/Letter, Lillyman to Legh, July 28, 1762 and WPL, Wrightington MSS/Eccleston Box/Old Coal Leases, Eccleston and Sutton/Eccleston v. Leigh *et al.,* May 19, 1762 and Eccleston v. Mackay and Leigh, 1763.
275. *Ibid.* and JRL, Legh MSS/Box 51/ Thoughts on the Dublin Trade from Sankey (n.d., 1763).
276. JRL, Legh MSS/Box 59/Letter, Mackay to Legh, Oct. 1, 1763.
277. JRL, Legh MSS/Box 51/ Thoughts on the Dublin Trade from Sankey (n.d. 1763).
278. JRL, Legh MSS/Box 51/Reply of R. Leigh to the proposals of Mr Mackay, Sept. 9, 1763. The scheme was to capture some of the Dublin market from the Lowthers of Whitehaven, whose coal 'is much complained of, not near so good as ours'. The capital cost would be £200 and running costs £140 a year to yield annual profits of £100. One of the reasons for Legh's refusal to become involved was that Mackay wanted two boats, one for his own colliery and one for that in which he was partnered by Leigh, whilst Gerard, Clayton and Legh were to be allowed only one each.
279. JRL, Legh MSS/Box 51/ St Helens Colliery, Jan. 2, 1765.
280. *Ibid.* and LplRO, *Lpl Gen Advt,* Nov. 1, 1771. St Helens colliery was not recorded after 1765.
281. JRL, Legh MSS/Box 43/Letters, Richardson to Orford, Jan. 4 (two versions), Feb. 10, Feb. 13 and Sept. 17, 1767 and July 31, 1768, and Richardson to Legh, Sept. 29, 1769 and Aug. 18, 1770.
282. JRL, Legh MSS/Box 56/Letter, Legh to Orford, Feb. 18, 1768.
283. JRL, Legh MSS/Box 56/Letter, Legh to Orford, April 28, 1768.
284. JRL, Legh MSS/Box 56/Letters, Legh to Orford, May 17 and 28 (two), 1768.
285. JRL, Legh MSS/Box 43/Letter, Richardson to Legh, Aug. 18, 1770.
286. WPL, Wrightington MSS/Eccleston Box/Old Coal Leases, Eccleston and Sutton/Eccleston v. Case, July 31, 1769 and appended note of Oct. 8, 1773.
287. The colliery closed through exhaustion and the equipment was sold. PRO, PL6/85/35 and LplRO, *Lpl Gen Advt,* July 14, 1769.
288. LRO, DDGe(E)/864.
289. LplRO, *Lpl Gen Advt,* Dec. 2, 1774.
290 Barker and Harris, *St Helens,* pp.46-7 and 113-15.
291. Mackay and Case agreed to buy 9750 tons a year for fifteen years from Gerard at 4/8d a ton and to resell it on the canal at that price, 4d a ton less than their own coal. Gerard could still sell coal to the slitting mill and copper works on his estate: the price agreed for the latter was 4d a ton less than a navigation price of 5/- to 6/-, or the same as the Mackay and Case price for his navigation coal. LRO, DDGe(E)/1399.

292. LplRO, *Lpl Gen Advt,* June 19 and July 3, 1778.
293. Harris, 'Copper industry', p.97.
294. Mackay's concession to the Ravenhead works seems to have been much more favourable than Gerard's to the Stanley works, a sign, perhaps, of his enthusiasm to acquire a further tied market for his coal. Gerard sold to the Stanley Company at 5d a ton less than navigation prices of 6/- a ton or over; 4d less than prices of 5/- to 6/-; 3d less than prices of 4/- to 5/- and at the canal price when it was less than 4/- a ton. The Ravenhead agreement stipulated a fixed price of 5/- per ton of '30 cwt neat pounds weight' – i.e. 1¼ tons of 20 cwt of 120 lb. Thus, Mackay's price was 4/- per common ton, which sold on the canal at 4/8d to 5/- in 1778. Gerard's relatively high price, agreed in 1772, was compensated somewhat by his provision at his own cost of a wagonway to the canal for the exclusive use of the Stanley Company. Harris, 'Copper industry', p.97 and LRO, DDGe(E)/864, 830 and 1399.
295. Hustler had financial control of the canal from its beginning to his death in 1790 and signed all the statements made by the proprietors: he was clearly the dominant person in its affairs. Blundell was responsible for the 'murky' dealings in which the Leeds and Liverpool Canal Company gained control of the Douglas Navigation, and therefore access to the coalfield. Harris, 'Canal controversies', pp.80 and 89; Hadfield and Biddle, *Canals,* pp.65-81 and Killick, 'Leeds and Liverpool Canal'.
296. JRL, Haigh MSS/Canals Box/Letter, Balcarres to Stanhope, 1789.
297. *Ibid.*
298. JRL, Haigh MSS/Canals Box/Letters, Balcarres to Stanhope, 1789 and n.d. (probably 1789). Balcarres to ? n.d. (two) and Balcarres to the Leeds and Liverpool Canal Company, April 28, 1788.
299. JRL, Haigh MSS/Canals Box/Letter, Hoare to Balcarres, Nov. 14, 1788.
300. In his initial proposal, Balcarres had pointed out that his reserves were 'in Extent . . . greater than All the Collieries put together that now supply Liverpool . . . they will find their way to Liverpool at the present prices, when it is impossible for Mr Bankes and Mr Standish to furnish them at that price'. JRL, Haigh MSS/Canals Box/Letter, Balcarres to Stanhope, 1789.
301. JRL, Haigh MSS/Canals Box/Letter, Balcarres to Stanhope, 1789. As a result of his experiences, Balcarres was 'very much cooled indeed on the subject'.
302. LRO, DP/175/p.64.
303. Standish (41) in 1788 and 1797, Shevington (155) in 1789, Ince (135) in 1791, Pemberton (169) in 1795 and Parbold (170) in 1798.
304. JRL, Haigh MSS/Canals Box/Letter, Balcarres to Greenwood, Mar. 9, 1790.
305. JRL, Haigh MSS/Canals Box/Lancaster Canal with Printed Papers, 1800 and enclosed letter, Haliburton to Balcarres.
306. *Ibid.* and JRL, Haigh MSS/Alexander, 6th Earl of Balcarres, Management of Mining Estates, 1818-22, pp.86-7.
307. *Ibid.* and JRL, Haigh MSS/Canals Box/Letters, Balcarres to Gregson, Oct. 1, 1801 and 1803.

## 8.  Conclusion

1. Nef, *Coal Industry* and Ashton and Sykes, *Coal Industry.* These estimates are discussed in Langton, 'Coal output', E. Kerridge, 'The coal industry in Tudor and Stuart England: a comment', *Econ. Hist. Rev.,* 2nd ser., 30 (1977), 340-2 and D. C. Coleman, 'The coal industry: a rejoinder', *Econ. Hist. Rev.* 2nd ser., 30 (1977), 343-5.

2. 1,211,050 and 763,200 tons respectively. Ashton and Sykes, *Coal Industry*, p.251.
3. 1,254,000 tons. Ashton and Sykes, *Coal Industry*, p.251.
4. The derivation of the St Helens figure is given on figure 44. The tonnage figures used for St Helens and Wigan are the output estimates for the areas in which they lay minus canal exports.
5. For a recent discussion of the role of competition in the development of the North Eastern coalfield see P. Cromar, 'The coal industry on Tyneside 1771-1800: oligopoly and spatial change', *Econ. Geog.*, 53 (1977), 80-93.
6. For a fuller elaboration of this idea and the journey times on which my multiplier is based see D. G. Janelle, 'Central place development in a time-space framework', *Professional Geographer*, 20 (1968), 5-10. Smiles observed that 'the locomotive . . . virtually reduced England to a sixth of its size'. S. Smiles, *The Life of George Stephenson* (London, 1904), p.vii.
7. For recent statements on this theme by non-geographers see A. Everitt, 'River and wold: reflections on the historical origins of regions and pays', *Jour. of Hist. Geog.*, 3 (1977), 1-20 and S. Thrupp, 'Comparative studies in the barnyard', *Jour. Econ. Hist.*, 35 (1975), 1-7.

## APPENDIX 1

1. Nef, *Coal Industry*, vol. II, app. C, pp. 374-7.
2. Especially at the Haigh cannel colliery. See WPL, Haigh Colliery Orders, end A, Orders for 1687, *passim* and end B, Orders for 1636, *passim*.
3. *Ibid.*, end B, fol.6v.
4. At Shevington in 1672 (LRO, DDHe/40/74); at Haigh in 1636 (WPL, Haigh Colliery Orders, end B, fol. 1r; and at St Helens in 1762 (WPL, Wrightington MSS/Eccleston Box/Old Coal Leases, Eccleston and Sutton/Eccleston v. Leigh *et al.* May 19, 1762).
5. JRL, Haigh MSS/Collieries Folder/Early Documents relating to Colliery Affairs/Account What Coals, Slate, &c.
6. KCC, 1/V/74.
7. LRO, DDGe(E)/849 and 1275; *cf.* LplRO, *Lpl Gen Advt*, May 24, 1771 and Aug. 20, 1773 where the basket and cwt of 120 lb are almost certainly used as equivalent measures; JRL, Haigh MSS/Canals Box/Report of the Lancaster Navigation Proprietors, Aug. 2, 1803 and WPL, Leigh Pemberton MSS/Deeds Reserving Rents and Coalmines/Title Deeds to Orrell Hall Estate/Lease, Atherton *et al.* v. Halliwell *et al.*, Mar. 25, 1756 and Colliery Account Books, 1771-6 (122 lb).
8. Wadsworth and Mann, *Cotton Trade*, pp.215n-216n.
9. WPL, Leigh-Pemberton MSS/Deeds Reserving Rents and Coalmines/Title Deeds to Orrell Hall Estate/Lease, Atherton *et al.* v. Halliwell *et al.*, Mar. 25, 1756.
10. The actual lease is not extant but it is evident that one containing this stipulation was made in 1771 from WPL, Leigh-Pemberton MSS/Letter, Leigh to Starkie, May 30, 1770 and Colliery Account Books, 1771-6.
11. WPL, Leigh-Pemberton MSS/Letter, Leigh to Starkie, May 30, 1770 and Colliery Account Books, 1771-6.
12. WPL, Leigh-Pemberton MSS/Articles of Agreement, Leigh v. Kingsley *et al.*, 1769.
13. Anderson, 'Blundells' collieries', p.109.
14. Anderson, 'Blundells' collieries', pp.124-5 and 133. After the improvement of

302 *Notes to pp.244-6*

underground haulage in the early nineteenth century the size of baskets was increased nearly fivefold.

15. LRO, DDL/832.
16. LRO, RCSt/6.
17. LplRO, *Lpl Gen Advt*, Jan. 27, 1775 and Nov. 1, 1776.
18. *VCH*, vol. II, pp.357-8. Best quality fuel was sold 'for 3d per 100 wt; that which is broken they sell for 3d a load which weighs 1200 wt'.
19. WPL, Leigh-Pemberton MSS/Articles of Agreement, Hodson *et al*. v. Blundell *et al*. Mar. 2, 1776.
20. LplRO, *Lpl Gen Advt*, Nov. 1, 1776.
21. LRO, DDWa/Ince Box/Ince, 1712-79/Articles of Agreement, Talbot v. Walmsley *et al.*, Nov. 3, 1779.
22. WPL, Euxton MSS/1/1.
23. JRL, Haigh MSS/Canals Box/Report of the Lancaster Canal Proprietors, Aug. 2, 1803.
24. Persons from outside the coalfield reported that coal was mined or sold by the cwt at collieries where the basket measure was used at Prescot in 1735 and Kirkless in 1751 (KCC, 1/V/74; and *VCH*, vol. II, pp.357-8). The measure used on the canals was the cwt, which was sold at the same price as baskets from the pitheads of supplying collieries, and the cwt and basket were obviously considered to be interchangeable terms by many colliery owners (see coal advertisments in LplRO, *Lpl Gen Advt*, May 24, 1771 and Aug. 20, 1773).
25. JRL, Haigh MSS/Canals Box/Report of the Lancaster Canal Proprietors, Aug. 2, 1803; LplRO, *Lpl Gen Advt*, June 5, 1772; Aug. 20, 1773; Dec. 2, 1774; Jan. 27, 1775 and Nov. 1, 1776, when 1 ton = 2400 lb (i.e. 20 cwt of 120 lb).
26. Harland, 'Shuttleworth accounts', pt 1, pp.105, 108, 109 and 519.
27. Bailey, 'Coalmining in Prescot', pp.1-20, *passim*.
28. Bailey, 'Coalmining in Prescot', p.2.
29. Nef, *Coal Industry*, vol. II, p.376.
30. At Smithils in 1593 (Harland, 'Shuttleworth accounts', pt 1, p.80) and at Liverpool in 1735 (Gibson, *Blundell's Diary*, p.212).
31. Aikin, *Description*. p.523.
32. Nef, *Coal Industry*, vol. II, p.376.
33. It was also used at Orrell in 1677 (WPL, Walthew Account Books/Debts, Legacies and Financial Expenses, p.5.
34. Crofton, 'Lancashire and Cheshire coalmining', p.44.
35. LRO, DDGe(E)/245.
36. Hawkes, 'Sir Roger Bradshaigh', p.3; LRO, DDCs/4/4; in the only Haigh colliery account to give details of the measure, penny baskets cost 10d per load and 'sail Cannel' at 3½d per basket cost 2/11d per load. JRL, Haigh MSS/Colliery Accounts, 1747-8 and Kirkless Colliery Account Book, 1750-6 and 1774-6 and An Account of Cannel got at Smallman's Hey Pit in Kirkless, June 29, 1758 (with 1774 account).
37. WPL, Haigh Colliery Orders/end A, fol.3r and LRO, DDGe(M)/94 and 96.
38. For example, for colliers' concessionary coal; for the depreciation of banked coal and for coal used about the colliery in smithy work, winter fires and, later in the period, engine fuel.
39. Coal was sold per number of individual baskets, no aggregative measures being used. LRO, DDBa/Coal Pitt Accompts, 1676-96.
40. *Ibid.* No direct statement concerning the composition of the load was made in the accounts, but it is apparent from a number of entries that it comprised eight baskets, e.g. one hundred baskets = twelve loads and four baskets, and eleven

loads and two baskets cost 11/3d at 1½d per basket.
41. Again no direct statement was made, but coal cost 2d per basket and 1/6d per load. LRO, DDBa/Colliery Accounts, 1766 and An Account of Sundry Trusts at the Higher Coal Pits, 1766.
42. LRO, DDBa/Colliery Accounts, 1766.
43. LRO, DDGe(M)/94-6.
44. The few collieries in this area are on the whole poorly documented but the score was used at Newburgh 1736-7 (LRO, DDK/2021/16) and at Skelmersdale (105) in the 1790s (CRO, EDC/5/1816-17, Rylance decd, 1816).
45. Wadsworth and Mann, *Cotton Trade,* pp.215n-216n; LRO, DDBl/54/12; DDGe(M)/94-6; DDHe/50/84 and 116/2 and DDK/2021/16; WPL, Leigh-Pemberton MSS/Articles of Agreement, Leigh v. Kingsley *et al.,* 1769 and Liverpool Canal Letter Book 1772, four loose sheets, Payments made by A. Leigh for Coal-Loading on the Douglas Navigation in 1740.
46. LRO, DDBa/Colliery Accounts, 1766.
47. LRO, DDBa/Accounts and Rentals/20 and Anderson, 'Blundells' collieries', p.109.
48. In the reckoning of coal sales total basket numbers were used. LRO, DDBa/An Account of Sundry Trusts at the Higher Coal Pitts, 1766; and Accounts and Rentals/20.
49. Twenty-one basket scores were used at collieries in Shevington (LRO, DDHe/40/74 and WPL, Wrightington MSS/Shevington Box/Lease, Hesketh v. Dicconson *et al.,* Jan. 29, 1731) and Wrightington (LRO, DDHe/50/84). Twenty-two basket scores were used to measure miners' concessionary coal at an Orrell leasehold colliery (WPL, Leigh-Pemberton MSS/Allowances for M. Halliwell). Twenty-three basket scores were used at collieries in Charnock Richard (PRO, FEC1/C/17 and LRO, DDTa/280/Articles of Agreement, Brooke v. Lancaster, June 12, 1716) and Wrightington (LRO, DDHe/55/15). Twenty-four basket scores were used at collieries in Adlington (JRL, Haigh MSS/A/5/56 – here called a 'long score'), Kirkless (JRL, Haigh MSS/Kirkless Colliery Account Book, 1750-6 and 1774-6; LRO, DDWa/Ince Box/Ince, 1712-79/Articles of Agreement, Talbot v. Walmsley *et al.,* Nov. 3, 1779 and WPL, Euxton MSS/1/1 and Leigh-Pemberton MSS/Articles of Agreement, Hodson *et al.,* Mar. 2, 1776), Orrell (PRO, DL1/503(1)), Ashton (LRO, DDGe(M)/1185) and Shevington (LRO, DDL/832). Twenty-six basket scores were used at certain Orrell collieries (LRO, DDWa/Coal, Leigh Estate, 1741-97/Lease, Leigh v. Longbotham, Jan. 20, 1775; PRO, DL1/503(1) and WPL, Leigh-Pemberton MSS/Deeds Reserving Rents and Coalmines, Orrell/Mr Porter's Account of Coal got at Hall Orrell).
50. CRO, EDC/5/1816-17, Rylance decd, 1816. A similar discrepancy between scores measured for sale and royalties occurred at Wrightington (55) (LRO, DDHe/50/84 and 116/2).
51. Bailey, 'Coalmining in Prescot',*passim;* KCC, 1/V/74; PRO, PL6/13/103 and DL4/118/10.
52. PRO, PL6/13/103.
53. PRO, DL4/118/10 and KCC, 1/V/74.
54. LRO, DDCa/1/47.
55. LplRO, *Wms Lpl Advt,* Feb. 16, 1759 and LplRO, Hq. 386.4100037, 1757.
56. LplRO, Holt and Gregson MSS/vol. X, p.177 and Harris, 'Copper industry', p.93.
57. LplRO, *Lpl Gen Advt,* June 5, 1772 and Dec. 2, 1774 in which the ton at four

canalside collieries was said to contain 20 cwt. In about 1790 Gregson
observed that the ton 'was then' 30 cwt in 1758. LplRO, Holt and Gregson
MSS/vol. X, p.177.
58. LplRO, *Lpl Gen Advt,* Jan. 17, 1772.
59. LplRO, *Lpl Gen Advt,* Jan. 27, 1775 and Nov. 1, 1776 and JRL, Haigh
MSS/Canals Box/Report of the Lancaster Canal Proprietors, 1803.
60. France, 'Papist estates', p.135; KCC, 1/V/74; LRO, DDGe(E)/849;
DDM/4/2, 5 and 7; DDWi/43/Copy of Case heard in Chancery, Case v. Moss *et
al.,* 1797; PRO, PL6/54/101 and 55/7; WPL, Wrightington MSS/Eccleston
Box/Old Coal Leases, Eccleston and Sutton/Eccleston v. Leigh *et al.,* May 19,
1762 and Eccleston v. Mackay and Leigh, 1763.
61. KCC, 1/V/74 and PRO, DL4/118/110.
62. KCC, 1/V/74 and LRO, DDK/295/5.
63. LRO, DDGe(E)/815 and DDM/4/13; WPL, Wrightington MSS/Eccleston
Box/Old Coal Leases, Eccleston and Sutton/Eccleston v. Mackay and Leigh,
1763; Eccleston v. Case, July 15, 1775 and Eccleston v. Orrell, Feb. 20, 1778.
64. WPL, Wrightington MSS/Eccleston Box/Old Coal Leases, Eccleston and
Sutton/Eccleston v. Mackay and Leigh, 1763; Eccleston v. Case, July 15, 1775
and Eccleston v. Orrell, Feb. 20, 1778.
65. LRO, DDGe(M)/1184.
66. LRO, DDBa/Coal Pitt Accompts, 1676-96.
67. JRL, Haigh MSS/Bradshaigh Accounts, 1744-54.

## APPENDIX 3

1. JRL, Haigh MSS/Bradshaigh Accounts, 1731-41 (which begins in 1725 and
ends in 1739), 1744-54, 1754-65 and 1765-74.
2. Wages consumed the largest proportion of any single mining cost; at Prescot in
1735 they accounted for c.forty per cent of mining expenses. KCC, 1/V/74.
3. For example see 1747 and 1754.
4. JRL, Haigh MSS/An Account of what Cannel has been got . . . Sept. 28, 1747
to Oct. 1, 1748; and Cannel, 1749 (Account Book for 1749-55).
5. See regulations concerning the size of the lumps to be produced in WPL, Haigh
Colliery Orders, *passim.*
6. The small cannel comprised 'tops', mined from the upper part of the seam and
the round cannel comprised 'bottoms' (cf. WPL, Euxton MSS/8/12). The
thickness of these two bands in the seam was variable, the 'bottoms' being,
apparently, usually thin or non-existent. After selling coal at all pits (three or
four each year) at 2/11d per load between 1748 and 1755 an entry of April 14,
1755 for Whalley Pit reads 'The Small Cannell advanc'd & the bottoms
enlarg'd which alters the Value of the Cannell from 2s. 11d. to 3s. 4d. P. Pitt
Load.' The duration of this new situation is unknown.
7. WPL, Haigh Colliery Orders, fol. 14r.
8. JRL, Haigh MSS/Collieries Folder/Early Documents relating to Colliery
Affairs/Account What Coals, Slate, &c.
9. JRL, Haigh MSS/Box EIII/Bundle II/Letter, Leigh to Bradshaigh, July 21,
1739.
10. See JRL, Haigh MSS/An Account of what Cannel has been got . . . 1747.
11. See JRL, Haigh MSS/Cannel, 1749, account for 1755.
12. LRO, DDBb/1/1.

**APPENDIX 4**

1. LRO, Land Tax Assessment Returns/Leyland, Salford and West Derby, 1782-99.
2. W. R. Ward, *The English Land Tax in the Eighteenth Century* (London, 1953), p.6.
3. The total of three cannel collieries entered for Haigh in 1797 and 1799 comprises the colliery of the Earl of Balcarres, counted as two, once for each year, and part of that of James Hodson, counted as half in each year.
4. CRO, EDC/5/1816-17, Rylance decd, 1816.
5. These 'men' were almost certainly hewers because they averaged only four or five a pit during the years when both pits and men were entered (see table 28).
6. Anderson, *Orrell Coalfield*, p.58.

# BIBLIOGRAPHY

**1. Manuscripts and newspapers**

BMus British Museum
   Add MSS Additional MSS

CLM Chetham Library, Manchester
   Piccope MSS

CRO Cheshire Record Office
   EDC/5 Consistory Court Papers

JRL John Rylands Library, Manchester
   Haigh MSS
   Legh MSS

KCC King's College Library, Cambridge
   Prescot MSS

LRO Lancashire Record Office

| | |
|---|---|
| DDAl | Alison of Park Hall MSS |
| DDBa | Bankes of Winstanley MSS |
| DDBb | Bryan Blundell MSS |
| DDBd | Pilgrim and Badgery MSS |
| DDBl | Blundell of Crosby MSS |
| DDCa | Cavendish of Holker MSS |
| DDCl | Clifton of Lytham MSS |
| DDCo | Walmsley-Cotham MSS |
| DDCr | Stanley of Crosse Hall MSS |
| DDCs | Cross of Prescot MSS |
| DDEl | Peace and Ellis MSS |
| DDFi | Finch of Mawdesley MSS |
| DDGe | Gerard of Ashton-in-Makerfield MSS |
| DDHc | Hornby Castle MSS |
| DDHe | Hesketh of Rufford MSS |
| DDHk | Hawkeshead-Talbot of Chorley MSS |
| DDHt | Houghton of Lowton MSS |
| DDK | Stanley, Earls of Derby MSS |
| DDKe | Kenyon of Peel MSS |
| DDL | Finch, Johnson and Lynn MSS |
| DDLm | Bootle-Wilbraham of Lathom MSS |
| DDM | Molyneux, Earls of Sefton MSS |
| DDMc | Machell MSS |
| DDSc | Scarisbrick of Scarisbrick MSS |
| DDSt | Weld of Stonyhurst MSS |
| DDTa | Tatton of Cuerden MSS |
| DDWa | Walmsley of Westwood MSS |

306

DDWi     Willis of Halsnead MSS
DDX     Miscellaneous Collections
DP     Purchased Documents
DX     Miscellaneous Documents
Hearth Tax Returns, 1673 (trans)
Land Tax Assessment Returns, 1782-99
Parish Registers and Bishops' Transcripts
QSP     Quarter Sessions Petitions
RCSt     Roman Catholic Records, Standish
Wills proved at Chester and Richmond

LplRO Liverpool Record Office
   Binns MSS
     *Gore's Lpl Advt*    *Gore's Liverpool Advertiser*
   Holt and Gregson MSS
   John Okill's Day Book
     *Lpl Gen Advt*    *Liverpool General Advertiser*
   Salisbury MSS
     *Wms Lpl Advt*    *Williamson's Liverpool Advertiser*

PPL Preston Public Library
   *Preston Guardian*

PRO Public Record Office
   DL1     Duchy of Lancaster Pleadings
   DL4     Duchy of Lancaster Depositions
   DL5     Duchy of Lancaster Decrees and Orders
   DL6     Duchy of Lancaster Decrees
   DL44     Duchy of Lancaster Special Commissions
   EDC     Hearth Tax Assessment Returns for Lancashire
   FEC     Forfeited Estates Commissions MSS
   PL6     Palatine of Lancaster Bills
   St Ch Procs Star Chamber Proceedings

StHRL St Helens Reference Library
   Accounts of the Supervisors of Parr Highways
   Parr Poor Law Papers
   Pleadings and Depositions taken at Lancaster concerning Sutton, 1547-1636 (trans)

WPL Wigan Public Library (now at Wigan Metropolitan Borough Record Office, Leigh)
   Account Book of Francis Sherrington
   Book of Common Council
   Bridgeman Ledger
   CL Court Leet Rolls
   Euxton MSS
   Haigh Colliery Orders
   KP Court of King's Pleas MSS
   Liverpool Canal Letter Book
   Leigh-Pemberton MSS
   Up Holland MS
   Walthew Account Book
   Wrightington MSS
Documents in the private possession of Mrs D. Bailey and Mr D. Anderson were kindly made available to me.

## 2. Books, articles and theses

Aikin, J. *A Description of the Country from Thirty to Forty Miles round Manchester* (London, 1795)

Agricola, G. (trans. H. C. and L. H. Hoover), *De Re Metallica* (1556; London, 1912)

Anderson, B. L. The attorney and the early capital market in Lancashire' in J. R. Harris (ed.), *Liverpool and Merseyside: Essays in the Economic and Social History of the Port and its Hinterland* (London, 1969)

Anderson, D. 'Blundell's collieries: the progress of the business', *Trans. Hist. Soc. of Lancs. and Chesh.*, 116 (1965), 69-116

*The Orrell Coalfield, Lancashire, 1740-1850* (Buxton, 1975)

Ashton, T. S. and J. Sykes, *The Coal Industry of the Eighteenth Century*, 2nd edn (Manchester, 1964)

Aspinall, A. *The Early English Trade Unions* (London, 1949)

Awty, B. W. 'Charcoal-ironmasters of Lancashire and Cheshire, 1600-1785', *Trans. Hist. Soc. of Lancs. and Chesh.*, 109 (1957), 71-125

*Bailey's northern directory* (1781)

Bailey, F. A. 'Coroners' inquests held in the manor of Prescot, 1746-89', *Trans. Hist. Soc. of Lancs. and Chesh.*, 86 (1934), 21-40

'The minutes of the trustees of the turnpike from Liverpool . . .', pt 1, 1726-53 and pt 2, 1753-89, *Trans. Hist. Soc. of Lancs. and Chesh.*, 88 (1936), 159-201 and 89 (1937), 31-91

'Prescot records, 1447-1660', *Trans. Rec. Soc. of Lancs. and Chesh.*, 89 (1937)

'Early coalmining in Prescot, Lancashire', *Trans. Hist. Soc. of Lancs. and Chesh.*, 99 (1947), 1-20

Baker, A. R. H. (ed.) *Progress in Historical Geography* (Newton Abbot, 1972)

Baker, A. R. H., J. D. Hamshere and J. Langton. 'Introduction' in *Geographical Interpretations of Historical Sources* (Newton Abbot, 1970)

Bankes, J. M. H. 'Records of mining in Winstanley and Orrell . . .', *Trans. Hist. Soc. of Lancs. and Chesh.*, 54 (1939), 31-65

'James Bankes and the manor of Winstanley, 1595-1617' *Trans. Hist. Soc. of Lancs. and Chesh.*, 94 (1942), 56-93

Barker, T. C. 'The Sankey Navigation: the first Lancashire canal', *Trans. Hist. Soc. of Lancs. and Chesh.*, 100 (1948), 121-55

'Lancashire coal, Cheshire salt and the rise of Liverpool', *Trans. Hist. Soc. of Lancs. and Chesh.*, 130 (1951), 83-103

Barker, T. C. and J. R. Harris. *A Merseyside Town in the Industrial Revolution: St Helens, 1750-1900* (Liverpool, 1954)

Beckmann, M. *Location Theory* (New York, 1968)

Berlin, I. 'History and theory: the concept of scientific history', *History and Theory*, 1 (1960-1), 1-33

Birch, A. 'The Haigh ironworks . . .', *Bull. of the John Rylands Library*, 35 (1953),.316-34

Blundell, M. *Blundell's Diary and Letterbook, 1702-82* (Liverpool, 1952)

Bowden, M. J. and W. A. Koelsch. 'Reviews in historical geography', *Econ. Geog.*, 46 (1972), 214-16

Bridgeman, G. T. O. 'The history of the church and manor of Wigan', *Chetham Society Publications*, N.S., 15-18 (1888, 1889 and 1890)

Brontë, Anne. *The Tenant of Wildfell Hall* (London, 1967)

Bulley, J. A. 'To Mendip for coal', *Proc. of Somerset Archaeol. and Nat. Hist. Soc.*, 97 (1952), 46-78 and 98 (1953), 17-54

Calvert, A. F. *Salt in Cheshire* (London, 1915)

Cartwright, J. J. *The Travels through England of Dr. Richard Pococke . . . during 1750, 1751 and Later Years,* Camden Society (2 vols., London, 1888)

Challinor, R. *The Lancashire and Cheshire Miners* (Newcastle, 1972)

Chandler, G. *Liverpool under James I* (Liverpool, 1960)

Cheetham, F. H. 'The church bells of Lancashire', pt 4, *Trans. Lancs. and Chesh. Antiq. Soc.,* 45 (1928), 89-123

Clegg, E. H. 'Some historical notes on the Wigan coalfield', *Trans. Inst. Min. Eng.,* 117 (1957-8), 784-801

Coleman, D. C. 'The coal industry: a rejoinder', *Econ. Hist. Rev.,* 2nd ser., 30 (1977), 343-5

Crofton, H. T. 'Lancashire and Cheshire coalmining records', *Trans. Lancs. and Chesh. Antiq. Soc.,* 7 (1889), 26-73

Cromar, P. 'The coal industry on Tyneside, 1771-1800: oligopoly and spatial change', *Econ. Geog.,* 53 (1977), 80-93

Deane, P. and W. A. Cole. *British Economic Growth 1688-1959* (Cambridge, 1964)

Defoe, D. *A Tour through the Whole Island of Great Britain* (Harmondsworth, 1971)

Derry, T. K. and T. I. Williams, *A Short History of Technology* (Oxford, 1960)

Duckham, B. F. *A History of the Scottish Coal Industry 1700-1815* (Newton Abbot, 1970)

Enfield, W. *An Essay towards the History of Liverpool* (Warrington, 1773)

Everitt, A. 'River and wold: reflections on the historical origins of regions and pays', *Jour. of Hist. Geog.,* 3 (1977), 1-20

Files, J. 'Mining with special reference to Lancashire', *Trans. Inst. Min. Eng.,* 92 (1936), 292-8

Finberg, H. P. R. Review in *Agri. Hist. Rev.,* 17 (1969), 77-8

Fischer, D. H. *Historians' Fallacies* (London, 1971)

France, R. S. 'Lancashire Papist estates in 1719', *Rec. Soc. of Lancs. and Chesh.,* 98 (1943) and 108 (1960)

Galloway, R. L. *Annals of Coalmining and the Coal Trade* (London, 1898)

Gibson, T. E. *Blundell's Diary: Selections from 1702-1728* (Liverpool, 1895)

Gilboy, E. W. *Wages in Eighteenth Century England* (Cambridge, Mass., 1934)

Gonner, E. C. K. 'The population of England in the eighteenth century', *Jour. Roy. Stat. Soc.,* N.S., 76 (1913-14), 216-96

Hadfield, C. and G. Biddle. *The Canals of North West England* (2 vols., Newton Abbot, 1970), vol. I

Hair, P. E. H. 'The Lancashire collier girl, 1795', *Trans. Hist. Soc. of Lancs. and Chesh.,* 120 (1968), 63-86

'Accidental death and suicide in Shropshire, 1780-1809', *Trans. Shropshire Archaeol. Soc.,* 59 (1969-70), 63-75

Hamilton, H. *The English Brass and Copper Trades* (London, 1926)

Harland, J. (ed.). 'The house and farm accounts of the Shuttleworths . . . 1582-1621', *Chetham Society Publications,* 35, 41, 43 and 46 (1856-8)

Harley, J. B. 'Changes in historical geography: a qualitative impression of quantitative methods', *Area,* 5 (1973), 69-74

Harris, J. R. 'Michael Hughes of Sutton: the influence of Welsh copper on Lancashire business, 1780-1815', *Trans. Hist. Soc. of Lancs. and Chesh.,* 101 (1949), 139-67

'The copper industry in Lancashire and North Wales, 1760-1815', unpubl. PhD thesis (Liverpool, 1952)

*The Copper King: A Biography of Thomas Williams of Llanidan* (Liverpool, 1964)

'Early Liverpool canal controversies', in J. R. Harris (ed.), *Liverpool and Merseyside: Essays in the Economic and Social History of the Port and its Hinterland* (London, 1969)

Harrison, W. 'Pre-turnpike highways of Lancashire and Cheshire', *Trans. Lancs. and Chesh. Antiq. Soc.*, 9 (1891), 101-31

Hart, J. F. 'Central tendency in areal distributions', *Econ. Geog.*, 30 (1954), 48-59

Harvey, D. *Explanation in Geography* (London, 1969)

'Conceptual and measurement problems in the cognitive behavioral approach to location theory' in K. R. Cox and R. G. Golledge (eds.), 'Behavioral problems in geography: a symposium', *Northwestern University Studies in Geography*, 17 (1969), 35-67

Hatcher, J. and T. C. Barker. *A History of British Pewter* (London, 1974)

Hawkes, A. J. 'Sir Roger Bradshaigh of Haigh, knight and baronet, 1628-84', *Chetham Society Publications*, N.S., 109 (1945)

Hickling, G. 'The coalmeasures of the Croxteth Park inlier', *Trans. Inst. Min. Eng.*, 50 (1915-16), 322-7

Hopkinson, G. G. 'The development of the South Yorkshire and North Derbyshire coalfield, 1500-1775', *Trans. Hunter Archaeol. Soc.*, 7 (1951-7), 295-319

Hoskins, W. G. *Local History in England* (London, 1959)

'Indexes to the wills and inventories now preserved in the court probate at Chester', *Rec. Soc. of Lancs. and Chesh.*, 2 (1879), 4 (1881), 15 (1887) and 18 (1888)

Isard, W. 'Notes on the use of regional science methods in economic history', *Jour. Econ. Hist.*, 20 (1960), 597-600

*General Theory: Social, Political, Economic and Regional* (Cambridge, Mass., 1969)

Jackman, W. T. *The Development of Transportation in Modern England* (2 vols., Cambridge, 1916)

Janelle, D. G. 'Central place development in a time-space framework', *Professional Geographer*, 20 (1968), 5-10

Jones, E. L. Review in *Econ. Hist. Rev.*, 2nd ser., 16 (1963-4), 571-2

Jones, R. C. B., C. H. Tonks and W. B. Wright. *The Geological Survey Memoir to the Wigan Sheet (no. 84)* (London, 1938)

*Journal of the House of Commons*, 19 (1718-21)

Kerridge, E. 'The coal industry in Tudor and Stuart England: a comment', *Econ. Hist. Rev.* 2nd ser., 30 (1977), 340-2

Killick, H. F. 'Notes on the early history of the Leeds and Liverpool canal'. *Bradford Antiquary*, N.S., 1 (1900), 169-238

'Lancashire itinerary of Dr. Kuerden, c.1695', in J. P. Earwaker (ed.), *Local Gleanings relating to Lancs. and Chesh.*, vol. I (1876), p.214

'Lancashire Quarter Sessions Records: Quarter Sessions Rolls, 1590-1606', *Chetham Society Publications*, N.S., 77 (1917)

'Lancashire Royalist composition papers, 1643-60', *Rec. Soc. of Lancs. and Chesh.*, 29 (1894)

Langton, J. 'The geography of the south west Lancashire mining industry, 1590-1799', unpubl. PhD thesis (2 vols., Wales, 1970)

'Coal output in south-west Lancashire, 1590-1799', *Econ. Hist. Rev.*, 2nd ser., 25 (1972), 28-54

'Potentialities and problems of adopting a systems approach to the study of change in human geography', *Progress in Geography*, 4 (1972), 127-79

'Landowners and the development of coalmining in South-west Lancashire, 1590-1799', in H. S. A. Fox and R. A. Butlin (eds.), *Change in the Countryside: Essays on Rural England, 1500-1900*, I. B. G. Special Publications, 10 (1979)

Lee, C. H. *Regional Economic Growth in the United Kingdom since the 1880s* (London, 1971)

Lösch, A. (trans. W. H. Woglom). *The Economics of Location* (New York, 1967)

Lerry, G. G. *The Collieries of Denbighshire – Past and Present* (Wrexham, 1946)

Long, W. H. 'Demand in space: some neglected aspects', *Papers Reg. Sci. Assoc.*, 27 (1971), 45-60

Ministry of Fuel and Power. *Northwestern Coalfields: Regional Survey Report* (London, 1945)

Moffitt, L. W. *England on the Eve of the Industrial Revolution*, 2nd edn (London, 1963)

Moller, A. W. R. 'The history of English coalmining, 1550-1750', unpubl. DPhil thesis (Oxford, 1923)

'Coal mining in the seventeenth century', *Trans. Roy. Hist. Soc.*, 4th ser., 8 (1925), 79-97

'The Moore rental', *Chetham Society Publications*, 12 (1847)

Moore, E. S. *Coal* (New York, 1955)

Morris, C. *The journeys of Celia Fiennes* (London, 1949)

Nef, J. U. *The Rise of the British Coal Industry* (2 vols., London, 1932)

Neft, D. S. *Statistical Analysis for Areal Distributions* (Philadelphia, 1966)

North, F. J. *Coal and the Coalfields in Wales* (Cardiff, 1926)

Olsson, G. Review in *Jour. Reg. Sci.*, 8 (1968), 253-5

Owen, G. *The Description of Pembrokeshire* (1603; London, 1892)

Parkinson, C. N. *The Rise of the Port of Liverpool* (Liverpool, 1952)

Parkinson, R. 'The Life of Adam Martindale written by Adam Martindale', *Chetham Society Publications*, 4 (1844)

Picton J. *Memorials of Liverpool* (2 vols., London, 1873)

Porteus, T. C. *History of Standish* (Wigan, 1927)

Postan, M. M. 'The maps of Domesday', *Econ. Hist. Rev.*, 2nd ser.,7 (1954-5), 98-100

Prince, H. C. 'Real, imagined and abstract worlds of the past', *Progress in Geography*, 3 (1971), 1-86

Richardson, H. W. *Regional Economics* (London, 1969)

Rodgers, H. B. 'The market area of Preston in the sixteenth and seventeenth centuries' *Geog. Studies*, 3 (1956), 46-55

Scott, N.M. 'Documents relating to coalmining in the Saltcoats district in the first quarter of the eighteenth century', *Scot. Hist. Rev.*, 19 (1922), 88-95

Scott, W. R. *Finance of English, Scottish and Irish Joint Stock Companies to 1720* (3 vols., Cambridge, 1910-12)

Shelley, R. J. A. 'Wigan and Liverpool pewterers', *Trans. Hist. Soc. of Lancs. and Chesh.*, 97 (1945) 3-15

Smiles, S. *The Life of George Stephenson* (London, 1904)

Smith, D. M. *Industrial Location: An Economic Geographical Analysis* (London, 1971)

Smith, R. S. 'The Willoughbys of Wollaton, 1500-1643, with special reference to early coalmining in Nottinghamshire', unpubl. PhD thesis (Nottingham, 1964)

Sprunt, W. C. 'Old Warrington trades and occupations', *Trans. Lancs. and Chesh. Antiq. Soc.,* 61 (1949), 167-9

Stone, L. 'An Elizabethan coalmine', *Econ. Hist. Rev.,* 2nd ser., 3 (1950-1), 97-106

Swanson, J. A. Review in *Econ. Hist Rev.,* 2nd ser., 22 (1969), 171

Thompson, A. *The Dynamics of the Industrial Revolution* (London, 1973)

Thrupp, S. 'Comparative studies in the barnyard', *Jour. Econ. Hist.,* 35 (1975), 1-7

Toynbee, A. *Lectures on the Industrial Revolution of the Eighteenth Century in England* (London, 1913)

Trott, C. D. J. 'The historical geography of the Neath region up to the eve of the industrial revolution', unpubl. MA thesis (Wales, 1964)

　'Coalmining in the borough of Neath in the seventeenth and early eighteenth centuries', *Morgannwg,* 13 (1969) 47-74

Tunnicliffe, W. *A Topographical Survey of the Counties of Staffordshire, Cheshire and Lancashire* (1787)

Tupling, G. H. 'The early metal trades and the beginnings of engineering in Lancashire', *Trans. Lancs. and Chesh. Antiq. Soc.,* 61 (1949), 1-35

　'The turnpike trusts of Lancashire', *Mems. and Procs. of the Manchester Lit. and Phil. Soc.,* 4 (1952-3), 39-62

Twemlow, J. A. *Liverpool Town Books,* vol. I, *1550-71* (Liverpool, 1918) and vol. II, *1571-1603* (Liverpool, 1935)

University of Liverpool, Department of Geography. 'Some aspects of the historical geography of Gloucester and Gloucestershire in the eighteenth century', second year students' field week report (mimeo., 1975)

　'Some aspects of the human geography of Liverpool in the late eighteenth century', second year students' field week report (mimeo, 1976)

Vance, J. E. *The Merchant's World: The Geography of Wholesaling* (Englewood Cliffs, 1970)

*Victoria History of the County of Lancaster (VCH)* (8 vols., London, 1906-1914)

Wadsworth, A. P. and J. de L. Mann. *The Cotton Trade and Industrial Lancashire, 1600–1728* (Manchester, 1931)

Ward, W. R. *The English Land Tax in the Eighteenth Century* (London, 1953)

Webber, M. J. *The Impact of Uncertainty on Locations* (Cambridge, Mass., 1972)

Weber, A. (trans. C.J. Freidrich). *Theory of the Location of Industries* (Chicago, 1929)

White, A. W. A. 'Sixty years of coalmining enterprise on the north Warwickshire estate of the Newdigates of Arbury', unpubl. MA thesis (Birmingham, 1969)

Willan, T. S. 'The navigation of the river Weaver in the eighteenth century', *Chetham Society Publications,* 3rd ser., 3 (1951)

Williams, J. E. 'Whitehaven in the eighteenth century', *Econ. Hist. Rev.,* 2nd ser., 8 (1955-6), 396-404

Wilson, C. *England's Apprenticeship* (London, 1966)

# INDEX

absenteeism, *see under* colliers
adit, 108
Adlington, collieries at, 72, 109, 121-2, 130, 249, 251, 270, 289, 303
Aikin, J., 245, 289, 290, 302
Amberswood Moss, 56, 101
anchorsmiths, 51, 52, 97
Anderson, D., 266, 284, 288, 294, 296, 301, 303, 305
Anglesey, copper ores of, 180
anthracite, 61
Arley, collieries at, 144, 251
Arley seam, 61, 99, 136, 162, 164, 178, 179-80, 182, 190, 224, 232, 291
Ashton family of Liverpool, 221, 283
Ashton, Henry, of Whiston, 55, 65
Ashton, T.S., 202, 203, 237, 263, 275, 278, 280, 283, 295, 300, 301
Ashton-in-Makerfield
    collieries at, 246, 247, 251, 255, 267, 297, *see also* Edge Green, Garswood and Park Lane
    lock and hingemakers at, 52, 274
Aspinall, A., 292, 294
Aspull, collieries at, 39, 40, 42, 43, 50, 53, 61, 66, 67, 69, 70, 71, 73, 74, 76, 79, 90, 109, 112, 122, 123, 126, 128, 158, 190, 200, 211, 214, 222, 224, 246, 249, 251, 255, 256-8, 263, 267, 270, 272, 285
Atherton, *see* Chowbent
auditors, *see under* collieries
Aughton, peat at, 56
Awty, B.W., 266, 274, 289, 290

Bailey, F.A., 245, 262, 264, 275, 289, 290, 291, 292, 293, 302, 303
Baker, A.R.H., 260
Balcarres, Earl of, 178, 219, 223, 230-2, 233, 235, 288, 297, 300, 305
Bankes family of Winstanley, 44, 56, 74, 125, 214, 217, 231, 271, 272, 300
Bankes, J.M.H., 267, 272
bankruptcy, *see under* coalmasters
barges, *see* flatts
Barker, T.C., 54, 266, 274, 275, 282, 283, 286, 290, 295, 297
Beckmann, M., 260, 261
bellows, 109, 292

Berry, Peter, of St Helens, 144, 220, 227
Bickerstaffe
    charcoal from, 275
    collieries at, 249, 251, *see also* Stanley Gate
    peat at, 56, 101, 181
Biddle, G., 285, 286, 289, 290, 300
Billinge
    nailors at, 51
    peat at, 101
Birch, A., 289, 290
Birch, Thomas, of Liverpool, 223
Birkacre
    colliery at, 126, 181, 251
    ironworks at, 177, 183, 275
Bispham, colliery at, 107, 251
Blackburn, coal from, 162
Blackburne family of Liverpool, 52, 100, 221
Blackrod
    collieries at, 147, 249, 251, 252, 255, 293, *see also* Arley
    colliers from, 116
    nailors at, 51
    parish registers of, 90-2
blacksmiths, 50, 51, 52, 96-7
Blaguegate, collieries at, 87, 101, 109, 121, 251
Blundell family of Little Crosby, 101, 105, 223
Blundell family of Liverpool, 148, 151-2, 223, 230-1, 246, 258-9, 284, 300
Blundell, M., 275
Bold family of Bold, 56
Bold, Samuel, of Wigan, 297
bonds of colliers, *see under* colliers
Bottlingwood, colliery at, 126, 181, 251, 276
Bowden, M.J., 260
Bradford, coalmasters from, 221-3, 224, 230-1, 233, 258, 297
Bradshaigh family of Haigh, 51, 74, 88, 123, 125, 137, 162, 178, 214, 216, 219, 271, 276
Bradshaw, James, of Parbold, 297
Bradshaw, Ralph, of Wigan, 297
brass industry, 52, 53-5, 58, 97, 98, 176-7, 179
brewing, 97, 100, 166, 173, 176
Bridgeman, Bishop, 262
Bridgeman, G.T.O., 262

## 314    *Index*

Bridgewater Canal, 171, 284
Bridgewater, Duke of, 172, 235, 236, 280
Briggs, William, of Preston, 297
Brock Mill ironworks, 177-8, 179
Browne, William, of Prescot and Whiston, 67, 74, 119, 271, 278
Bulley, J.A., 264, 273
Burgh, colliery at, 178, 181, 225, 251
Byrom family of Parr, 123

Calvert, A.F., 267
canals, *see* coal transportation by water, flatts, Bridgewater Canal, Lancaster Canal, Leeds and Liverpool Canal and Sankey Navigation. *See also river navigations*
cannel, 40-1, 60, 68, 71, 78, 79, 80, 99, 109, 122, 125, 130, 136, 161, 162, 164, 175, 178, 182-3, 190, 194, 230, 231, 233, 253-4, 267, 281, 285, 291, 298, 301, 304, 305
    fire-damp in, 109, 190-1
    nature of, 54, 62, 79, 254
    pits, cost of, 64; depth of, 64, 66-7; production of, 88, 148, 150, 151; value of, 256-7
    price of, 52, 60, 79, 104, 128, 165, 172, 224, 254, 272, 288, 298, 302, 304, *see also* Aspull, Haigh, Ince and Kirkless
capital, *see* collieries, cost of
Carr Mill ironworks, 98, 178, 179
cartels, *see under* coalmasters
Cartwright, J.J., 290, 291
Case family of Liverpool and Red Hazels, 100, 123, 127, 139, 141, 144, 152, 189, 215, 220, 221, 225-6, 229, 232, 233, 235, 280, 283, 298, 299
Case, Richard, of Whiston, 55
Catterall family of Crooke, 272
Chadwick family of Birkacre, 98, 128, 177-8, 179, 183, 219, 220, 223, 275, 276
Challinor, R., 199, 269, 293
Chandler, G.A., 266
charcoal, 101, 275
Charnock Richard, collieries at, 108, 112, 123, 126, 130, 190, 249, 251, 270, 276, 303
Cheshire
    as coal market, 78, 165, 166, 183, 227,\234
    coal supplies of, 49-50, 100, 106, 107, 110, 167, 184, *see also* Weaver Navigation
    collieries in, 172 *see also* Neston
    iron industry of, 51, 97
    salt industry of, 55, 58, 99-100, 106, 127, 166, 173, 224
Chorley, 48
    canal at, 183
    coal buyers of, 181-2
    collieries at, 251, 252
    colliers from, 270
    ironworks near, 98
    population of, 96
Chorley, John, of Prescot, 226, 297

Chowbent, nailors at, 51-2, 96-7
Civil War, effect on mining, 48, *see also* Jacobites
Clarke family of Liverpool, 215, 223, 290
Clayton, Sarah, of Liverpool, 123, 143, 221, 227-9, 232, 280, 299
Clegg, E.H., 288
coal
    consumers, domestic, 6, 14-15, 16, 17, 27, 28, 45-50, 57, 60, 94-6, 102, 165-6, 181-3, 238, *see also* households; industrial, 15-16, 17, 18, 27, 28, 50-4, 55, 58, 96-100, 102, 123, 176-81, 216-17, 225, 229, 236, 238-9
    credit sales of, *see* coal, trust sales of
    demand for, 5-11, 13-17, 18, 19, 26, 28, 29, 30, 45-61, 94-105, 161-89, *see also* demand
    exploration for, 55, 57, 58, 67, 107-8, 267, 268
    exports of, 16, 17, 27, 28, 54-5, 58, 100-1, 143, 161-76, 225, 229, 236, 285, 286-7, 288, *see also* Dungeon Point, Liverpool and Tarleton
    markets for, 5, 7, 10, 16, 17, 21, 30, 46, 49-50, 54, 55, 57-61, 79-80, 94-105, 128-31, 158, 161-89, 224-36, 238-9, 298-9, *see also* Cheshire, Dublin, Fylde, Liverpool, Preston, St Helens, Warrington and Wigan
    pits, *see also* cannel pits; cost of, 40-1, 42, 64, 109, 122, 125; depth of, 19, 63-4, 66-7, 108-10, 112, 130, 267-8, 276, 291; production of, 88, 150, 151, 213, 263-4; value of, 256-7
    prices of, 7, 8, 9, 10, 13, 15, 16, 17, 18, 19, 20, 21, 28, 45, 77, 78, 100, *see also* cannel, prices of; pithead, 58-60, 80, 102-4, 128, 129, 130, 143, 165, 168, 184-7, 201, 202-3, 224, 225, 227-31, 234, 235, 236, 246, 261, 264, 281, 290, 293, 298, 299, 302, 303; market, 60, 100, 106, 163, 165, 168-9, 171-3, 180, 187, 224, 230-1, 226, 229, 235, 239, 267, 291
    royalties of, *see* collieries, rents of
    seams, 17, 18, 28, 40, 54, 61-2, 64, 107-10, 189-90, 224, 225, 226, 229, 239, 267, 291, *see also* coal, types of, Arley seam, cannel, outcrops and smith coal
    transportation, cost of, 5, 6, 7, 8, 10, 16, 17, 21, 22, 27, 29, 45, 57, 78, 105-7, 129, 130, 183, 235, 239; by road, 21, 22, 25, 30, 45-6, 55, 60, 105-7, 165, 167, 169, 174-6, 179, 193, 275, 289, *see also* turnpikes; by wagonway, 174-5, 188, 215, *see also* wagonways; by water, 100-2, 105-7, 129, 135, 161-76, 179, 233, 284-5, 286, 287, 288, 289, 296, *see also* coal, exports, canals *and river navigations*
    trust sales of, 123, 216-17, 281, 283, 296

types of, 14, 15, 18, 26, 28, 54, 61-2, 79, 99, 128-9, 136, 162, 164, 165, 168, 172, 178-9, 179-81, 181-2, 187, 190, 224-5, 229, 230-1, 234, 239, 287-8, 289, 290, 297, *see also* anthracite, Arley seam, cannel, coking coal and smith coal

weights and measures of, 40, 80, 117, 118, 152, 167, 169, 172, 201-3, 243-8, 278, 284, 286-7, 290, 293, 301-4

coalmasters

bankruptcies of, 123, 127, 220, 221, 226, 229

cartels of, 143, 144, 228, 229, 230-1, 236

credit sales by, *see under* coal

competition between, 15, 16, 20-3, 25-6, 29, 78-9, 128-30, 167, 169, 171, 172-3, 183, 224-36, 240-1, 261, *see also* coalmasters, price cutting of, monopoly and oligopoly

freeholders as, 19, 24, 25, 26, 75, 219, 220-1

gentry as, 15, 24-5, 64, 74-9, 123-5, 127, 217-19, 221-2, 228

knowledge of, 10, 23-5, 29

leaseholders as, 19, 24, 25, 74-7, 80, 124, 127, 219-24, 280

litigation between, 39-40, 64, 85

mobility of, 126-7, 131, 219, 280

mortgages of, 123, 219, 220, 221, 280, 297

motives of, 7, 8, 10, 17, 23-5

partnerships of, 75, 76-7, 125-6, 219-24, 230, 234, 236, 240, 281, 297, 297-8

perceptions of, 23-4

price cutting of, 16, 17, 26, 78, 79, 129, 168, 225, 227-9, 232, 233, 235

profits of, *see under* collieries

Quakers as, 223

*see also under* Bradford and Liverpool

coalmeasures, 17, 60, 61, 108, 189, *see also* lower coalmeasures

Cobham family of Prescot, 99, 127, 139, 220, 225-6, 281

coking coal, 61, 178, 182, 225, 289

Cole, W.A., 274

Coleman, D.C., 300

collieries, *see also* coal pits and wayleaves

auditors of, 68, 69, 71, 72, 112, 123, 253, 270, 278, 284, 295; wages of, 69-70, 117, 205

on commons, 262-3

costs of, 5, 6, 8, 9, 18-23, 29, 64, 65-7, 73-4, 108, 112, 117-18, 120-3, 129, 187, 193, 212-17, 220, 233, 235, 267, 268, 269, 271, 275, 279, 295, 296, 304

demand orientation of, 58, 60, 106, 113, 164-5, 166-7, 170-1, 174, 184, 189, 232-3

drainage of, 19, 25, 64-7, 73, 74, 108-12, 191-2, 193, 214-15, *see also* drainage engines and soughs

duration of, 15, 28, 39-40, 57, 79-80, 85,

92, 102, 136-7, 140-1, 147-8, 154, 193, 239, 272, 273

engineers at, 199

fire-damp in, 109, 110, 123, 190-1, 200, 228, 279-80

flooding of, 67, 74, 109, 130, 151, 189, 191-3, 194, 228, 275-6

landsale, 187, 232

lessees of, *see under* coalmasters

location pattern of, 4, 8-10, 15, 16, 17, 18, 19, 20, 21, 22, 23, 26, 28, 30-1, 35-44, 83-9, 93, 102-5, 110, 113, 128, 135-60, 163-4, 166-7, 174, 183-9, 224, 227, 230

market areas of, 7-8, 10, 18, 24, 78, 261, *see also* coal, consumers and coal, markets

numbers of, 35-9, 43, 79, 83-5, 92, 135-7, 139-40, 145-7, 283-4

outputs of, 7-11, 18, 19, 20-2, 28, 40-4, 60, 73, 79, 88-9, 102, 108, 110, 113, 121, 122, 123, 128, 137-9, 141-5, 148-53, 167, 184, 220, 233, 234-5, 245, 253-4, 256-9, 272, 273, 281, 282, 283, 284; seasonality in, 43, 45-6, 265

profits of, 5, 8, 9, 11, 17, 18, 20-1, 23, 24, 72-4, 76, 120-3, 124-5, 128, 139, 201, 212-13, 216, 227, 230-1, 232, 233, 235, 236, 256, 258, 272, 279, 280, 282, 283, 296

receipts of, 5, 43, 73-4, 121, 123, 212-13, 253-4, 264, 279-80, 281

rents of, 19, 41, 42, 74, 76, 77, 80, 121, 125-6, 137, 144, 244, 245, 261, 271, 273, 280

sequestration of, 74, 76, 123-4, 271, 280

sites of, 18, 28, 64-7, 110-13, 193

or urban copyholds, 35-6, 39, 42

ventilation of, 109, 191, 292

workers at, *see* colliers

colliers

absenteeism of, 151, 204-5, 294

bonds of, 68, 71, 194, 269, 295

children as, 195, 197

communities of, 72, 211-12, 240

families of, 72, 114, 116-17, 119-20, 195-8, 200, 210-11, 270, 273, 292, 293

fines of, 69, 70-1, 203, 264, 294

journeys to work of, 71-2, 115, 211-12, 270, 271, 295-6

life-styles of, 43, 71, 119, 120, 199-200, 204-5, 211-12, 278, 293

literacy of, 199, 270, 293

mobility of, 22, 25, 71-2, 113-17, 119, 131, 197-9, 204, 210, 212, 235, 240, 276, 277, 280, 292

numbers of, 22, 25, 29, 40-2, 90-3, 113, 144, 155-9, 194-5, 256-7, 263, 284, 292, 305

Poor Law and, 71, 115-16, 198-9, 270, 277-8, 293, *see also* paupers

productivity of, 70, 88, 194, 203, 240, 258-9, 278, 284
recruitment of, 72, 111, 117-18, 195-8, 210, 213, 240
shortage of, 71, 194
specialisation of, 67-9
wages of, 22, 29, 70-1, 117-19, 200-11, 235, 240, 245, 254, 269, 278, 293, 295, 303
women as, 69, 71, 270
work of, 68, 71, 72, 80, 117, 191, 195, 196, 199, 200, 202-5, 209, 243, 264, 270, 276, 278, 293, 295
*see also* collieries, auditors of and collieries, engineers at
commons, *see* collieries, on commons
competition, *see* coalmasters, competition between
copper industry, 98, 152, 166, 171, 173, 179, 180-1, 217, 221, 229, 295, 299, 300
copperas, 50
Coppull
collieries at, 158, 251, 255, *see also* Birkacre
colliers of, 115
costs, *see under* collieries
credit sales, see coal, trust sales of
Crofton, H.T., 267, 302
Cromar, P., 301
Croston, 96
Croxteth, collieries at, 108, 187, 189, 251, 268
Croxteth Park outlier, 61, 105, 107-8
Culcheth, Richard, of Orrell, 297
Cumberland, coal exports from, 54, 55, 172
*see also* Whitehaven
Curr, John, 296

Dagnall, Charles, of Eccleston, 182, 187, 191, 220, 290, 296
Dalton, colliery at, 251
Deane, P., 274
Defoe, Daniel, 272, 274
Delaval, Sir J.H., 215
demand schedules for coal, 8-9, 10, 13-15, 21, 261 *see also* coal, demand for
Derby, Earl of, 46, 56, 77, 101, 124, 125, 137, 144, 182, 219, 221, 223, 226, 231, 283, 298
Derry, T.K., 269
Dicconson family of Wrightington, 112, 125, 126, 219, 223, 244, 297
Ditton, 55, 266
Douglas, river, 36, 79
navigation of, 45, 107, 108, 132, 136, 154, 158, 160, 161-5, 166, 170, 174, 175, 176, 178, 217, 223, 224, 233, 234, 235, 244, 247, 285, 286-7, 300
valley of, 64, 66, 146, 160, 184, 193
water power of, 51, 52, 177, 184, 187
drainage engines, 64-5, 73, 110, 112-13, 120-1, 122, 123, 125, 151, 191-2, 214-15, 226, 239, 269, 276, 279, 281,

292, *see also* horse power, steam engines, water power and windmills
Dublin, coal supplies of, 161, 162, 228, 285, 299, *see also* Ireland
Dungeon Point
coal exports from, 55, 60, 106, 107, 129, 165, 183
road to, 55, 131
saltworks at, 100, 127, 298
duty on sea-borne coal, 162, 285
Duxbury, colliery at, 251, 255, *see also* Burgh
Dwight, John, 52

Earle, William, of Liverpool, 223
Eccleston near Chorley, colliery at, 249
Eccleston near Prescot, collieries at, 42, 64, 123, 129, 144, 150, 151, 182, 191, 201, 202, 217, 251, 252, 255, 263, 267, 289, *see also* Gillars Green
Eccleston family of Eccleston, 77, 125, 217
economies of scale in mining, 20-3
Eden, Robert, of Prescot, 226, 297
Edge Green, collieries at, 89, 109, 125, 251, 273, 276
Eltonhead family of Sutton, 45
engineers, *see under* collieries
Everitt, A., 301

family reconstitution, 195-7, 277, 292, *see also* colliers, families of
faults, 62, 74, 108, 281
feedback, 27, 30
Files, J., 47
Finberg, H.P.R., 260
fire-damp, *see under* cannel and collieries
firewood, 56, 101, 275
Fischer, D.H., 260
flatts, 143, 162, 163, 164, 168, 215, 285, 295, 296, 299
Fletcher, Ralph, of Prescot, 42
Flintshire, coal exports of, 54, 172
flooding of collieries, *see under* collieries
folding, 62, 108
France, R.S., 276
Frodsham, 99, 106, 165
Furness, iron industry of, 98
Fylde, coal supplies of, 128, 281, 285

Galloway, R.L., 266, 272
Garswood, collieries at, 89, 102, 106, 112, 143, 145, 152, 190, 212, 213, 219, 221, 227-9, 251, 252, 273, 287, 290
gentry, *see under* coalmasters
geomorphology, influence on colliery location of, 19, 29, 66-7, 108, 110, 111, 113, 123, 193, 239
Gerard family of Garswood, 66, 76, 125, 143, 219, 221, 223, 227-9, 232, 267, 274, 280, 297, 299, 300
Gerard family of Ince, 39, 50, 56, 66-7, 70, 76, 79, 101, 122, 217, 272, 280
German, John of Orrell, 223

Gibson, T.E., 274, 275
Gilboy, E.W., 209, 294, 295
Gildart, James, of Liverpool, 144, 189, 221, 225, 295
Gillars Green, collieries at, 129, 137, 140, 213, 251, 282, 289
glacial drift, 62-3, 108
glass industry, 50, 52, 61, 97, 98-9,100, 127, 137, 152, 176, 179-80, 181, 220, 221, 229, 274, 281
Golborne
  collieries at, 146, 252, 255, *see also* Edge Green
  peat at, 56
Gonner, E.C.K., 94, 265, 274
Goodicar, John, of Prescot, 39
Goose Green, collieries at, 42, 249, 269
Gorsuch family of Gorsuch, 272
Great Sankey, ironworks at, 98
Gregson, Matthew, of Liverpool, 171, 172
gunsmiths, 97

Hadfield, C., 285, 286, 289, 290, 300
Haigh
  canal at, 183
  collieries at, 40, 41-2, 46, 54, 61, 64, 66, 67, 68-9, 70, 71, 72, 74, 77, 79, 85, 88, 90, 101, 107, 109, 112, 115, 116, 117, 123, 125, 126, 128, 137, 139, 144, 145, 146, 147, 148, 149, 150, 158, 164, 175, 178, 181, 190, 200, 201, 202-3, 209, 211, 214, 219, 220, 222, 223, 224, 230-2, 243, 244, 245, 246, 247, 249, 251, 253-4, 255, 256-8, 264, 267-8, 269, 272, 276, 278, 279-80, 285, 289, 291, 292, 295, 296, 298, 301, 305, *see also* Arley
  industries at, 51, 52, 97-8, 178, 179, 219, 224, 231
  roads to, 105, 267
Hair, P.E.H., 200, 203, 293
Halebank, 55, *see also* Dungeon Point
Hall, Edward, 98
Halliwell family of Wrightington, 112, 126, 127, 219-20
Hamilton, H., 266
Hamshere, J.D., 260
Hardcastle, Thomas, of Bradford, 223
Hardshaw, collieries at, 282, *see also* Windle
Harland, J., 265, 303
Harley, J.B., 262
Harrington family of Huyton, 123, 276
Harris, J.R., 274, 282, 285, 290, 295, 297, 300
Harrison, James, 283
Harrison, W., 265
Harrock Hill, colliery at, 108, 110, 130, 251
Harvey, D., 3, 260, 261, 262
Hatcher, J., 54, 266
Hatton, Thomas, 162, 297
Hawkes, A.J., 265, 266, 271, 302
Haydock, collieries at, 106, 109, 112, 136,

143, 145, 150, 155, 172, 174, 175, 176, 190, 191, 194, 211, 217, 219, 221, 225, 227-9, 251, 272, 273, 276, 279, 287, 288, 289, 291, 292, 295, *see also* Laffack
  woods at, 56
hearth taxes, 46-8, 96
Hesketh family of Rufford, 76, 146, 217
Heskin, 79, 128
  colliery at, 249
Hindley,
  collieries at, 42, 64, 65, 66, 73, 76, 222, 249, 252, 263, 269
  nailors at, 51
  peat at, 56
  population of, 95
hingemaking, 52, 97, 98, 274
Hodson, James, of Wigan, 256, 305
Homle family of Up Holland, 125, 137, 148, 151, 152, 164, 214, 216-17, 223, 246, 258-9, 284, 291
Holt, Roger, of Wigan, 161
Hopkinson, G.G., 264
horse power, 65, 66, 112, 121, 174, 192, 214, 226, 268-9, 279, *see also* drainage engines
Hoskins, W.G., 265
Houghton, John, of Prescot, 297
households, coal consumption of, 46, 94, 181
Howarden, Thomas, of Wigan, 191, 296
Hustler, John, of Bradford, 221-2, 223, 230-1, 297, 300
Hustler, William, of Bradford, 223, 297
Huyton, collieries at, 75, 100, 110, 123, 129, 251, 252, 255, 276, 291

Ince
  collieries at, 146, 151, 158, 190, 217, 222, 251, 252, 300, *see also* Kirkless
  peat at, 56-7, 101
inventories, 51, 57
Ireland, coal exports to, 54, 55, 58, 172, *see also* Dublin
iron industry, 50, 51, 52, 97-8, 101, 176-9, 181, 231, 275, 289, 299
Isard, W., 23, 260, 262

Jackman, W.T., 296
Jackson family of Orrell, 192, 220, 223
Jacobites, 123, 125, 219, *see also* collieries, sequestration of
Janelle, D.G., 301
Jarratt, John, of Bradford, 223, 297
Johnson, Sir Thomas, of Liverpool, 100
Jones, E.L., 260
Jones, R.C.B., 261, 288

Kerridge, E., 300
Killick, H.F., 286, 300
King's College, Cambridge, 77, 127, 137, 225, 226, 262, 279, 280-1
Kirkby, 267, 268
Kirkless, collieries at, 137, 146, 147, 148,

150, 152, 174, 190-1, 192, 200, 201,
    202-3, 204, 208, 211, 212, 213, 222,
    224, 244, 245, 246, 251, 252, 254, 291,
    294, 298, 302, 303
Knowsley,
    collieries at, 89, 127, 251
    peat at, 101
Koelsch, W.A., 260

labour market, 210, *see also* colliers,
    recruitment of
Lace, Ambrose, of Liverpool, 223
Laffack, collieries at, 143, 178, 194, 225, 227,
    251, 252, 289
Laithwaite, Raphe, of Wigan, 67
Lancaster Canal, 183, 219, 224, 231-2, 235,
    244, 247, 284, 289
Lancaster, coal supplies of, 162
Lancaster, Robert Patrick, of Coppull, 112,
    126, 278
land tax, 146-8, 152, 194, 203, 219, 255-9,
    283-4, 297
Langton family of Hindley, 76, 123, 280
Langton, J., 260, 262, 264, 271, 281, 284,
    300
Lathom
    collieries at, 68, 102, 108, 130, 147, 230,
        251, *see also* Blaguegate, Newburgh
        and Westhead
    colliers of, 115, 119
    peat at, 56, 101
law suits, *see under* coalmasters
Layton, Philip, of Prescot, 262
Leafe family, glassmakers, 98-9, 127
Lee, C.H., 260
Leeds and Liverpool Canal, 135, 145, 146,
    147, 148, 152, 153, 155, 158, 161, 165,
    166, 169-73, 174-5, 179, 183, 184, 187,
    188, 191, 215, 217, 220, 221-3, 224-5,
    230-2, 235-6, 238, 239-40, 244, 247,
    254, 286-7, 287-8, 290, 296, 300
Legh family of Lyme Park, 143, 219, 221,
    223, 227-9, 232, 282, 289, 299
Leigh, Alexander, of Wigan, 161-2, 163, 164,
    296, 297
Leigh, Thomas, of Warrington, 144, 220,
    227, 228, 299
Lerry, G.G., 273, 279
Leyland Mill ironworks, 178
Liverpool
    canal proprietors of, 170, 230-1
    coal exploration at, 55, 267
    coal exports of, 54-5, 100, 143, 171, 225,
        229, 236, 238, 287
    as coal market, 55, 58, 78, 100, 131, 152,
        158, 160, 162, 163, 165-73, 175, 183,
        184, 217, 225, 227, 236, 238, 239
    coalmasters from, 127, 178, 179, 220-1,
        222-3, 258
    coal supplies of, 106-7, 108, 110, 127, 129,
        143, 165-73, 180, 232, 272, 286-7,
        287-8, 288, 297, 300

communications to, 105-6, 107, 234, 239,
    285, *see also* Leeds and Liverpool
    Canal, Mersey river, Sankey
    Navigation and turnpikes
corporation of, 54, 55, 57
industries of, 52, 98-9, 100, 127, 137,
    165-7, 171, 173, 176, 180, 298
peat at, 57
population of, 48, 55, 95, 96, 165-6, 171,
    182, 238
slave trade of, 221
location theory, 5-11, 12, 13, 23, 25, 30-1, *see
    also* Lösch, A. and optimum location
    patterns
lockmaking, 52, 98
Lofthouse, John, of Liverpool, 223, 297
London, 56, 161, 162, 221, 238
Long, W.H., 13, 261
Longbotham, John, of Hargreaves, 151, 221,
    283
longwall mining, 152, 194, 235, 240, 279
Lösch, A., 5, 6, 11, 12, 13, 17, 18, 19, 20, 22,
    23, 25, 26, 28, 61, 78, 232, 233, 261
lower coalmeasures, 61-2, 64, 189, 193,
    291-2
Lowther, Sir James, 162, 299
Lowton, peat at, 56


Mackay, John, 143, 144, 181, 189, 221,
    228-9, 299, 300
Makin family of Prescot, 98, 127, 139, 189,
    220, 225-6, 281, 297
Mann, J. de L., 279, 301
manors, 14
    rights in, 47, 55, 57, 66, 265
    courts of, 56, 101, 275
markets for coal, *see under* coal
Martlu, Peter, of Lathom, 67
Mascall, William, of Frodsham, 106
Master, Legh, of New Hall, 220, 227
Mersey, river and estuary
    coal trade, 55-6, 98-100, 162, 298, *see also*
        Liverpool, coal exports of, Sankey
        Navigation and Weaver Navigation
    industries on, 98-100, 102, 103, 106, 131,
        176, 180, *see also under* Liverpool and
        Warrington
    navigation of, 98-9, 106, 131, 165, 168,
        173
methodology of economic history and
    historical geography, 1-5, 12, 13-14,
    20-2, 23-4, 25, 26, 30-1, *see also* systems
    model of mining industry
Middlewich, 100, 106
migration of colliers, *see* colliers, mobility of
Millstone Grit, 61, 62, 107, 161
Moffitt, L.W., 96, 274
Moller, A.W.R., 264, 266, 267, 275
Molyneux family of Hawkley, 270
Molyneux, Viscounts, 46, 77, 123, 189,
    213-14

monopoly, 6, 7, 16, 181, 183, 229, 232, 233, 235, 236
Moore family of Liverpool, 57
mortgages, *see under* coalmasters
mosslands, 36, 56-7, 63-4, 101-2, 108, 130, *see also* peat
Mostyn, coal exports of, 54, 55

nailors, 51-2, 96-8, 178
Nef, J.U., 35, 43, 45, 46, 68, 237, 245, 262, 264, 265, 267, 268, 269, 270, 273, 295, 300, 301, 302
Neft, D.S., 262
Neston, colliery at, 172
Newburgh
  collieries at, 87, 108, 110, 111, 112, 123, 130, 251, 272, 275-6, 279, 303
  canal at, 160, 170
Newton, peat at, 56
North, F.J., 272
Northwich, 99, 100, 106, 107, 176, 217

ochre, 108, 276
Ogle family of Whiston, 42, 76, 77, 123
oligopoly, 16, 17
Olsson, G., 3, 260
optimum location patterns, 5-6, 8-9, 10, 11, 30, 261
Ormskirk
  peat at, 56, 57, 182
  population of, 48, 96
Orrell
  collieries at, 40, 42, 50, 53, 64, 70, 71, 72, 79, 109, 114, 115, 127, 128, 136, 137, 144, 145, 146, 147, 148, 150, 151-3, 155, 170, 173, 175, 188, 190, 192, 193, 194, 197, 201-3, 209, 211, 212, 213, 214, 215, 217, 219, 220, 222, 223, 224, 229, 230-1, 244, 245, 246, 249, 251, 252, 255-9, 267, 268, 278, 285, 287, 289, 291, 292, 296, 298, 302, 303
  colliers of, 115, 194-5, 201, *see also* Up Holland, colliers of
Orrell family of Orrell, 272
Orrell, James, 220
outcrops of coal seams, 62, 64, 66-7, 108, 110
  workings at, 64, 118, 279
output of coalfield, 35, 43, 58, 60, 93, 154-5, 159-60, 238-9, 264, *see also under* collieries
Owen, George of Henllys, 267

Parbold
  collieries at, 153, 189, 222, 252, 255, 291, 300
  colliers of, 115
parish registers, 90-2, 95, 113-14, 115-16, 135, 155-9, 195-8, 274, 277, 295
Park Lane, collieries at, 67, 123, 137, 214, 215, 220, 222, 251, 263, 285, 291, 295
Parkinson, C.N., 266

Parr
  collieries at, 71, 76, 89, 92, 101, 123, 143, 144, 145, 146, 155, 176, 181, 191, 211, 221, 227-9, 251, 252, 255, 267, 270, 272, 283, 297, *see also* Laffack
  colliers of, 115-16, 198-9, 200, 277-8
  iron ore at, 98, 289
  roads in, 275
Parrott, Stanier, 127
Patten, Thomas, of Warrington, 98
paupers, 115-16, 198-9, 277, *see also* colliers, Poor Law and
peat, 56-7, 78, 101-2, 181, 182
Pemberton, collieries at, 40, 136, 144, 146, 147, 153, 155, 158, 170, 175, 188, 190, 191, 211, 214, 222, 231, 251, 252, 255, 269, 270, 283-4, 285, 295, 300, *see also* Goose Green
Pennant, Thomas, 158
Pennington, Nicholas, of Wigan, 39, 42, 67
pewter industry, 52, 53-5, 58, 97, 98, 176-7
Picton, J., 266
pinmaking, 97
pit props, 122, 279
plague, 265
Platt, Peter, of Wigan, 39, 42
Pococke, Richard, 181
Poole, Josiah, of Liverpool, 127
Poor Laws, *see under* colliers
population, of South West Lancashire, 46-8, 57, 60, 61, 94-6, 131, 181, 182, 184, 238, 265, *see also* Chorley, Liverpool, Prescot, Preston, St Helens, Warrington and Wigan
Porteus, T.C., 274, 275, 276
Postan, M.M., 260
Potter, Thomas, 125
pottery industry, 52, 99, 100, 101-2, 127, 166, 176, 179, 274
Prescot
  collieries at or near, 39, 40, 43, 45, 59, 60, 64, 65, 72, 89, 99, 106, 110, 113, 115, 118, 119, 120, 121, 122, 127, 129, 132, 137-9, 141, 144, 145, 150, 151, 163-4, 167-8, 172-3, 174, 189, 190, 191, 197, 204, 213, 215, 216, 220, 225-6, 243, 245, 246-7, 249, 251, 252, 262, 263, 272, 273, 276, 279, 281, 283, 288, 290, 291, 295, 302, 304
  industries of, 52, 99, 184
  population of, 48, 96, 182
  roads to, 105-6, 107, 175-6, 184, 234
Preston, 48, 52, 95, 108
  canal at, 183
  as coal market, 60, 107, 128, 134
  coalmaster from, 127
  coal supplies of, 108, 130, 131, 162, 163, 184, 224, 232, 272, 281, 289, 290, 298
  population of, 96, 182
Priestley, Joseph, 160
Prince, H.C., 261
profits, *see under* collieries

Rainford
    collieries at, 87, 105, 251, 252, 255
    peat at, 56, 101, 181
    potteries at, 52, 99, 101-2, 105, 274
Ravenhead.
    collieries at, 144, 145, 152, 154, 172, 181,
        189, 194, 210, 211, 221, 224, 225,
        229, 251, 252, 282, 291, 292, 300, see
        also Thatto Heath
    industries at, 179-81
receipts, see under collieries
Ribble, river, 46, 132, 137, 161, 162
Richardson, H.W., 260
Rigby, John, of Shevington, 39, 76
roads, see under coal, transportation of
Roberts, E., of Liverpool, 221
Rodgers, H.B., 266
royalties, see collieries, rents of
Russell, John, of Portway, 97-8

St Helens
    canal to, see Sankey Navigation
    as coal market, 238
    collieries at, 136, 137, 144, 215, 220, 227,
        228, 245, 251, 252, 283, 288, 301, see
        also Eccleston near Prescot,
        Hardshaw, Parr, Ravenhead, Sutton
        and Windle
    colliers of, 113-17, 119-20, 195-200,
        210-12, 277, 292
    growth of, 179, 235
    industries of, 99, 179-81, 239
    population of, 181, 182, 238
    turnpike to, 175, 184, 234
salt industry, 50, 55, 99-100, 166, 173, 221,
    281, 287, 298, see also under Cheshire
Salt, George, 127
Sankey Navigation, 135, 140, 141, 143, 147,
    148, 152, 153, 154, 158, 162, 165-9,
    170, 171, 172, 173, 175, 176, 184,
    187-8, 189, 190, 224, 225, 227-30, 233,
    234, 235, 247, 287, 288, 296, 298
Scarth Moss, 56, 57
Scott family of Wigan, 53
Scott, N.M., 280
Scott, W.R., 274
Sefton, search for coal at, 57, 67, 267, 268
Shelley, R.J.A., 266
Sherrington, Francis, 263
Shevington, collieries at, 39, 42, 64, 65, 69,
    70, 73, 74, 76, 88, 109, 112, 123, 125,
    126, 127, 130, 146, 152-3, 155, 217,
    219, 222, 223, 230, 231, 243, 244, 245,
    251, 252, 255, 263, 268, 271, 272, 278,
    280, 300, 301, 303
Shuttleworth family of Smithils, 45, 46, 49,
    57, 272
Skelmersdale, collieries at, 147, 148, 183,
    187, 193, 197, 201, 202-9, 213, 231,
    246, 251, 252, 255, 258, 290, 294, 303
Smiles, Samuel, 301
Smith, D.M., 311

smith coal, 54, 67, 78, 79, 80, 99, 128, 130,
    136, 162, 165, 172, 173, 175, 189, 190,
    193, 224, 225, 281, 288, 291, 298
Smithils, coal supplies of, 46, 49, 78, 302
soap industry, 50, 173
soughs, 19, 25, 65-7, 73-4, 108, 109, 112,
    120, 121, 122, 123, 125, 151, 193, 214,
    216, 267, 268, 269, 271, 276, 279, 292
South Sea bubble, 107
Southworth family of Hindley, 65-6
Speakman, Joseph, 220
Sprunt, W.C., 266, 274
Staffordshire
    coal from, 49-50, 55, 100, 106, 165
    coalmaster from, 109
    iron industry of, 51, 97-8
Standish
    collieries at, 117, 121, 123, 130, 137, 146,
        147, 148, 150, 152, 155, 158, 170,
        175, 188, 190, 197, 201, 202, 204-5,
        217, 222, 223, 230, 231, 244, 251,
        255, 256-7, 282, 295, 296, 298, 300
    colliers of, 115
    nailors at, 51
    parish registers of, 90-2
    population of, 95-6
Standish family of Standish, 217, 231, 300
Stanley family of Aughton, 56
Stanley Gate, colliery at, 78, 272
starch industry, 50
steam engines, 110, 113, 121, 127, 143, 144,
    150, 151-2, 160, 164, 178, 180, 191-2,
    193, 214-15, 226, 233, 235, 239, 240,
    279, 284, 292, 296
Stock, James, of Ashton, 214, 215, 220, 291
Stone, L., 263, 264
sugar industry, 50, 98, 100, 166, 173, 217
supply schedules for coal, 8-9, 10, 20-1
Sutton,
    collieries at, 40, 45, 64, 66, 73, 89, 106,
        123, 144, 155, 211, 244, 249, 251,
        255, 263, 269, 273, 289, 291, see also
        Ravenhead and Thatto Heath
    industries of, 52, 98-9, 127, see also
        Ravenhead and St Helens
    peat at, 56, 101
Swanson, J.A., 260
Sykes, J., 202, 203, 237, 263, 275, 278, 283,
    295, 300, 301
systems model of mining industry, 9-13,
    27-30, 262

tanning bark, 56
Tarbock,
    coal to, 266
    collieries at, 89, 109, 110, 116, 120, 121,
        123, 127, 137, 141, 144, 194, 201,
        202, 215, 217, 251, 272, 278, 279,
        280, 284, 291, 292, 294, 295
    road at, 55
Tarbuck family of Windle, 215, 220, 227, 229
Tarleton, 158, 160, 162, 285, 286-7

Thatto Heath, collieries at, 88, 89, 93, 99, 106, 110, 112-13, 122, 127, 129, 141, 144, 154, 251, 272, 278, 282
Thompson, A., 260
Thrupp, S., 301
Tonks, C.H., 261, 288
towns of South West Lancashire, 46-55, *see also* Chorley, Liverpool, Ormskirk, Prescot, Preston, St Helens, Tarleton, Warrington and Wigan
Toynbee, A., 94, 274
transportation, *see under* coal
Trott, C.J.D., 264, 266
Tupling, G.H., 266, 274, 275
turbary, 56-7, 101
turf, *see* peat
Turner, Richard, of Middlewich, 106, 127
turnpikes, 107, 129, 131, 171, 173, 175-6, 183, 193, 224, 232, 235, 275, 289, *see also* coal, transportation
Twemlow, J.A., 266

Ulverston, coal to, 162
Up Holland
  collieries at, 146, 153, 155, 175, 188, 190, 193, 211, 222, 223, 224, 230, 251, 252, 255
  colliers of, 113-17, 119-20, 195, 196-8, 210-12, 277, 278, 292
  industries at, 50, 274

Vance, J.E., 260
Victoria History of the County of Lancaster, 274, 280, 203

Wadsworth, A.P., 275, 301
wages, *see under* collieries, auditors of, and colliers
wagonways, 143, 193, 230, 288, 296, 299
Walton, coal to, 272
Ward, W.R., 256, 305
Warren, Samuel, of Liverpool, 223, 287-8
Warrington
  as coal market, 55, 60, 98, 100, 165, 166, 180
  coal supplies of, 162, 175, 176, 272
  industries of, 51, 52, 97, 98, 99, 127, 180, 290
  population of, 48, 95, 96, 182
  transport routes to, 98, 176, 180
water power, 51, 52, 65, 66, 67, 98, 110-13, 121-2, 177-8, 192, 214, 269, 276
wayleaves, 227, 282, 289
Weatherby family of Whiston, 271
Weaver Navigation, 55, 98, 99-101, 106-7, 131, 165, 166, 170, 173, 179, 183
Webber, M.J., 261
Weber, A., 5, 260
weights and measures, *see under* coal
West, Thomas, of St Helens, 179, 181, 220
Westhead, collieries at, 87, 88, 121, 122, 251, 255, 272, 279

Whiston, collieries at, 39, 40, 42, 55, 60, 65, 67, 71, 74, 76, 89, 100, 106, 110, 113, 115, 120, 121, 122, 127, 139, 141, 144, 147, 152, 167-8, 174, 175-6, 189, 190, 191, 214, 219, 221, 225-6, 247, 249, 251, 252, 255, 266, 269, 272, 279, 281, 283, 291
White, A.W.A., 264, 265
Whitehaven, 55, 162, 264, 279, 299
whitesmiths, 51, 97, 98, 172
Whittle-le-Woods, 96
Wigan
  as coal market, 78, 96-9, 100, 128, 129, 224, 238-9
  collieries at or near, 39, 42, 43, 58, 60, 64, 67, 107, 109, 125, 127, 132, 146, 156, 169, 222, 244, 245, 246, 249, 251, 262, 275, *see also* Bottlingwood
  colliers of, 71, 115, 155-6, 198, 211-12, 277, 295
  industries of, 51, 52-5, 58, 97, 98, 176, 178, 179, 180, 234, 239, 274
  parish registers of, 90-2
  roads to, 105, 107, 267, 275
  waterways to, 107, 136, 161, 162, 165, 169-74, 285, 287, *see also* Douglas Navigation and Leeds and Liverpool Canal
Willan, T.S., 265, 275
Williams, J.E., 279
Williams, T.I., 269
Williamson, John, of Liverpool, 220, 227, 297
Willis family of Liverpool and Halsnead, 183, 221, 298
wills, 50-1, 96-7, 274
Wilson, C., 267
Windle
  collieries at, 140, 141, 144, 146, 152, 153, 155, 181, 182, 211, 215, 221, 227, 229, 244, 249, 251, 252, 255, 256-8, 273, 289, 291, 292, *see also* Hardshaw and St Helens
  iron ore at, 289
windmills, for colliery drainage, 110, 113, 121
Winstanley
  collieries at, 40, 42, 43, 46, 64, 65, 67, 68, 69, 70, 71, 72, 73, 74, 77, 78, 79, 85, 88, 117, 121, 125, 126, 128, 136, 145, 150, 151, 153, 162, 173, 175, 182, 194, 201, 202-3, 204, 208, 211, 213, 214, 215, 216, 217, 222, 224, 230, 231, 246, 247, 249, 251, 255, 263, 267, 268, 271, 278, 285, 292, 294, 295, 298
  colliers of, 115, 116
  nailors of, 51
  woodland at, 56
Winstanley, James, 162, 297
wire industry, 52
Withington, Nicholas, of Chowbent, 51-2

Workington, 172
Wright, W.B., 261, 288
Wrightington
  collieries at, 65, 74, 76, 90, 93, 109, 112,
    123, 126, 130, 189, 251, 272, 275,

276, 303, *see also* Harrock Hill
  colliers of, 115, 116, 119-20, 277, 278

Yarrow, river, 36, 177, 178, 193
Yorkshire, iron from, 51